D0215056

ADAPTIVE GENETIC VARIATION IN THE WILD

ADAPTIVE GENETIC VARIATION IN THE WILD

Edited by Timothy A. Mousseau
Barry Sinervo
John A. Endler

New York Oxford

Oxford University Press

2000

Oxford University Press

Oxford New York
Athens Auckland Bangkok Bogotá Buenos Aires Calcutta
Cape Town Chennai Dar es Salaam Delhi Florence Hong Kong Istanbul
Karachi Kuala Lumpur Madrid Melbourne Mexico City Mumbai
Nairobi Paris São Paulo Singapore Taipei Tokyo Toronto Warsaw

and associated companies in
Berlin Ibadan

Published by Oxford University Press, Inc.,
198 Madison Avenue, New York, New York 10016

Library of Congress Cataloging-in-Publication Data
Adaptive genetic variation in the wild / edited by Timothy A. Mousseau,
Barry Sinervo, John A. Endler.
p. cm.
Includes bibliographical references and index.
ISBN 0-19-512183-X
1. Variation (Biology) 2. Population genetics I. Mousseau,
Timothy A. II. Sinervo, Barry. III. Endler, John A., 1947–
QH401.A395 2000
576.5'4—dc21 99-13844

9 8 7 6 5 4 3 2 1

Printed in the United States of America
on acid-free paper

Preface

Two of the great mysteries of biology yet to be explored concern the distribution and abundance of genetic variation in natural populations and the genetic architecture of complex traits. The two are intimately entwined by their relationship to natural selection and evolutionary history, and the key to their disclosure lie in studies of wild organisms in their natural environments.

This volume and its associated symposium (held at the Ecological Society of America meetings in Albuquerque) were premised on the ideal that a fundamental objective of evolutionary ecology is to predict organismal, population, community, and ecosystem response to environmental change. It is generally acknowledged that anthropogenic influences will likely lead to rapid environmental change in the coming decades (e.g., elevated CO_2, global warming, increased UV, shifts in global climate); a prerequisite for ecological predictions concerning population and ecosystem response to environmental change is knowledge of the genetic basis of traits likely to be under selection in a dynamically changing environment.

The central thesis of this collective work is that the expression of genetic variation is modulated and shaped by the action of natural selection in the natural environment. Thus, it is only within the context of an organism's ecology that novel insights concerning the tempo and mode of evolution can be elucidated. Further, it is not sufficient to simply explain past patterns of adaptation, although this is a worthy and informative first step. The utility of evolutionary ecology will rest on its ability to generate testable and useful predictions concerning organismal response to selection.

The chapters contained in this volume represent the leading edge of studies concerning the adaptive significance of genetic variation in natural populations. These studies employ a wide variety of techniques to test for genetic variation and its consequences for adaptation. These techniques range from classical demography to the use of molecular markers to reveal patterns of relatedness that can then be used to infer the genetic basis for traits associated with adaptation. The hope is that these studies may serve as templates for future research in this area.

I gratefully thank my co-editors Barry Sinervo and John Endler, and the many contributors to this book for their efforts. In addition, I would like to acknowledge and thank the following individuals and organizations for their support of this venture: The Ecological Society of America, Kirk Jensen and Lisa Stallings at Oxford University Press, The National Science Foundation for providing a collegial atmosphere in which to edit manuscripts, and my wife, Heather Preston, for indulging my need to complete this exercise on weekends, nights, and early mornings.

Columbia, South Carolina T.A.M.

Contents

Contributors

John A. Endler Department of Ecology, Evolution and Marine Biology, University of California, Santa Barbara, CA 93106 USA

Derek J. Girman Romberg Tiburon Center, San Francisco State University, Tiburon, CA 94920, and Center for Population Biology, University of California, Davis, CA 95616 USA

Peter R. Grant Department of Ecology and Evolutionary Biology, Princeton University, Princeton, NJ 08544-1003 USA

B. Rosemary Grant Department of Ecology and Evolutionary Biology, Princeton University, Princeton, NJ 08544-1003 USA

Ary A. Hoffmann School of Genetics, La Trobe University, Bundoora, 3083 Australia

Lukas F. Keller Department of Ecology and Evolutionary Biology, Princeton University, Princeton, NJ 08544-1003 USA

Keli Landau Department of Biology, University of Southwestern Louisiana, Lafayette, LA 70504-2451 USA

Susan J. Mazer Department of Ecology, Evolution and Marine Biology University of California, Santa Barbara, CA 93106 USA

Daniel E. Meade Department of Ecology, Evolution and Marine Biology, University of California, Santa Barbara, CA 93106 USA

Susan Mopper Department of Biology, University of Southwestern Louisiana, Lafayette, LA 70504-2451 USA

Timothy A. Mousseau Department of Biological Sciences, University of South Carolina, Columbia, SC 29208 USA

Ruedi G. Nager Division of Environmental and Evolutionary Biology, University of Glasgow, Glasgow G12 8QQ, Lanark, Scotland

Kermit Ritland Department of Forest Sciences, University of British Columbia, Vancouver, BC V6T1Z4 Canada

Beren W. Robinson Department of Zoology, University of Guelph, Guelph, Ontario N1G 2W1, Canada

Dolph Schluter Department of Zoology and Center for Biodiversity, The University of British Columbia, 6270 University Boulevard, Vancouver, BC, V6T 1Z4, Canada

Barry Sinervo Department of Biology, University of California, Santa Cruz, CA 95064 USA

Thomas B. Smith Department of Biology, San Francisco State University, San Francisco, CA 94132, and Center for Population Biology, University of California, Davis, CA 95616 USA

Arie J. van Noordwijk Netherlands Institute of Ecology, NL-6666 ZG Heteren, Netherlands

Peter Van Zandt Department of Biology, University of Southwestern Louisiana, Lafayette, LA 70504-2451 USA

ADAPTIVE GENETIC VARIATION IN THE WILD

1

Quantitative Genetic Variation in Populations of Darwin's Finches

PETER R. GRANT AND B. ROSEMARY GRANT

Some populations of organisms are much more variable than others in genetic characteristics and quantitative phenotypic traits. Theories explaining genetic variation focus on a balance between mutation on the one hand and stabilizing selection and drift on the other. Additional factors are needed to explain why some populations are more variable than others. These factors include introgression of genes and nonstabilizing forms of selection (directional and diversifying). A long-term field study of Darwin's Finches on the Galapagos island of Daphne Major shows that differential introgressive hybridization is partly responsible for the higher levels of additive genetic variance in the medium ground finch (*Geospiza fortis*), an ecological generalist species comprising specialist phenotypes, than in the cactus finch (*G. scandens*), a specialist species. The two species hybridize, rarely, and in addition, the medium ground finch hybridizes with the small ground finch (*G. fuliginosa*). There is little or no fitness loss in the hybrids and backcrosses. Additive genetic variances of the medium ground finch and cactus finch are approximately at equilibrium. Equilibrium implies that the medium ground finch is subjected to stronger forces of selection than the cactus finch, balancing the stronger genetic input.

Evolutionary implications of these findings are considered in the context of an apparently rapid adaptive diversification of this closely related group of species. High levels of additive genetic variation have facilitated the evolutionary transition of one species to another, subject to constraints arising from character correlations. The inferred role of selection in the past is based on the demonstrated role in the present; microevolutionary responses to directional selection have been observed and measured in the medium ground finch population. Transformations of body size in the past have been achieved more easily than transformations of shape. Reconstruction of those transformations suggests that, given current genetic parameters, certain directions of evolution were more likely to have occurred than their opposites. Thus the study of quantitative variation in nature is of value in illuminating and interpreting evolution in the past as well as being of interest in its own right as a phenomenon to be explained.

1.1 Genetics and Ecology

Why do populations vary so much in quantitative traits such as the flowering phenology of plants, the body size of animals, and the dormancy period of copepod eggs? Some of the variation is attributable to effects of age, sex, and growth conditions, but even when these are accounted for there still remains a large amount of variation.

Two classes of answers, genetic and ecological, can be given to this question. Both are needed for a comprehensive understanding of quantitative variation in nature (P. R. Grant and Price 1981). The genetic answer is that mutation is balanced by selection and/or drift (Kimura 1965, Bulmer 1971a, 1989, Lande 1976, Barton and Turelli 1989, Barton 1990; see a recent review of genetic factors, Roff 1997). Polygenic variation is continually augmented by new mutations (M). These are reassorted by recombination and may have sex-specific effects that tend to preserve variation (Wayne et al. 1997). Variation is depleted by stabilizing selection (*S*), which occurs either directly on the trait or on one or more correlated traits (Barton 1990, Keightley and Hill 1990, Kondrashov and Turelli 1992, Caballero and Keightley 1994, Mackay et al. 1995); in the latter case selection on the trait of interest is only apparent. Variation neither increases nor decreases over a moderate time span because these opposite tendencies are assumed to be equal and hence balanced.

A simple theory of balance by mutation and selection alone (*M-S*) is conceptually sound but quantitatively inadequate (Mackay et al. 1995). Strong stabilizing selection removes polygenic variation slowly, and environments are not constant for long enough to allow selective removal to proceed far (Gavrilets and Hastings 1995; see also Lynch and Hill 1986, Houle 1989). Genetic drift (D), however, augments the selective loss of variation (Bulmer 1972, Lynch and Hill 1986, Houle 1989, Caballero and Keightley 1994, Gavrilets and Hastings 1995). Mutation is thus balanced by selection plus drift (*M-SD*). This is the baseline for examining particular factors in detail and for seeking explanations for heterogeneity in the amount of variation maintained in populations. The upper part of figure 1.1 schematically represents the concept of balance.

Because populations in nature are open, usually large, and rarely inbred to a pronounced degree, consideration must be given to two further processes that elevate the standing level of polygenic variation. The first is introgression (*I*) of alleles, here construed broadly to mean genetic input from either conspecific (Bulmer 1971b, Slatkin 1978, 1987) or heterospecific sources (P. R. Grant and Price 1981, B. R. Grant and P. R. Grant 1989a, P. R. Grant 1994, Phillips 1996). Interconnectedness of populations through gene flow is widespread in nature; therefore a balanced theory of variation should include introgression (*MI-SD*). The lower part of figure 1.1 illustrates the essential ingredients of *MI-SD* theory. The second process is a conversion of nonadditive to additive genetic variation that sometimes occurs during bottlenecks in populations which fluctuate in size (Bryant et al. 1986, Goodnight 1987, 1988, Willis and Orr 1993, Bryant and Meffert 1995, 1996, Cheverud and Routman 1996). It is not yet clear how important and prevalent this process is in nature. Recent theoretical and laboratory work suggests that additive-by-additive epistasis can suppress or release additive genetic variance according to the frequency of alleles at the interacting gene loci (Goodnight 1987, Routman and Cheverud 1997).

The ecological answer to why quantitative traits are so variable is that environments are diverse and heterogeneous in time and space. When environmental conditions fluctuate strongly, causing populations with overlapping generations to fluctuate in size, genetic vari-

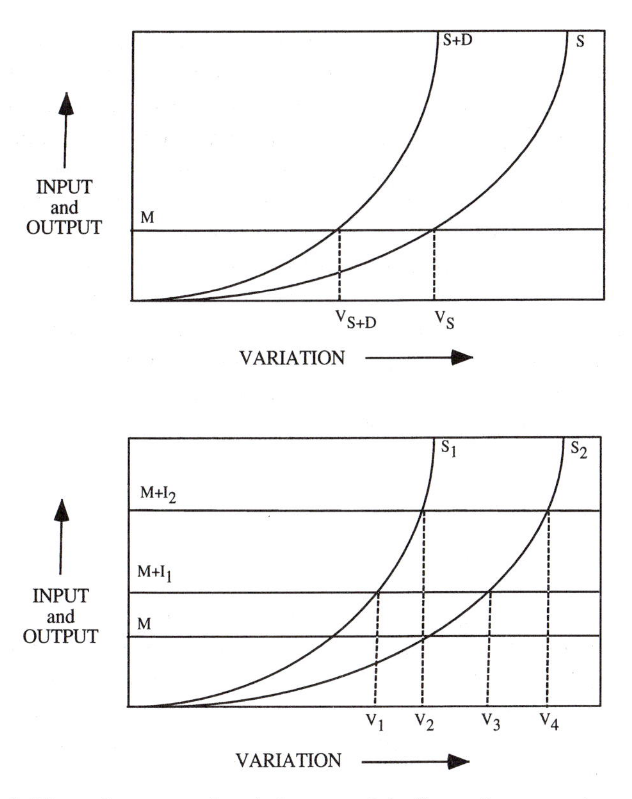

Figure 1.1 The maintenance of variation: a model of input from mutation and introgression balanced by output from selection and drift. (Top) An equilibrial level of variation is determined by a constant rate of input of genetic novelty through mutation balanced by selection alone (S) or selection combined with drift ($S + D$). At constant or occasionally low population sizes, where drift is likely to be most effective, variation is relatively low ($V_{S+D} < V_S$). (Bottom) Two levels of constant introgression (I_1, I_2) and two levels of selection intensity (S_1, S_2) are added to the scheme above to show their joint determination of equilibrial levels of variation. Drift, not shown, could be added equally or unequally to the selection functions. Modified from P. R. Grant and Price (1981).

ation can be sustained at a high level (Ellner and Hairston 1994, Sasaki and Ellner 1997). Heterogeneity in space is likely to be even more effective in sustaining high levels of genetic variation (Via and Lande 1985, 1987, Gillespie and Turelli 1989), especially for sessile organisms. However, heterogeneity in time and space is not necessary for the maintenance of large trait variation providing that the environment is sufficiently rich in resources to allow partitioning of the ecological niche by different, partially specialized phenotypes in the population (Van Valen 1965, Roughgarden 1972). In all these cases of environmental control over the level of expressed variation, the agency responsible for the match between environmental variation and trait variation is natural selection — directional, stabilizing or diversifying. In addition, sexual selection of these same three types occurs on variation in traits that are used in the acquisition of mates (Pomiankowski and Møller 1995, Rowe and Houle 1996).

Most empirical investigations of quantitative trait variation in animals have been conducted with inbred lines of laboratory mice and *Drosophila* and have focused on the estimation of mutational variance and its relationship to fitness (e.g., Keightley et al. 1993, Caballero and Keightley 1994, Caballero et al. 1995, Mackay et al. 1995). Environmental heterogeneity has been generally ignored (but see Mackay 1981). In contrast, field studies have been few and brief and have concentrated on ecological factors relevant to trait variation. Genetic factors have been generally ignored. In this chapter we review the results of a long-term field study of Darwin's finches on the Galápagos Islands designed in part to understand why some populations vary much more than others in beak and body size traits (Lack 1945, 1947, Bowman 1961, P. R. Grant 1986) in terms of both genetic and environmental factors. We first discuss the ecology of morphological trait variation, then the estimation of quantitative genetic parameters and the question of whether observed variation is in a state of balance between enhancing and depleting forces, and finally the evolutionary significance of variation in the context of speciation.

1.2 Ecology of Darwin's Finches

Our study of Darwin's finches began in 1973. The immediate stimulus for our research was an article by Van Valen (1965) in which he argued that variation may be adaptive; different members of a population with different phenotypes have different fitnesses when exploiting the environment in the same way, or similar fitnesses when doing so in different phenotype-dependent ways. This was not an original or surprising idea when applied to discontinuous variation (e.g., Ludwig 1950, Levene 1953), such as shell color polymorphism in snails (Cain and Sheppard 1954, Johannesson et al. 1997) or flower color polymorphism in many plant species (e.g., Levin and Kerster 1967). It was novel, however, when applied to continuous variation, such as body size, in an explicitly ecological context.

Van Valen indirectly tested the idea (1965) by comparing beak size variation in populations of bird species with broad ecological (feeding) niches and in populations of relatives with narrower niches in different environments. The comparative tests yielded statistical evidence in support of the hypothesis: populations with broad niches, generally on islands, were more variable in beak dimensions than narrow-niche relatives, living generally in species-rich continental communities.

There are different ways in which a population may have a niche of broad dimensions (Van Valen 1965, Soulé and Stewart 1970, Van Valen and Grant 1970, Roughgarden 1972, Rothstein 1973a, Feinsinger and Swarm 1982, Werner and Sherry 1987, Taper and Case 1992). The population might be composed of identical generalists or different specialists. For organisms in general, some degree of different specialization might be associated with differences in age and growth stage (Polis 1984, Werner and Gilliam 1984), sexual dimorphism in adults (Selander 1966, Rothstein 1973a), or behavior unrelated to morphology (Werner and Sherry 1987). Clones of some asexual species may also differ in their ecological niches, with the result that the population niche is the sum of its clonal parts (Vrijenhoek 1984).

An obvious question for us was whether adults of the same sex fed in different, specialized ways on islands where the ecological niche was broad, as expected under Van Valen's hypothesis, or whether they fed in a similar generalist manner. In attempting to answer this question, we encountered difficulties in obtaining reliable field data in two of the situations examined by Van Valen (P. R. Grant 1979a, 1979b). This led us to look elsewhere, and we chose Darwin's finches on the Galápagos Islands as the most suitable species for a detailed field study. We chose these populations because, first, from the work of Lack (1945, 1947) and Bowman (1961), we knew that some populations were unusually variable in beak dimensions (fig. 1.2) and, separately, some populations apparently had broader niches than others. Second, the birds were known to be tame and easily observed. We sought an island where we could make two types of comparisons: among different phenotypes within a population and between populations (in the same environment) that differed in levels of phenotypic variation. The first comparison could give us insights into absolute variation and the second into relative variation.

Finches on the island of Daphne Major were selected for detailed observations because the birds were abundant and the island was small enough (~0.34 ha) to facilitate resightings (P. R. Grant et al. 1976). Two species have sizeable breeding populations on this island: *Geospiza fortis*, the medium ground finch, and *G. scandens*, the cactus finch. They display contrasting levels of variation in bill dimensions (table 1.1). With reference to the frequency distributions of the coefficients of variation in figure 1.2, *G. scandens* on Daphne would be classified as a typically varying species in beak depth and in beak length, whereas the population of *G. fortis* on this island would be classified as an unusually variable species in both dimensions (neither population is included in the figure). *G. fortis* varies more in bill depth than *G. scandens* by as much as 65–75%.

1.2.1 Morphological variation and feeding ecology

Geospiza fortis is a granivorous generalist, and *G. scandens* is a cactus specialist. Our first task was to uniquely band, measure, and then quantify feeding observations of a large number of individuals in each population to determine the diets of different phenotypes. Timed observations revealed that *G. fortis* individuals with different beak sizes fed in different ways. In the dry season, when food supply is likely to be limiting and most mortality occurs, they fed on food items of different size and hardness in proportion to their beak sizes. Moreover, the efficiency they displayed in cracking seeds varied depending on beak size (P. R. Grant et al. 1976, P. R. Grant 1981a, Price 1987), as has been found for other seed-eating finches elsewhere (e.g., T. B. Smith 1987, 1993, Benkman 1993, Diaz 1994, Benkman and Miller 1996). This generalist population is therefore made up of different specialists to some degree. Specialists are not discretely different in diets; rather, the diets of some of them do not span the full dimensions of the population's niche. Price (1987) estimated the between-phenotype component of the population's niche width to be 11%.

G. scandens fed in a more uniform manner on cactus products (nectar, pollen, and seeds) and on other seeds to a lesser extent at this time (Boag and Grant 1984a, Millington and Grant 1983, B. R. Grant and P. R. Grant 1996a, 1998, P. R. Grant and B. R. Grant 1996a). All members of this population are specialized in similar ways.

Additional studies of marked and measured birds revealed that fitness of *G. fortis* individuals was related to beak size under stressful conditions. Under the drought conditions prevailing in 1976–77, birds with large beak size (and body size) survived better than those with small beaks, apparently as a result of a feeding efficiency advantage on large and hard seeds that increasingly dominated the seed supply due to differential consumption of the small seeds. In contrast, as the island dried out after an exceptionally severe El Niño event in

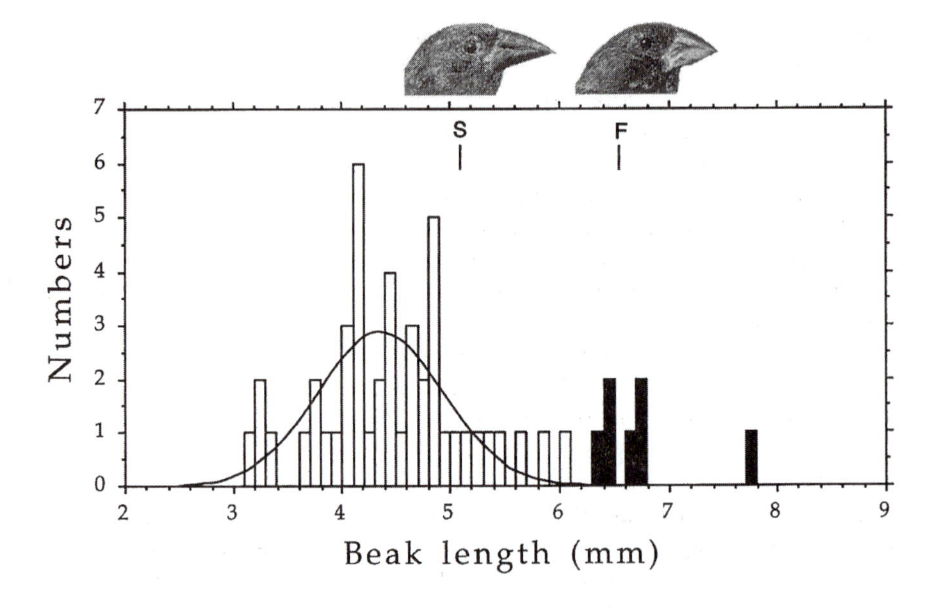

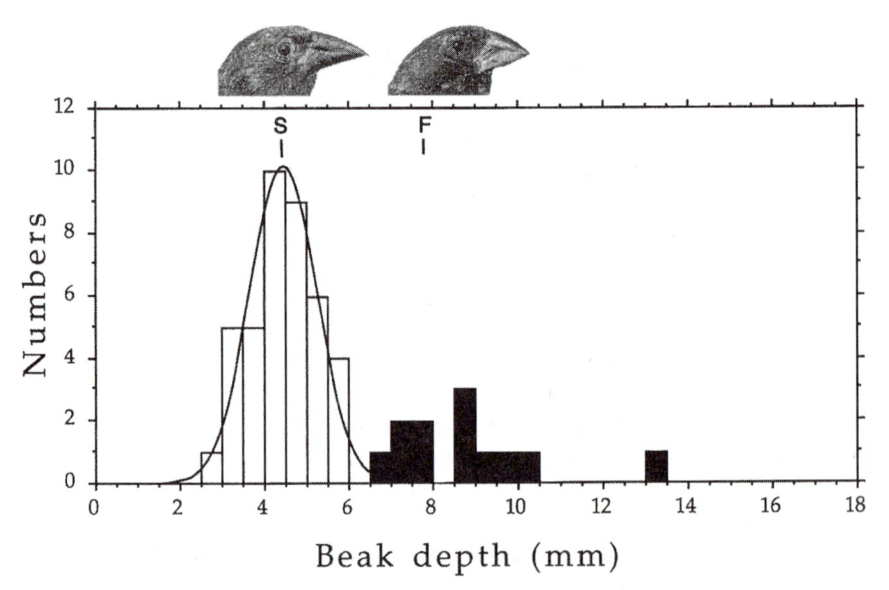

Figure 1.2 Frequencies of coefficients of variation of beak length and beak depth for adult males in populations of *Geospiza* species. Based on data in table 61 of Bowman (1961). Normal curves have been fitted by eye, and values beyond the normal ranges have been highlighted in black. An average value on both axes for the typically variable populations is about 4.5, which is in line with average values for continental emberizine species (related buntings and sparrows, B. R. Grant and P. R. Grant (1989a); for a broader survey of avian coefficients, see Fitzpatrick (1997). The Daphne population of *G. fortis*, indicated by F in the figure, is 40.5% more variable than *G. scandens* (S) in bill length and 75.8% more variable in bill depth.

Table 1.1 Coefficients of phenotypic variation for the two main finch populations on Daphne Major island

	G. fortis		*G. scandens*	
	Males	Females	Males	Females
Mass (g)	10.90	10.24	8.20	8.86
Wing length (mm)	3.23	3.12	2.90	3.07
Tarsus length (mm)	3.82	4.04	3.53	3.66
Bill length (mm)	7.04	6.81	5.80	6.03
Bill depth (mm)	8.39	8.28	4.81	5.02
Bill width (mm)	6.84	6.58	4.62	4.55
n	798	636	317	238

Only those born (hatched) on the island, or who bred, are included.

1982–83, small-beaked birds in this population survived better than large-beaked birds. The apparent reason in this case was a sharp decrease in the large and hard seeds that are rewarding for the large-beaked birds: the low-growing plants (*Tribulus cistoides*) that produced the largest seeds were smothered by other vegetation in 1982–83. The plant producing the second largest seed (*Opuntia echios*) also suffered high mortality and depressed reproductive output.

The less varying *G. scandens* suffered lower mortality than *G. fortis* during the first event. There was a relatively weak selective advantage to large body size during the first episode (Boag and Grant 1984b), with little or no independent directional selection on beak traits, unlike the situation with *G. fortis*. Also unlike *G. fortis*, beak and body size traits of *G. scandens* were subject to stabilizing selection (P. R. Grant and Price 1981, Price et al. 1984a; note that negative signs of the selection coefficients are missing in table 4 of Price et al. 1984a). Higher *G. scandens* mortality occurred in the second event as a result of adverse effects of heavy rain on the *Opuntia* cactus. All members of the population were affected by the reduction in their major food source approximately equally: mortality was random with respect to morphological traits.

Summarizing to this point, *G. fortis* is morphologically and ecologically variable. Feeding skills of individuals vary in relation to beak size. Relative fitness of individuals of a particular phenotype varies in both direction and magnitude according to environmental conditions. Natural selection occurs episodically, and it oscillates in direction, with cumulative effects over the long term perhaps equivalent to a weak but persistent stabilizing selection in the short term. Selection in opposite directions at different times of a life cycle (Price and Grant 1984; see also Schluter and Smith 1986, B. R. Grant and P. R. Grant 1989a, 1989b) may contribute to a cumulative stabilizing selection function on each generation. *G. scandens*, on the other hand, is both morphologically and ecologically less variable. Relative fitnesses of different phenotypes vary much less in this species, although stabilizing selection does occur.

All of these findings are consistent with the view that ecological factors control the level of morphological variation in a population by permitting or constraining increases in variation. Do they do more than this? Do they actually enhance population variation by giving an advantage to individuals furthest from the mean, at least under some conditions? Is selection stabilizing yet relaxed, or is it diversifying (Van Valen 1965, Rothstein 1973a, 1973b)?

Diversifying selection would be revealed as curvilinearity in the fitness function, as determined by multiple regression (Lande and Arnold 1983, Price and Boag 1987) or cubic spline (Schluter 1988, Schluter and Nychka 1994) techniques. There is some evidence for enhanced fitness of the smallest *G. fortis* individuals when in 1976–77 large size was selectively favored (P. R. Grant and Price 1981, Schluter et al. 1985, Price and Boag 1987), but the effect is not statistically demonstrable (Schluter 1988). Integration of different forms of natural and sexual selection occurring within a short period of time (e.g., Price et al. 1984a) could yield a single diversifying selection function on beak and body size traits. A thorough analysis of selection on all cohorts has not yet been performed, but comparison of means and variances in successive years does not indicate any overlooked linear or nonlinear selection event.

Elsewhere, evidence of disruptive selection has been presented for tarsus length in tree swallows (*Tachycineta bicolor*) by Wiggins (1991) and for beak size in African seedcracker finches (*Pyrenestes*) by T. B. Smith (1993; Smith and Girman, this volume). Both studies used cubic splines to depict fitnesses in relation to morphology. Diversifying (disruptive) selection appears to be rare, although it may be prevalent yet generally weak and overlooked as a result of difficulties of detection. The seedcracker case is exceptional in another respect: the frequency distribution of beak sizes in the population is strongly bimodal (Smith and Girman, this volume). Bimodality may be more common in other taxa, particularly fish (Robinson and Schluter, this volume).

1.2.2 Morphological variation and competition

In some quantitative genetic models of competitive character displacement (Taper and Case 1985, Doebeli 1996), trait means diverge at the cost of genetic variance: variance declines as a result of competition, in accordance with expectation from Van Valen's hypothesis (Van Valen 1965). Therefore, in the absence of competitors the variance of a species may increase, released from the constraints of competitors. In fact, the evidence for "variance release" in populations on small islands with few or no competitors is exceedingly slim (Feinsinger and Swarm 1982, P. R. Grant et al. 1985, Schoener 1986). Lack of evidence for variance release is difficult to interpret. Resources are likely to be less diverse on small islands, nullifying the effect of the absence of competitors. Moreover, the expected increase in variance would be caused presumably by an accumulation of mutations and so would require a long time. The variance is not expected to be large (e.g., see Doebeli 1996) and would be opposed by drift. This short discussion of competition highlights the inadequacy of considering morphological variance solely in terms of ecological factors (resources and competitors). One needs to know the genetic sources of variation and not just the ecological factors that permit or constrain variation.

1.3 Genetics

1.3.1 Heritable variation

Phenotypic variation is enhanced by spatial and temporal environmental variation and by genetic variation. In turn, genetic variation is enhanced by mutation, recombination, and introgression (figure 1.1). At the outset we knew nothing about these factors. Our most important question was whether phenotypic variation reflected an underlying genetic variation, as would be necessary for the adaptive argument of Van Valen to apply, or whether it resulted from the effects of strong environmental variation in space and time on a genetically uniform or near-uniform population (Rothstein 1973a). In 1973 there were no studies of heritable morphological variation of wild birds to give us insight into how this question would be answered.

In 1976 we began a breeding study of the finches on Daphne (Boag and Grant 1978, Boag 1983). Offspring were banded as nestlings and recaptured and measured when growth had finished or nearly finished. In many cases the nestlings' parents, identified by observing them incubating the eggs (female parent only) or feeding the nestlings, had been previously banded and measured. We were therefore able to estimate the heritabilities of the measured morphological traits by regressing the average of the offspring measurements on measurements of each of the parents separately or on the average of the measurements of both parents.

Heritabilities (h^2) for *G. fortis* (22 families) were found to be unusually high from midparent regressions, ranging from a minimum of 0.53 for wing length to a maximum of 0.95 for bill width (Boag and Grant 1978). In other words, h^2 estimates for some traits were close to the theoretical maximum of 1.0 in the absence of measurement error. The average was 0.67. The calculations were repeated with a new set of breeding and morphological data in 1978 (Boag 1983). Results were similar, with a slightly higher mean of the individual trait heritabilities calculated from midparent regressions (0.74). Regardless of the particular value of separate or combined estimates, the conclusion is clear: these traits appear to have a large amount of additive genetic variation, as indicated by these statistical associations between offspring and parents.

High values of *G. fortis* heritabilities are confirmed by the results of analyses of 5 more years of data (table 1.2). Beak heritabilities are consistently > 0.5 and are usually higher than body mass, wing, and tarsus heritabilities. Estimates for each of the years are not strictly independent because some birds bred in more than one year. The 1976 and 1991 estimates are entirely independent, however. The important point is the consistency of the estimates across 16 years. This is illustrated with bill depth in figure 1.3. Precision of the estimates could be improved by using a family weighting according to sample size (Falconer 1989), which has not been done.

The first attempts to estimate heritabilities for *G. scandens* were hampered by small sample sizes. It appeared that heritable variation was low or lacking in this species (Boag 1983). This turned out to be wrong when larger samples were assembled (Price et al. 1984a, P. R. Grant and B. R. Grant 1994). Table 1.2 gives heritability estimates for this species in the 3 years when sample sizes of families reached 20 or more. Heritabilities are similar in this species and *G. fortis* but tend to differ in the beak dimensions. Thus all beak traits in *G. fortis* are highly heritable to approximately the same degree, whereas in *G. scandens* beak length variation is more highly heritable than variation in the other two dimensions.

Because heritability estimates are influenced by sample sizes, which vary strongly among years, we also calculated heritabilities for composite samples of these two species (table 1.3). Examples are given in figure 1.4. In addition, a principal components analysis (PCA) was performed to reduce the six measured variables to a smaller number of synthetic traits, and these were then subjected to heritability analysis. Results are shown in table 1.3.

Heritabilities of the six measured traits averaged 0.70 for *G. fortis* and 0.50 for *G. scandens*. The heritabilities of an overall body size factor (PC1) and a shape factor (PC2) are

Table 1.2 Heritabilities calculated from midoffspring–midparent regressions

	1976	1978	1981	1983	1984	1987	1991	Mean
				G. fortis				
Mass	1.044****	0.645***	0.595****	0.270*	0.450***	0.640****	0.573****	0.602
Wing length	0.751	0.567***	0.496****	0.381***	0.408**	0.713****	0.627****	0.563
Tarsus length	0.299	0.666***	0.301*	0.246*	0.471***	0.549****	0.534****	0.438
Bill length	0.651**	0.770****	0.579****	0.664****	0.564****	0.791****	0.847****	0.695
Bill depth	0.776****	0.843****	0.644****	0.747****	0.569****	0.695****	0.828****	0.729
Bill width	0.894****	0.926****	0.669****	0.700****	0.500****	0.714****	0.811****	0.745
Families	22	21	51	64	66	101	99	
Offspring	25	82	99	201	102	467	220	
				G. scandens				
Mass	—	—	0.398	0.450***	—	0.736***	—	0.528
Wing length	—	—	0.672***	0.396***	—	0.361	—	0.476
Tarsus length	—	—	0.578*	0.458***	—	0.214	—	0.397
Bill length	—	—	0.997****	0.621****	—	0.970****	—	0.863
Bill depth	—	—	0.681*	0.271	—	0.713*	—	0.555
Bill width	—	—	0.582**	0.242	—	0.298	—	0.357
Families	<20	<20	37	68	<20	25	<20	
Offspring			62	151		65		

Statistically significant differences from zero at $^*p<.05$, $^{**}p<.01$, $^{***}p<.005$, and $^{****}p<.001$.

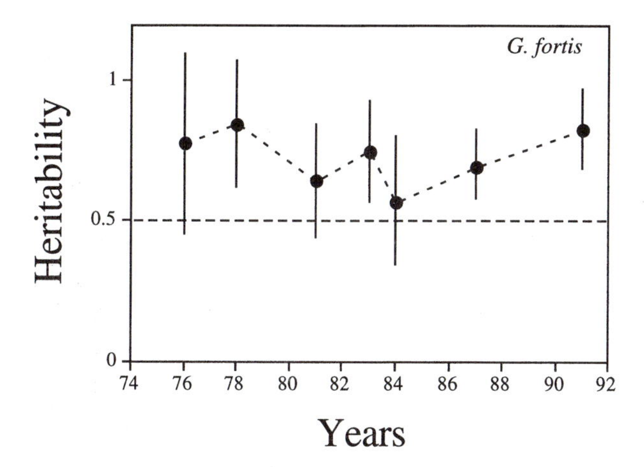

Figure 1.3 Consistency in the estimates of *G. fortis* heritability of bill depth. Vertical bars show 95% confidence limits on the estimates from midoffspring–midparent regressions (table 1.2).

> 0.50 in both species. The second component can be understood from the factor loadings to express large bill depth and bill width in relation to bill length. In *G. scandens,* PC3, which gives a large weighting to wing length in relation to the other variables, is also heritable. It is noteworthy that the heritabilities of shape factors apply to relatively little phenotypic variation. After the removal of a size factor (PC1) accounting for 54.3% of the total multivariate variation in *G. scandens* and a shape factor (PC2) accounting for a further 14.9%, there still remains variation in a third factor (10.2%) that is heritable.

There are genetic and nongenetic reasons to expect a nonlinear relationship between parent and offspring trait measurements (Gimmelfarb and Willis 1994). We have not performed the appropriate analysis using polynomial functions to detect departures from linearity, but inspection of individual scatter plots gave no indication of nonlinearities (see also fig. 1.4). Nonlinearity for genetic reasons might be quite small (Turelli and Barton 1994).

These heritabilities are surprisingly high in view of the low (<100) genetically effective sizes of the populations (ignoring gene flow; P. R. Grant and B. R. Grant 1992a). Theoretical models that make different assumptions agree that populations this small should exhibit heritabilities of <0.25 (Houle 1989, Caballero and Keightley 1994).

1.3.2 Other species

These results are likely to be generally true for Darwin's finches. Heritable variation has been estimated in two other populations of ground finches (table 1.3). Heritabilities for five of six traits of *G. magnirostris* on Daphne are apparently similar in magnitude to those of *G. fortis* and *G. scandens*. A breeding population of *G. magnirostris* only became established on Daphne in 1982–83 (P. R. Grant and B. R. Grant 1995a), and the sample size is still too small to yield reliable estimates of heritable variation. Only one heritability is statistically significant. For *G. conirostris* on the island of Genovesa, all heritabilities exceed 0.50 and are sim-

Table 1.3 Heritabilities (h^2) for four species of Darwin's finches estimated from midoffspring–midparent regressions

Trait	*G. fortis*		*G. scandens*		*G. magnirostris*		*G. conirostris*	
	h^2	SE	h^2	SE	h^2	SE	h^2	SE
Mass (g)	0.68	0.06	0.50	0.09	-0.37	0.84	0.69	0.20
Wing length	0.64	0.07	0.49	0.09	0.46	0.47	0.58	0.18
Tarsus length	0.49	0.07	0.43	0.08	0.81	0.20	0.82	0.16
Bill length	0.85	0.04	0.80	0.07	0.26	0.49	0.66	0.20
Bill depth	0.75	0.05	0.55	0.10	0.64	0.37	0.69	0.11
Bill width	0.78	0.05	0.39	0.10	0.64	0.59	0.77	0.16
Mean	0.70		0.53		0.49		0.70	
PC1	0.76	0.05	0.60	0.09	0.71	0.35	0.81	0.18
PC2	0.63	0.07	0.52	0.10	-0.30	0.68	0.67	0.17
PC3	—	—	0.42	0.09	0.67	0.42	0.63	0.16
Families	207		158		5		57	
Offspring	674		412		13		82	

All heritability estimates differ from zero ($p<<.01$) except for *G. magnirostris*; in this species only tarsus length has a significant heritability ($p<.05$). Note principal components (PCs) are not strictly comparable among species. In all cases PC1 represents overall size. PC2 has a bill shape element in all cases; factor loadings on bill length differ consistently in sign from loadings on bill depth and bill width. There is no consistency in the factor loadings on PC3. Percent variance explained by the first two components in *G. fortis* is 69.6 and 11.0, respectively; the third component makes no significant contribution in this species. In the other species the percent variance is approximately the same, about 60% for PC1, 12% for PC2, and 10% for PC3.

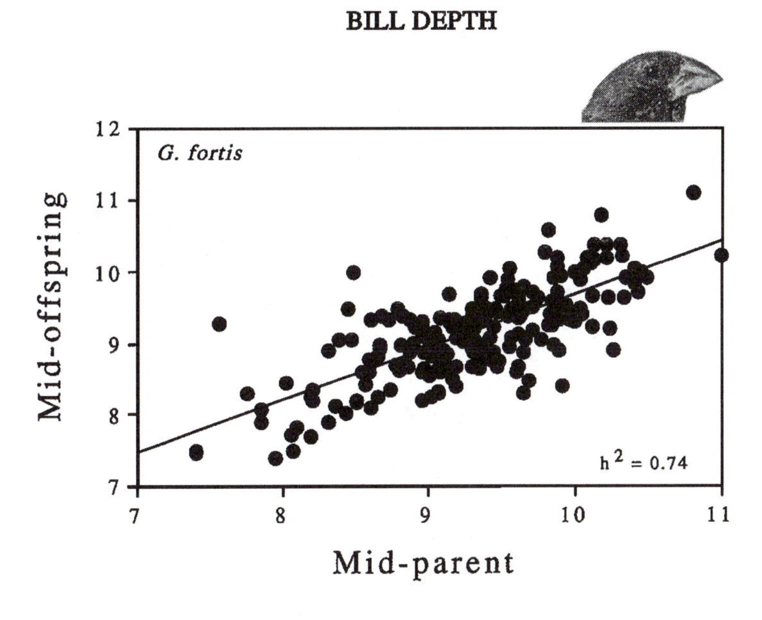

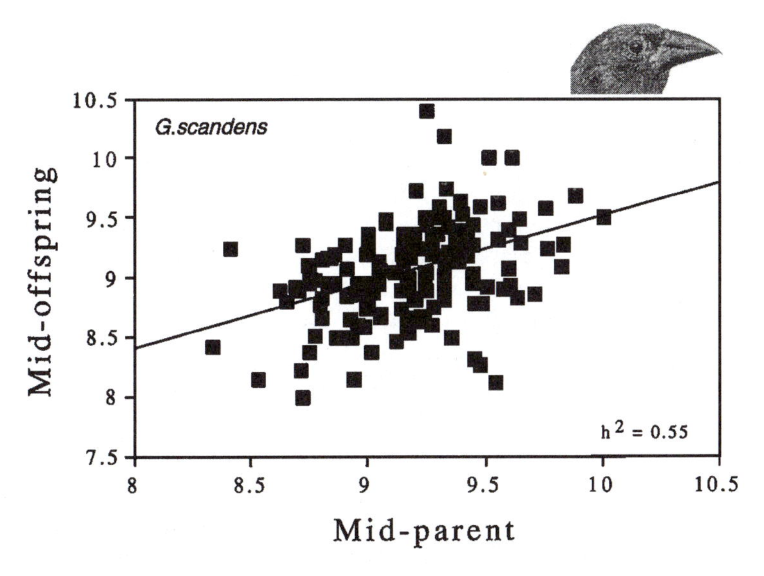

Figure 1.4 Family resemblance indicates strongly heritable variation in *G. fortis* and *G. scandens* on Daphne Island; data based on the combined samples from all years.

ilar to those of *G. fortis*. Heritabilities of the first three components of a principal components analysis are also well above 0.50 (table 1.3). A fifth species, *G. difficilis* on Genovesa, is also likely to display heritable morphological variation as day 8 nestling size (mass) is strongly correlated with midparent (adult) size (P. R. Grant 1981b). Heritable variation in tree finch species and the warbler finch has not been assessed.

Table 1.4 Morphological heritabilities of *Geospiza* species and seven other passerine species

Species	Tarsus	Bill Length	Bill Depth	Bill Width	Source
Melospiza melodia	0.76	0.37	0.98	0.56	Smith and Dhondt (1980)
Parus caeruleus	0.55[a]	—	—	—	Dhondt (1982, 1991)
Parus major	0.55[a]	0.49	0.71	0.68	Dhondt (1991), van Noordwijk (1987)
Parus montanus	0.61	—	—	—	Thessing and Ekman (1994)
Ficedula hypoleuca	0.50	—	—	—	Alatalo and Lundberg (1986)
Ficedula albicollis	0.49[b]	0.44	0.30	0.39	Merilä and Gustafsson (1996)
Tachycineta bicolor	0.75	—	—	—	Wiggins (1989)
Hirundo rustica	0.27	—	—	—	Møller (1987)
Acrocephalus arundinaceus	0.63[a]	—	—	—	Hasselquist et al. (1995)
Geospiza fortis	0.49	0.85	0.75	0.78	Table 1.3
Geospiza scandens	0.43	0.80	0.55	0.39	Table 1.3
Geospiza magnirostris	0.81	0.26	0.64	0.64	Table 1.3
Geospiza conirostris	0.82	0.66	0.69	0.77	Table 1.3
Mean others	0.57	0.43	0.66	0.54	
Mean *Geospiza*	0.64	0.64	0.66	0.64	

[a] The average of several estimates from single-parent regressions; possibly inflated by maternal effects in *Acrocephalus arundinaceus.*

[b] The average of several estimates from midoffspring–midparent regressions.

Beyond the archipelago, heritabilities of some of these morphological traits have been estimated for a few populations of passerine birds (table 1.4). The only other emberizine population that has been studied is the song sparrow (*Melospiza melodia*) on Mandarte Island off the coast of British Columbia, Canada. Heritabilities of the same six traits varied from 0.04 to 0.98 (Smith and Zach 1979, Smith and Dhondt 1980, Schluter and Smith 1986). Estimates vary among species (table 1.3; Boag and van Noordwijk 1987, Larsson 1993, Merilä and Gustafsson 1996); nevertheless, there is a general tendency for weight and wing length to have the lowest heritabilities and bill depth and width to have the highest, as in Darwin's finches.

1.3.3 Genetic sources of bias

Because the heritability estimates were obtained under field conditions and not from a controlled breeding program under aviary conditions, they are subject to bias, and we must consider the major sources of potential bias. True heritabilities may be higher than apparent heritabilities if parentage has been incorrectly determined by observations alone. In 11 other emberizine species, DNA fingerprinting has revealed that extrapair copulations have resulted in extrapair fertilizations, ranging from 5% in the corn bunting (Hartley et al. 1993) to 55% in the reed bunting (Dixon et al. 1994). Incorrectly assigned paternity is therefore common, whereas incorrectly assigned maternity is rare among birds in general (Westneat and Webster 1994). As a result, offspring resemble their mothers more than their putative fathers (Alatalo and Lundberg 1984, Møller 1987, Payne and Payne 1989, Møller and Birkhead 1992; but see Dhondt 1991, Gebhardt-Henrich and Nager 1991, Thessing and Ekman 1994).

The slightly lower heritabilities for *G. scandens* than for *G. fortis* (table 1.3) are expected from the lower genetically effective size of the *G. scandens* population (P. R. Grant and B. R. Grant 1992a) because heritabilities increase with an increase in effective population size (Lynch and Hill 1986, Houle 1989, Caballero and Keightley 1994). Nevertheless, the difference can be alternatively explained as the result of a higher frequency of incorrect identification of *G. scandens* parents. A combination of direct and indirect evidence argues against this possibility.

Incorrectly assigned parentage is a small source of bias in *G. scandens*. On the basis of variation at eight microsatellite DNA loci, Petren et al. (1999a) found that 8% of 152 offspring were not sired by males attending the nest, whereas the mother was correctly identified in all cases. Assuming that the two biological parents genetically contribute equally to the determination of trait size in their offspring, regressions of offspring measurements on putative male parent measurements should be slightly lower on average than regressions on female parent measurements when a few of the true male parents have been misidentified. This was found to be the case in 1983 but not in 1981, the only other year in which sample sizes of families for single-parent regressions exceeded our arbitrary minimum criterion of 20 (table 1.5). In 1981 male parent regressions were much higher than female parent regressions, probably due to the vagaries of sampling, as the standard errors of the heritability estimates are large and broadly overlapping.

The situation in *G. fortis* is different. Molecular determinations of parentage in *G. fortis* are lacking, so we have to rely on single-parent regressions to assess the possibility of a bias from misidentified parents (table 1.5). These have been calculated for the 7 years in which family sizes exceed 20, and table 1.5 shows the results for just the 2 years when sample sizes of *G. scandens* were also sufficient. In 1976 and 1981 (table 1.5) there was no consistent difference between the two sets of regressions in *G. fortis*. In 1987 and 1991 male-parent regressions were somewhat lower than female-parent regressions, and in 1978, 1983 (table 1.5), and 1984 the differences were pronounced, although not significant by ANCOVA. These last results may have arisen from misidentification of the fathers. Extrapair fertilizations are the likely cause and may be more frequent in this species than in *G. scandens*, though not in all years. Another possibility is that female-parent regressions are inflated by maternal effects (Boag 1983, Price 1998). Maternal effects have been invoked to explain the higher female-parent regressions in the great reed warbler in the near absence of extrapair fertilizations (Hasselquist et al. 1995).

Taken at face value, and assuming no misidentified female parents through egg dumping or strong maternal effects, results of comparing single-parent regressions in *G. fortis* imply that heritabilities are sometimes underestimated from offspring–midparent regressions. Correcting this error would increase, not decrease, the small differences between the estimates of heritabilities in *G. fortis* and *G. scandens.*

1.3.4 Environmental sources of bias

An opposite bias giving inflated estimates of true heritable variation may arise for environmental reasons. If offspring and parents share environments, as well as genes, their morphological resemblance may be due partly to common environmental factors experienced during development to full adult size. Experimental studies with a variety of passerine birds have demonstrated the enduring effects of dietary factors during growth on the size of traits in

Table 1.5 Heritability estimates ($h^2 \pm$ SE) calculated from male parent and female parent regressions and their differences (D) expressed as a percentage of the female parent heritabilities

	1981					1983				
	Male		Female			Male		Female		
	h^2	SE	h^2	SE	D	h^2	SE	h^2	SE	D
G. fortis										
Mass	0.648	0.200	0.842	0.158	**-23.0**	-0.094	0.172	0.746	0.146	**-112.6**
Wing length	0.460	0.220	0.436	0.162	**+5.5**	0.340	0.194	0.428	0.160	**-20.6**
Tarsus length	0.378	0.228	0.276	0.196	**+37.0**	0.298	0.180	0.514	0.168	**-42.0**
Bill length	0.734	0.176	0.794	0.146	**-7.6**	0.552	0.174	0.876	0.148	**-37.0**
Bill depth	0.754	0.230	0.900	0.126	**-16.2**	0.618	0.198	0.996	0.128	**-37.9**
Bill width	0.886	0.230	0.796	0.164	**+13.1**	0.414	0.164	0.800	0.124	**-48.2**
Families	57		54			83		70		
Offspring	107		103			220		237		
G. scandens										
Mass	0.606	0.252	0.210	0.378	**+188.6**	0.452	0.160	0.388	0.246	**+16.5**
Wing length	0.742	0.312	1.024	0.290	**-27.5**	0.450	0.196	0.448	0.226	**+0.4**
Tarsus length	0.602	0.300	0.506	0.364	**+19.0**	0.702	0.216	0.440	0.200	**+60.0**
Bill length	1.382	0.238	0.452	0.382	**+205.8**	0.816	0.170	0.600	0.238	**+ 36.0**
Bill depth	0.420	0.298	0.666	0.318	**-36.9**	0.032	0.200	0.780	0.210	**-95.9**
Bill width	0.756	0.210	0.376	0.356	**+101.1**	0.264	0.184	0.586	0.238	**-55.0**
Families	38		42			69		60		
Offspring	63		67			155		152		

Table 1.6 Experimental assessment of genotype × environment correlations: regression slopes of offspring tarsus length on true or foster parent tarsus length

	Control	Experimental		
Species	True	True	Foster	Source
Melospiza melodia	1.01*	0.76*	-0.06	Smith and Dhondt (1980)
Tachycineta bicolor	0.75*	0.50*	0.16	Wiggins (1989)
Parus caeruleus	—	0.61*	-0.37	Dhondt (1982)
Parus montanus	0.62*	0.67*	-0.03	Thessing and Ekman (1994)
Sturnus vulgaris	0.49*	0.43*	0.14	Smith and Wettermark (1995)
Ficedula hypoleuca	0.53*	0.50*	0.04	Alatalo and Lundberg (1986)
Ficedula albicollis	0.51*	0.59*	-0.28	Gustafsson and Merilä (1994)

*Significantly different from 0 ($p < .05$ or lower).

adults (Boag 1987, Boag and van Noordwijk 1987, Gebhardt-Henrich and van Noordwijk 1991, Richner et al. 1989, Merilä 1997; see also Price 1985, 1998). A population may occupy a heterogeneous environment, with some parts more favorable for raising offspring than others. If parents breed in the same type of habitat as the one in which they were raised, then their offspring, subject to the same environmental factors during growth as themselves, are likely to realize their growth potential to the same degree and thus resemble them.

There are two reasons for believing that inflation of heritabilities caused by possible genotype–environment correlations are trivial and ignorable. First, *G. fortis* do not show a tendency to breed in the same type of habitat as their natal habitat (Boag and Grant 1978, Boag 1983). Second, experimental manipulation of broods of other species has not changed the estimates of heritable morphological variation (table 1.6). The experimental technique is to break possible genotype–environment correlations by randomly reassigning clutches or nestlings to nests (i.e., cross-fostering). Such experiments have failed to detect an effect of rearing environment on the size of adult traits in the offspring (Smith and Dhondt 1980, Dhondt 1982, Alatalo and Lundberg 1986, Wiggins 1989, H. G. Smith 1993, Gustafsson and Merilä 1994, Thessing and Ekman 1994, Smith and Wettermark 1995; but see Ricklefs and Peters 1981). However, these experiments do not exclude the possibility of genotype–environment correlations arising from parental influences during early embryogenesis.

Even when there are no genotype–environment correlations, the estimation of heritability may be affected by spatial and temporal variation in environmental conditions. Heritability of size traits may be underestimated when offspring-rearing conditions are poor (van Noordwijk 1984, Price 1985, 1991, Gebhart-Henrich and Marks 1993), and pooling samples from years of contrasting conditions can have the same effect (Hõrak and Tammaru 1996). Nevertheless, Merilä and Gustafsson (1996) found little temporal or spatial variation in the estimates of heritable variation in collared flycatchers (*Ficedula albicollis*). Similarly, we have found little temporal variation in the estimates for Darwin's finches (table 1.2 and fig. 1.3).

In other organisms heritabilities have been found to increase with environmental stress (Ward 1994; but see also Blows and Sokolowski 1995), either because new genes are active under these conditions or because the mean environmental contributions to the trait increase (Ward 1994) while the variance decreases. Annual variation in heritabilities in Darwin's finch is too small to suggest a direct relationship with the environment. Variation in heritabilities does not match the known variation in density and dry-season feeding conditions (Grant and Grant 1996b). For example, in the years of highest density, heritabilities were generally high in one (1976) and low in the other (1984) (table 1.2 and fig. 1.3). Food was plentiful in both years (B. R. Grant and P. R. Grant 1993, P. R. Grant and B. R. Grant 1996a). When little rain falls and the vegetation responds to it weakly, either breeding fails to take place or, if it does occur, the growth conditions for young finches are so poor that most of them die before they are old enough to be captured, measured, and included in samples (Price and Grant 1984, Price 1985). As a result, heritabilities can only be estimated in productive years and are relatively unvarying (table 1.2).

1.3.5 A check on the estimates by prediction

Selection on heritable variation of a trait produces an evolutionary response (Falconer 1989). We compared expected with observed evolutionary responses to selection as a means of checking the validity of our heritability estimates.

Directional natural selection on the *G. fortis* population on Daphne occurred in 1976–77, and selection coefficients were calculated (Price et al. 1984b, P. R. Grant and B. R. Grant 1995b). Evolutionary responses to selection on the six morphological traits were calculated as the difference between means of each trait before selection in 1976 and in the next generation produced in 1978 (Boag 1983, P. R. Grant and B. R. Grant 1995b). Observed responses were found to be close to those predicted from the direct effects of selection on each trait (selection gradients; Lande 1979), trait heritabilities, and genetic correlations between traits (fig. 1.5). Estimated and realized heritabilities were statistically indistinguishable (P. R. Grant and B. R. Grant 1995b). Similar results were obtained when the calculations were repeated for the 1984–86 selection episode (fig. 1.5). We conclude that heritabilities are estimated with reasonable accuracy by the method of regressing offspring measurements on parental measurements in unmanipulated populations of Darwin's finches.

1.3.6 Genetic variation in G. fortis *and* G. scandens

Heritabilities are proportions of phenotypic variance that are attributable to additive genetic variance. For all six traits, *G. fortis* heritabilities are higher than *G. scandens* heritabilities (table 1.3). The differences are much smaller than the differences in phenotypic variation (table 1.1). To compare genetic variation in the species, we use the coefficient of additive genetic variation (i.e., [square root of the additive genetic variance × 100]/the mean; P. R. Grant and Price 1981, Houle 1992). This comparison reveals whether the differences between species in observed heritable variation are due entirely to differences in additive genetic variation, differences in residual variation (nonadditive genetic and environmental), or some combination of the two.

Table 1.7 shows that *G. fortis* is much more variable than *G. scandens*, both genetically and phenotypically, especially in body size (mass), bill depth, and bill width. The species differ little in their coefficients of residual variation, except in bill depth. One component of the residual variation, measurement error, also differs little between the species: as a proportion of the total phenotypic variation, measurement error is 0.06–0.21 for *G. fortis* traits and

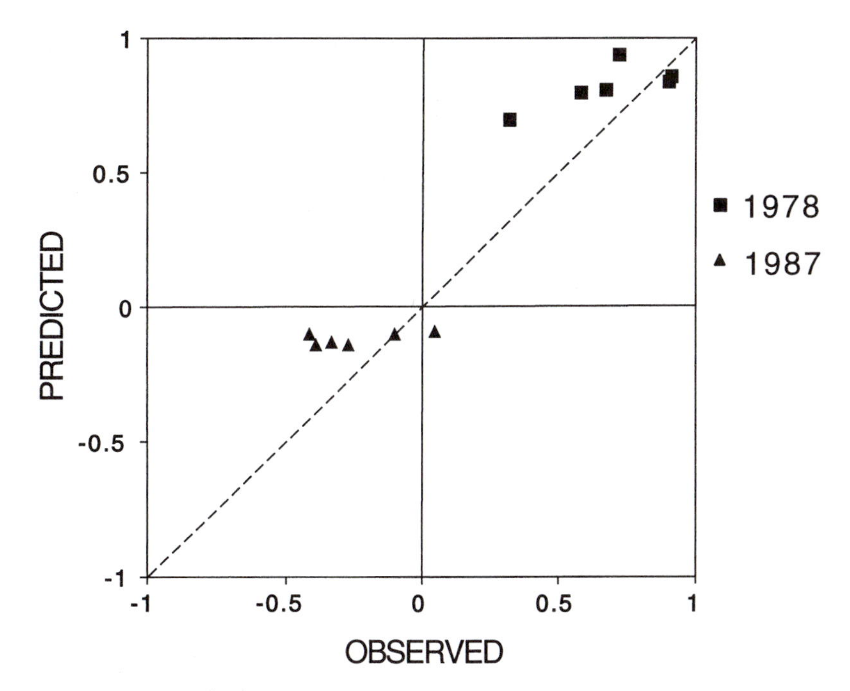

Figure 1.5 Prediction of evolutionary response in *G. fortis* from directional selection on heritable variation in six morphological traits (from P. R. Grant and B. R. Grant 1995b).

0.11–0.29 for *G. scandens* traits (P. R. Grant and B. R. Grant 1994). Measurement error has been minimized in this study by averaging repeated measurements of the same individuals. Nonadditive genetic variation (dominance and epistasis) has been detected by joint-scaling tests. It appears to be small, though possibly confounded by uncontrolled sources of environmental variation (P. R. Grant and B. R. Grant 1994).

1.3.7 Effects of introgression on genetic variation

We now return to the question of what genetic factors are responsible for the maintenance of a particular level of quantitative trait variation in a population and for the differences between the species. Rates of mutation, segregation, and recombination are unknown. They are only now becoming amenable to direct study in wild populations of birds through the use of molecular techniques (e.g., see Yauk and Quinn 1996) — techniques that will soon be applied to avian quantitative trait loci. We assume that the two species of Darwin's finches do not differ in these genetic processes (P. R. Grant and Price 1981). Introgression, on the other hand, can be estimated directly from quantified observations of breeding birds. Differences between species in rates and sources of introgression could be responsible for the differences in standing variation (figure 1.1). We examine this possibility by considering two sources: conspecific gene flow and hybridization.

Table 1.7 Coefficients of phenotypic (CV_P), additive genetic (CV_A), and residual (CV_R) variation of six traits of *G. scandens* (158 families) and *G. fortis* (207 families)

	CV_P		CV_A		CV_R	
	G. scandens	*G. fortis*	*G. scandens*	*G. fortis*	*G. scandens*	*G. fortis*
Mass	8.41*	12.62	5.94	10.38	5.96	7.17
Wing length	3.22**	3.54	2.26	2.83	2.89	2.12
Tarsus length	3.78	4.41	2.44	3.07	2.84	3.16
Bill length	6.47**	8.13	5.80	7.51	2.86	3.13
Bill depth	4.62	9.89*	3.43	8.55	3.10	4.96
Bill width	4.53	7.76*	2.82	6.83	3.54	3.67

The same families of *G. scandens* and *G. fortis* are used here as in table 1.3. Most phenotypic distributions are weakly skewed, negatively, as a result of hybridization.

Significantly skewed midparent distributions, *$p < .05$) or **$p < .01$). In addition, *G. scandens* offspring distributions of wing length (*) and bill length (**) measurements are skewed.

1.3.7.1 Conspecific gene flow Conspecific gene flow has not been estimated directly but appears to be low and annually variable. In some years immigrants arrive on Daphne Major after the breeding season, presumably from the neighboring large island of Santa Cruz. All but a few birds disappear before or at the start of the following breeding season. For example, in 1977, a year of unsuccessful breeding and little or no molting, several young *G. fortis* in new plumage appeared on the island (Boag and Grant 1984b). Some of them were captured, banded, and measured and were found to have the large dimensions typical of Santa Cruz birds. None stayed to breed in 1978. Large influxes were recorded in 1983–85, during and after an exceptional El Niño event (P. R. Grant and B. R. Grant 1995a). In 1990 all but one *G. scandens* and all *G. fortis* on the island had been banded, and no extra birds lacking bands were on the island (and breeding) in 1991, so in this year we can be certain that no immigration of breeders took place.

Immigrant *G. scandens* are smaller on average than the residents, whereas *G. fortis* immigrants are larger on average than residents. None of the suspected *G. scandens* immigrants have bred on Daphne, but a few of the *G. fortis* immigrants have. Estimating their numbers is difficult because only the largest can be distinguished from the residents. To estimate the minimum number of those that have bred, we have compared the measurements of birds banded as nestlings on the island with those captured and banded as adults from 1981 onward and which subsequently bred (before 1981 the majority of breeding birds had been banded as adults). The adult-banded group contains possible immigrants as well as several residents. The criterion for identifying immigrants is stringent. Immigrants of past generations that bred would have produced large offspring, and these would tend to be the largest members of the resident population. Their effect is to make it less likely that new immigrants can be identified. Immigration (and breeding) may be biased in a size-selective manner, with small birds that resemble the residents staying to breed more frequently than large birds. These, too, would not be identified as immigrants.

None of the females in the adult-banded sample of *G. fortis* ($n = 42$) can be identified as immigrants, from either univariate measurements or PC scores. The four largest birds, lying beyond three standard deviations above the mean PC 1 score, were all born (hatched) on the island. In contrast, the two largest males were banded as adults and may have been immigrants. An ellipse of 3 standard deviations around the joint PC 1 and PC 2 mean encloses all but 3 males of a sample of 41 captured as adults. We consider these three to be probable immigrants.

Addition of the three identified immigrants to the combined sample of the remaining 341 breeding males of *G. fortis* increases phenotypic variances of all traits. The increases range from 4.4% for tarsus length to 20.8% for weight. On one hand, these are minimum estimates of the effects of immigrants because we have not included other probable immigrants. On the other hand, these others, being morphologically indistinguishable from residents, would probably make only a small additional contribution to variances, even if they are large on average.

Genetic effects of immigration are not as large as might be expected from these calculations. Although all three males bred, the offspring of two of them ($n = 7$) failed to survive to breed. The third male (no. 5110) was known to have bred 14 times with 6 females in 11 years, and to have produced 19 fledglings, of which 5 survived to breed. Addition of these families to the composite sample of *G. fortis* families lacking them had trivial effects on heritability estimates. Apart from weight, heritabilities of all traits were affected at the third decimal place, and were lower, not higher, when the immigrant family was included. Similarly, coefficients of additive genetic variation were scarcely affected by inclusion of these birds. Our overall conclusion is that conspecific gene flow into the population of *G. fortis* has little effect on quantitative trait variation in the residents.

1.3.7.2 Hybridization Interbreeding between *G. fortis* and *G. fuliginosa* and between *G. fortis* and *G. scandens* was first detected in 1976 (P. R. Grant and Price 1981, Boag and Grant 1984b, Price et al. 1984a, P. R. Grant 1986). The production and breeding of hybrids and backcrosses has been monitored since then (P. R. Grant and B. R. Grant 1992b, 1997a, 1997b, P. R. Grant 1993). Interbreeding has occurred repeatedly but rarely, involving less than 5% of pairs in any one year (P. R. Grant 1993). Short-term effects of hybrids and backcrosses on phenotypic and additive genetic variances of the six morphological traits have been estimated by first calculating variances for the samples of *G. fortis* and *G. scandens* alone, adding the hybrids and backcrosses to the samples, and then recalculating the variances (P. R. Grant and B. R. Grant 1994). These estimates have been recalculated with slightly expanded sample sizes. Table 1.8 gives the changes in coefficients of variation.

1.3.7.3 Effects on G. scandens Two F_1 families (*G. fortis* × *G. scandens*), five first-generation backcross families, and two second-generation backcross families were added to the total sample of 149 *G. scandens* families lacking known hybrids and backcrosses (i.e., hybrids and backcrosses constituted about 5% of the total, which approximates their incidence in the population). All offspring bred with *G. scandens* and were therefore members of the *G. scandens* population (P. R. Grant and B. R. Grant 1997b).

Effects of hybridization on *G. scandens* are summarized as follows. As expected, coefficients of additive genetic variation of the six traits increased, on average by 15%. Residual variances (environmental and nonadditive genetic) also increased, though generally by a smaller amount, and as a result heritabilities increased to a small extent. *G. scandens* is more variable phenotypically as a result of hybridizing with *G. fortis* than it would be in the absence of hybridization.

Table 1.8 Effects on heritabilities (h^2) and coefficients of variation (CV) of including F_1 hybrids and backcrosses in samples of *G. scandens* and *G. fortis*.

	h^2	CV_P	CV_A	CV_R
G. scandens				
Mass	+33.9	+16.4	+32.2	+4.1
Wing length	+3.1	+15.0	+16.7	+43.0
Tarsus length	+3.3	+10.3	+10.2	+8.5
Bill length	+11.2	+22.4	+29.2	+2.7
Bill depth	-1.4	+3.6	+2.7	+4.8
Bill width	-9.0	+6.2	+1.0	+10.7
G. fortis				
Mass	+19.2	+37.9	+50.6	+18.9
Wing length	+40.0	+30.1	+53.9	+5.9
Tarsus length	+29.3	+15.4	+31.0	+4.8
Bill length	+7.8	+22.1	+26.8	+2.5
Bill depth	+11.0	+26.9	+33.8	+11.4
Bill width	+10.1	+28.4	+34.7	+12.0

See table 1.7 for the inclusive samples. Subscripts to CV are P, phenotypic; A, additive; and R, residual.

If species differ phenotypically because they are genetically constituted differently, then, as a result of hybridization, additive genetic variances should increase the most in those traits for which the species differ most. This is in fact observed; additive genetic variances (see P. R. Grant and B. R. Grant 1994), coefficients of additive genetic variation, and heritabilities increase the most in body mass and bill length (table 1.8). Coefficients increase the least, and heritabilities apparently decrease, in those traits (bill depth and width) for which the species differ least.

The magnitude of the effects of hybridization on heritable variation in *G. scandens* is likely to be underestimated because some of the individuals classified as *G. scandens* at the beginning of the study may have been backcrosses, and possibly F_1 hybrids (P. R. Grant 1993). The more extensive pedigrees of *G. fortis* allow us to effectively eliminate this problem.

1.3.7.4 Effects on G. fortis For a sample of *G. fortis* in the absence of F_1 and backcrosses, we analyzed heritable variation in the 167 families for which all 4 grandparents were known (i.e., banded). To this sample we added all available hybrid families; 12 F_1 families, 13 families involving first-generation backcrosses to *G. fortis*, and another 15 second-generation backcross and other families which involved descendants of interbreeding (see P. R. Grant 1993). Among the "other" category were two families in which one of the grandparents was a *G. fortis* × *G. scandens* F_1 hybrid and the other was a *G. fortis* × *G. fuliginosa* F_1 hybrid. Hybrid and backcross individuals constituted about 20% of the total, which approximates their incidence in the population in recent years. All bred with *G. fortis* or their hybrids; none bred with *G. scandens*.

As in *G. scandens*, the effect of hybridization was to increase phenotypic, additive genetic, and residual variances (P. R. Grant and B. R. Grant 1994). Heritabilities increased as a result of a disproportionate increase in the additive genetic variances. The magnitude of the effect varied among traits. Increases in heritability, for example, were much larger for body mass, wing length, and tarsus length than they were for the three bill traits. On average, heritabilities for these traits increased from about 0.6 to 0.7. Coefficients of additive genetic variation increased by 0.25 to 0.50 (table 1.8). Hybridization had a larger influence on variation in *G. fortis* than in *G. scandens*, partly because the frequency of hybridization was three or four times greater (P. R. Grant 1993), and partly because *G. fortis* receives genes from two markedly different species, whereas *G. scandens* receives genes from only one. The generally small differences between the species in coefficients of residual variation are scarcely affected by the addition of hybrids. This may be the result of environmental variance decreasing and nonadditive genetic variance increasing by about the same amounts.

Effects of hybridization on variances might be slightly underestimated by the inclusion of unidentified second- and third-generation backcrosses in the sample of *G. fortis* families alone. Nonetheless, there might be a small bias in the opposite direction attributable to misidentified paternity (see above). Correcting this bias might then reduce the difference between the slopes of midparent regressions with and without the hybrids. Most of the male parents died before our blood sampling program for paternity identification began, so it is impossible to correct the bias, if it exists. There is no such bias in the samples of hybrids. In fact, slopes of the regressions on male parents are higher than the corresponding slopes of female parent regressions for four of the six traits, though the differences are individually small. We repeated the calculations of the effects of introgression on variances using female parent regressions instead of midparent regressions and obtained almost identical results.

In conclusion, hybridization, unlike conspecific gene flow, substantially increases coefficients of additive genetic variation. Effects are larger in *G. fortis* than in *G. scandens*, and as a result hybridization increases the interspecific difference in additive genetic variation. In terms of figure 1.1 (lower panel), equilibrial values for *G. fortis* variation (V_2 or V_4) are larger than the equilibrial values for *G. scandens* (V_1 or V_3) as a result of a higher level of genetic input: $M+I_2>M+I_1$. Without knowing the selection functions on the two species, it is not possible to be more specific about the difference in variation between the species. However, an indirect argument (below) suggests that the equilibrial levels of variation are V_1 and V_4.

1.3.8 Selection–introgression balance

The absence of hybridization would not eliminate the differences between the species, as can be seen by reducing the coefficients in table 1.7 by the factors given in table 1.8. For example, the coefficient of additive genetic variation of *G. fortis* bill depth in the absence of hybrids and backcrosses (6.39) is almost twice as large as the comparable coefficient of *G. scandens* (3.34). It is also larger than the coefficient of the *G. scandens* sample that includes hybrids (3.43). Thus, hybridization has a short-term effect of increasing variation, but differential introgression in the short term is not large enough to explain why the Daphne *G. fortis* population is more variable than *G. scandens*.

Failure of contemporary hybridization to fully account for the species differences in trait variation leaves open several possibilities. The remaining difference may simply reflect the greater cumulative (storage) effects of past hybridization in *G. fortis* than in *G. scandens*. Our assumption of equal mutation rates may be wrong; those rates may be higher in the more variable *G. fortis* than in the less variable *G. scandens* as a result of hybridization (e.g., P. R.

Grant and B. R. Grant 1994). Another contributing factor might be drift. Effects of drift are likely to be more pronounced in *G. scandens*, the smaller population, than in *G. fortis* (Price et al. 1984a). Perhaps most important, effects of stabilizing selection on a given level of variation may be weaker in *G. fortis* (S_2 in fig. 1.1) than in *G. scandens* (S_1) (P. R. Grant and Price 1981), in which case the equilibrial coefficient of *G. fortis* in figure 1.1 is V_4 and the equilibrial coefficient of *G. scandens* is V_1. Removal of the short-term effects of hybridization reduces the variation of *G. fortis* from V_4 to a lower value, perhaps in the vicinity of V_3, but above the equilibrial level determined by mutation alone as a result of storage effects from past hybridization. This lower value is still much larger than the equilibrial value of *G. scandens* when it receives genetic input from introgression. Therefore, the *G. scandens* equilibrial value is more likely to be V_1 than V_3, indicating a stronger regime of selection at a given level of variation on this species than on *G. fortis*.

Relaxed selection on ecological generalists with broad niches such as *G. fortis* was the theme of an earlier analysis of relative variation (Rothstein 1973a, 1973b), but hybridization complicates the picture. If there is a balance between introgression (and mutation) and selection (and drift), the high introgression to the *G. fortis* population must be balanced by a high selective depletion of the variation at equilibrium. Under these conditions it is the generalist *G. fortis* and not the specialist *G. scandens* that should experience the greater effects of selection. But is there a balance between genetic input and output in either species?

As shown by the data in table 1.9 (see also fig. 1.6) coefficients of additive genetic variation have been approximately stable over 15 years. An apparent decrease in *G. fortis* variation early in the study coincided with a strong directional selection event in 1976–77 when fitness was close to being an exponential function of beak size and body size (Boag and Grant 1984b, Price and Boag 1987, Schluter 1988). An apparent increase in variation after 1984 coincided with recruitment of hybrids and backcrosses (P. R. Grant and B. R. Grant 1992b, B. R. Grant and P. R. Grant 1993), even though this followed another, but weaker, directional selection event (Gibbs and Grant 1987). The apparent changes are not statistically significant, even though possibly biologically real. It is clear that coefficients of additive genetic variation remain consistently higher than those of *G. scandens*. Taken together, these results indicate an approximate balance between genetic input and output in each of the two populations. The balance is struck at different levels in the two species. This implies that the more variable *G. fortis* not only receives genetic input at a higher rate than *G. scandens* but that it loses genetic input at a higher rate than *G. scandens*.

1.4 Evolution

Darwin's finches are model organisms for the study of two evolutionary phenomena: the maintenance of quantitative variation in populations and the adaptive diversification of species (P. R. Grant and Price 1981, Price et al. 1984a, P. R. Grant 1994, P. R. Grant and B. R. Grant 1996b, 1997c). One provides the basis for the other. Additive genetic variation is the raw material from which new species are formed. We link the two with some brief remarks on the evolutionary potential of populations that differ in levels of additive genetic variation and covariation.

The transformation of *G. fortis* to either *G. scandens*, *G. magnirostris,* or *G. fuliginosa* has been reconstructed by using genetic variances and covariances for *G. fortis* and the

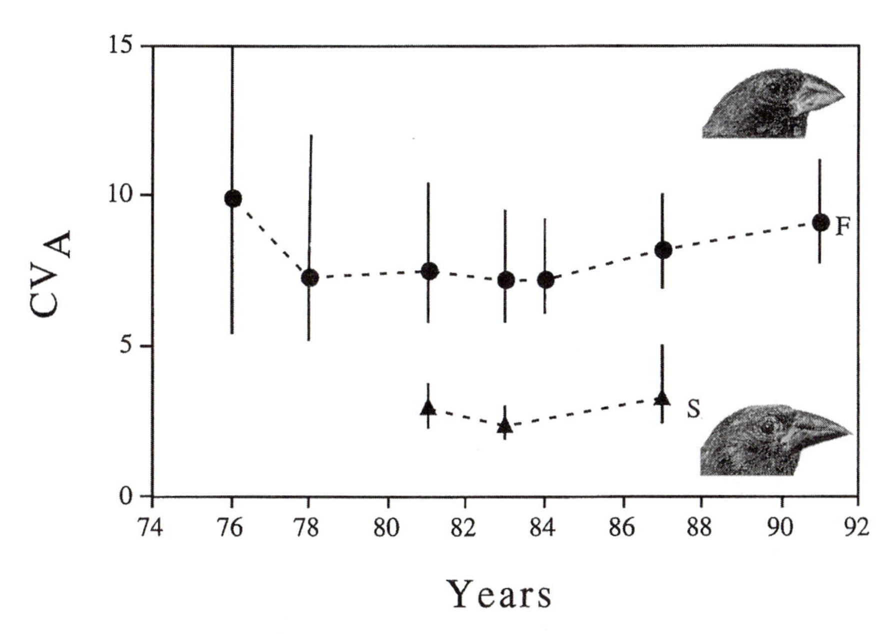

Figure 1.6 Equilibrial genetic variation in bill depth for *G. fortis* (F) and *G. scandens* (S). Mean coefficients of additive genetic variation are shown with 95% confidence limits. Confidence intervals are asymmetrical. Asymmetry is most pronounced where sample size of families is < 25, as in the case of *G. fortis* in 1976 (22 families, upper confidence limit = 34.3).

known morphological differences between *G. fortis* and each of the other species (Price et al. 1984a). With these genetic and phenotypic values, the minimum cumulative or net forces of selection required to effect the transformation can be estimated (Lande 1979, Price et al. 1984a, Schluter 1984). Genetic covariances and correlations between *G. fortis* traits are uniformly strong and positive. This implies (P. R. Grant 1981b, 1983), and analysis shows (Price et al. 1984a), that transformations involving primarily overall size require less selection than those involving primarily shape; from the *G. fortis* starting point the *G. magnirostris* and *G. fuliginosa* morphologies can be reached more easily than can the *G. scandens* morphology. Size transformation occurs readily because there is a large amount of genetic variance for size. Shape changes, in contrast, tend to be opposed by the strong genetic correlations between traits, and in this sense encounter genetic resistance (Schluter 1984, 1996). In other species, negative genetic correlations between traits, even weak ones, can act as a more severe hindrance to evolutionary increases or decreases in both traits (Merilä et al. 1994, Björklund 1996).

We present the results of a new analysis in three steps to make the same basic points. The approach is to find the residual shape change in standard deviation units remaining after size changes have been accomplished in the transformation of one species to another. A low value of residual shape change implies that the transition from one species to another has been largely accomplished by selection on size alone. A high value implies that selection on shape itself is needed. In this we are not attempting to reproduce the actual transitions in both size

Table 1.9 Coefficients of additive genetic variation (CV_A) ± standard error (SE) (see Sokal and Rohlf 1981)

	1976	1978	1981	1983	1984	1987	1991
				G. fortis			
Mass	9.44 ± 2.03	7.67 ± 1.68	8.22 ± 1.16	4.64 ± 0.58	6.93 ± 0.86	10.30 ± 1.04	10.68 ± 1.09
Wing length	2.21 ± 0.47	2.29 ± 0.50	2.23 ± 0.31	1.81 ± 0.23	2.19 ± 0.27	2.90 ± 0.29	2.69 ± 0.27
Tarsus length	2.38 ± 0.51	3.42 ± 0.75	2.34 ± 0.33	2.12 ± 0.27	3.17 ± 0.39	3.39 ± 0.34	2.94 ± 0.30
Bill length	6.83 ± 1.46	5.18 ± 1.13	6.04 ± 0.85	5.54 ± 0.69	5.93 ± 0.73	7.56 ± 0.76	7.85 ± 0.79
Bill depth	9.96 ± 2.14	7.29 ± 1.60	7.51 ± 1.06	7.22 ± 0.91	7.22 ± 0.89	8.20 ± 0.82	9.13 ± 0.93
Bill width	7.76 ± 1.66	5.61 ± 1.23	6.08 ± 0.85	5.51 ± 0.69	5.02 ± 0.62	6.61 ± 0.66	7.35 ± 0.74
				G. scandens			
Mass	–	–	4.03 ± 0.66	4.72 ± 0.57	–	7.60 ± 1.53	–
Wing length	–	–	2.08 ± 0.34	1.90 ± 0.23	–	1.45 ± 0.29	–
Tarsus length	–	–	2.39 ± 0.39	2.58 ± 0.31	–	3.40 ± 0.68	–
Bill length	–	–	4.43 ± 0.73	4.35 ± 0.53	–	4.81 ± 0.96	–
Bill depth	–	–	2.98 ± 0.49	2.38 ± 0.29	–	3.27 ± 0.65	–
Bill width	–	–	2.75 ± 0.45	2.32 ± 0.28	–	2.23 ± 0.45	–

and shape, nor are we using a known, statistically well-supported phylogeny, which was not available (Stern and Grant 1996) when this chapter was written (but see Petren et al. 1999b). Rather, we use the calculations to illustrate the fact that some morphological transformations are much more easily accomplished than others, perhaps more likely to have occurred than others (Schluter 1996), and possibly to have occurred in less time. In doing so we assume that the genetic covariance matrix changes relatively little during the transformations. This may not be correct if selection oscillates in direction in a more extreme manner than we have observed (Gibbs and Grant 1987, B. R. Grant and P. R. Grant 1989b) and is truncational (Shaw et al. 1995).

The three core species for which we have genetic and phenotypic data are *G. fortis*, *G. scandens,* and *G. conirostris*. Using equal numbers of families for a correlation-based principal components analysis (table 1.10), we characterize structural size variation among pairs of these species as the first component. Between 63.0% and 84.9% of the total variation is accounted for in each analysis by PC1, and in all analyses the six morphological traits have approximately equal loadings, which justifies using it as a measure of overall size. The second principal component accounts for most of the remaining variation, 7.8–25.7%. In each analysis, loadings on PC 2 for bill depth and bill width are large and positive, whereas the loading for bill length is negative, and loadings for the other traits are generally small. Therefore, PC 2 is mainly a beak-shape factor, varying from pointed to blunt (see Boag 1983, B. R. Grant and P. R. Grant 1989a, for results of similar analyses).

Next we calculate the heritabilities of PC1 and PC2 scores for *G. fortis*, *G. scandens,* and *G. conirostris* and genetic correlations between them (table 1.10). Even though the two axes of variation are uncorrelated in the combined data of species pairs, the individual species do sometimes display positive correlations. Table 1.10 gives the interesting result that heritable variation for the size and shape factors and the additive genetic correlation between them vary strongly among the species. Third, we calculate the net forces of selection on the size factor (alone) involved in each of the six possible (reciprocal) species transformations using

Table 1.10 Heritability (h^2) of size (PC1) and shape (PC2) factors and the genetic correlation (r_A) between them

	PC1		PC2		
Species	h^2	SE	h^2	SE	r_A
G. scandens	0.457	0.191	0.429	0.197	-0.382
G. fortis	1.037	0.153	0.797	0.156	-0.095
G. conirostris	0.407	0.122	0.452	0.121	-0.690
G. fortis	1.024	0.151	0.678	0.151	-0.903
G. scandens	0.480	0.196	0.983	0.245	0.802
G. conirostris	0.704	0.204	0.686	0.157	0.242
G. fortis	1.034	0.155	0.273	0.174	0.184
G. fuliginosa					
G. magnirostris					

In the first 3 analyses, 33 or 34 families of each species were included: father, mother, and a single offspring. In the last analysis the same families of *G. fortis* were included, together with 100 *G. fuliginosa* and 100 *G. magnirostris* individuals from Daphne Major in lieu of families.

these genetic parameters and the phenotypic distances in standard deviation units of principal components scores. Shape change arising from selection solely on size can now be calculated from these measures of selection and genetic variation and covariation; it is the product of the net forces of selection on size, the square root of the heritability of size, and the genetic correlation between size and shape (Lande 1979, Price et al. 1984a, Schluter 1984). Unidirectional transformations to *G. fuliginosa* and *G. magnirostris* can be reconstructed by using *G. fortis*, the most similar species, as the starting point.

Figure 1.7 illustrates the transformations. Three points can be made. First, transformation of species differing little in shape is accomplished by selection on a size factor alone in two ways. Either there is little heritable variation in the shape factor and a low genetic correlation between size and shape, and hence no correlated effect of selection on size (e.g., *G. fortis*, *G. fuliginosa,* and *G. magnirostris*), or the indirect component of selection closes the gap in shape between the species' morphologies, even in the absence of selection on shape (*G. conirostris* and *G. scandens*). The negative sign on the residual shape variation for the *G. scandens* to *G. conirostris* transition indicates that the correlated effect on shape has been too strong. Genetic facilitation rather than genetic resistance has caused an overshoot of the shape target.

Second, species that differ most in shape are the least interconvertible. Values marked with an asterisk in the bottom figure show those cases where selection on size results in a correlated change in shape in the opposite direction, so that the difference between species in shape becomes magnified as selection decreases the difference in size. This raises the possibility that in these cases one extant species did not give rise to another; instead, a nonintermediate common ancestor gave rise to both. If the ancestor was either smaller or larger than both of the extant species and differed from both in shape in the same way, each transformation could have proceeded with little genetic resistance. Suppose, for example, that the common ancestor of *G. fortis* and *G. conirostris* was smaller and with a more pointed beak than both (see fig. 1.7, upper left PC panel) and with the same axis of covariation. Each of the extant species could be derived from the ancestor by a change in both size and shape, with only a small departure from the axis of covariation, the magnitude depending on the particular ancestral starting point.

Third, within a pair of species, one direction of change is easier than the reciprocal change. Thus a transition from *G. scandens* to *G. fortis* results in a smaller negative effect on shape than does a transition from *G. fortis* to *G. scandens*. The difference arises from the lower heritability of shape and the lower genetic correlation between size and shape in *G. scandens* than in *G. fortis*. In other words, genetic resistance is weaker in the former case.

These results have an obvious bearing on phylogenetic questions. Schluter (1984) performed a selection gradient analysis to estimate phylogenetic relationships among all of the Darwin's finch species. He used minimum selective distances between species, based on the *G. fortis* genetic variance–covariance matrix and assuming that the shorter the distance, between two species the more closely related they are. The assumption of similar genetic structure among species now has a stronger foundation: genetic matrices for *G. fortis*, *G. scandens* (P. R. Grant and B. R. Grant 1994), and *G. conirostris* (B. R. Grant and P. R. Grant 1989a) are similar; moreover, the genetic matrix for *G. fortis* has remained largely unchanged from 1976 to 1991 (unpublished observations). The phylogenetic relationships bore a strong resemblance to the original, nonquantitative phylogenetic tree constructed by Lack (1947) on the basis of plumage and size and shape variation and to another tree based

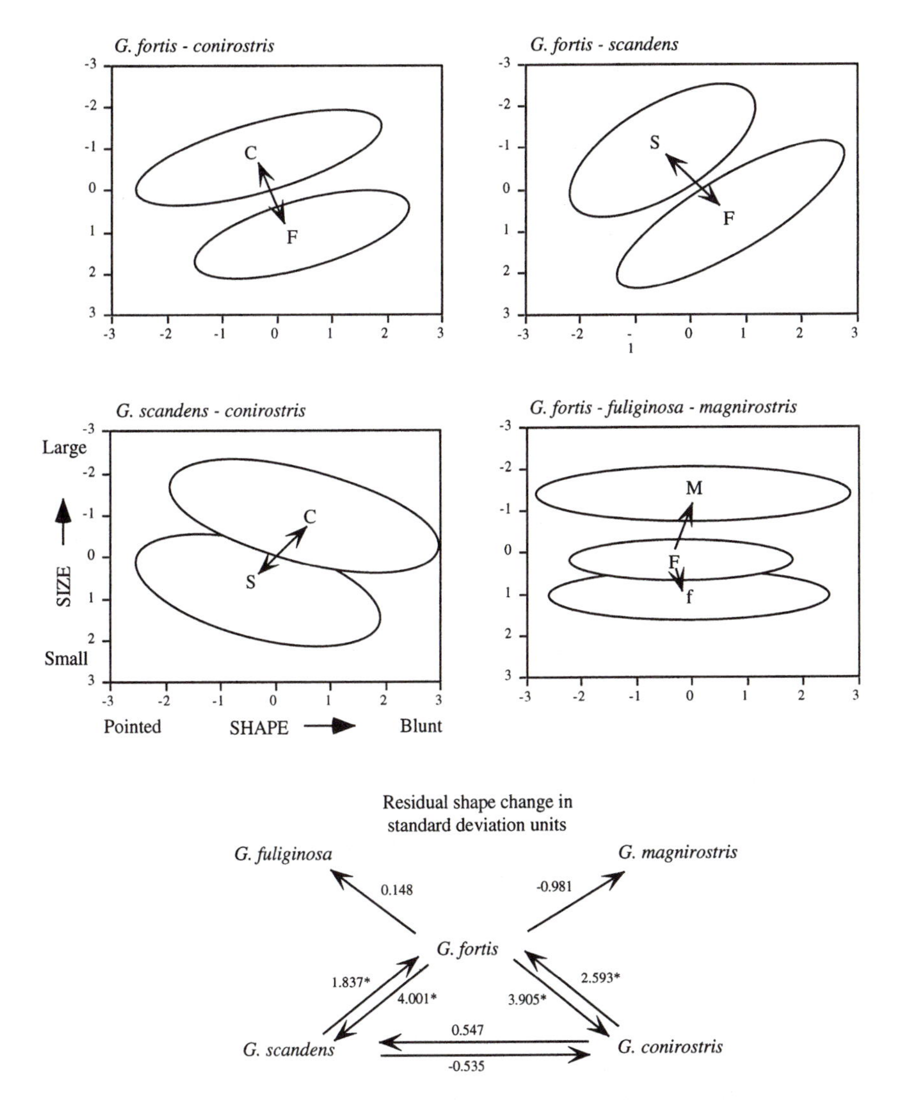

Figure 1.7 Evolutionary transitions along body size (PC1) and beak shape (PC2) axes (top), and residual shape changes in standard deviation units after the size changes alone have been effected (bottom). A negative sign indicates that change in shape associated with a change in size has gone too far, and residual shape change must be in the reverse direction. An asterisk indicates that the difference in shape between two species has increased as a result of a change in size. In upper principal components plots, note the reversal of the y-axis; as a result, bivariate associations are of opposite sign to the correlations in table 1.10.

on allozyme variation (Yang and Patton 1981). These last two trees, in turn, have been shown to resemble a tree based on microsatellite DNA variation (Petren et al. 1999b). Similarities provide support for the assumption that minimum selective distances between species reflect genealogical relationships.

The main implication of these findings is that important aspects of the evolutionary history of a group of species can be recovered from a knowledge of phenotypic and genetic variation in adaptive, quantitative characters.

1.5 Generality?

We have taken advantage of a convenient situation in the Galápagos Islands to investigate causes of quantitative variation. Some populations of Darwin's Finches are shown to be genetically more variable than others, partly as a result of different levels of introgressive hybridization without reduced fitness, and partly as a result of different, niche-based selection regimes. The question arises as to whether the results of a study of one group of organisms in a single location are broadly generalizable. We think they are generalizable for three main reasons.

First, Darwin's Finches are not unique in being highly variable. Natural populations vary in phenotypic (Yablakov 1974) and quantitative genetic variation (Mousseau and Roff 1987, Pomiankowski and Møller 1995), sometimes to a strong degree, and the range of variation encompasses the variation exhibited by Darwin's finch populations. For example, Pomiankowski and Møller (1995) found coefficients of additive genetic variation for naturally selected, quantitative traits to vary from 0.77 to 14.23 in a survey of 22 characters of 19 species (cf. tables 1.7 and 1.9). An even larger range of variation of coefficients of additive genetic variation, 0.13–90.50, was found for 37 secondary sexual traits in 30 species. In general, these coefficients are higher than those for naturally selected traits, although not all coefficients are comparable because the traits are scaled differently. Reasons for the difference between the two sets of coefficients have been debated (Pomiankowski and Møller 1995, Rowe and Houle 1996). The question of why each of the groups of coefficients considered separately should vary so much has not been addressed. The concept of a balance between elevating and depleting forces (fig. 1.1) could provide a useful starting point to investigate the role of special factors such as biased mutation (Pomiankowski and Møller 1995) and genetic or nongenetic condition dependence of the traits (Rowe and Houle 1996).

Second, natural selection on ecologically significant traits in nature has been reported numerous times in a wide variety of animal and plant taxa (e.g., Endler 1986). Most reported cases of natural selection are directional, as in our study of Darwin's finches. Directional selection usually reduces the phenotypic variance (Bulmer 1971a, Schnol and Kondrashov 1993). Stabilizing selection, an important ingredient of MS-ID theory, has been detected in nature comparatively rarely, and diversifying selection even more rarely, perhaps as a result of detection difficulties.

The third reason for believing that our results are generalizable is that populations in nature are open to immigration to varying degrees. The exchange of genes between neighboring conspecific populations is widespread and noncontroversial. Genetic effects are rarely measured directly, however. They are usually inferred from population genetic structure (e.g., Slatkin 1987), and the estimates of these effects vary broadly (Neigel 1997). Exchange of genes between species is less well known, although it is potentially more important in elevating levels of additive genetic variation within populations. The increasing number of studies reporting introgressive hybridization in a wide variety of animal taxa (P. R. Grant 1998) is showing that its evolutionary importance has been underestimated. It would be worth examining the role of introgressive hybridization in the maintenance of variation in sexually selected traits.

Acknowledgments We thank the Galápagos National Parks Service and the Charles Darwin Research Station for administrative and logistical support of our research on the Galapagos Islands, and the National Science and Engineering Research Council of Canada and the National Science Foundation of the USA for financial support. Lukas Keller, Trevor Price, Barry Sinervo, Dolph Schluter, and an anonymous reviewer gave valuable comments on the manuscript.

References

Alatalo, R. V., and A. Lundberg. 1984. High frequency of cuckoldry in pied and collared flycatchers. Oikos 42:41–47.

Alatalo, R. V., and A. Lundberg. 1986. Heritability and selection on tarsus length in the pied flycatcher (*Ficedula hypoleuca*). Evolution 40:574–583.

Barton, N. H. 1990. Pleiotropic models of quantitative variation. Genetics 124:773–782.

Barton, N. H., and M. Turelli 1989. Evolutionary quantitative genetics: how little do we know? Annu. Rev. Genet. 23:337–370.

Benkman, C. W. 1993. Adaptation to single resources and the evolution of crossbill (*Loxia*) diversity. Ecol. Monogr. 63:305–325.

Benkman, C. W., and R. E. Miller. 1996. Morphological evolution in response to fluctuating selection. Evolution 50:2499–2504.

Björklund, M. 1996. The importance of evolutionary constraints in ecological time scales. Evol. Ecol. 10:423–431.

Blows, M. W., and M. B. Sokolowski. 1995. The expression of additive and nonadditive genetic variation under stress. Genetics 140:1149–1159.

Boag, P. T. 1983. The heritability of external morphology in Darwin's ground finches (*Geospiza*) on Isla Daphne Major, Galápagos. Evolution 37:877–894.

Boag, P. T. 1987. Effects of nestling diet on growth and adult size of zebra finches (*Poephila guttata*). Auk 104:155–166.

Boag, P. T., and P. R. Grant. 1978. Heritability of external morphology in Darwin's finches. Nature 274:793–794.

Boag, P. T., and P. R. Grant. 1984a. Darwin's finches (*Geospiza*) on Isla Daphne Major, Galápagos: breeding and feeding ecology in a climatically variable environment. Ecol. Monogr. 54:463–489.

Boag, P. T., and P. R. Grant. 1984b. The classical case of character release: Darwin's finches (*Geospiza*) on Isla Daphne Major, Galápagos. Biol. J. Linn. Soc. 22:243–287.

Boag, P. T., and A. J. van Noordwijk. 1987. Quantitative genetics. Pp. 45–78 *in* F. Cooke and P. A. Buckley, eds., Avian Genetics. Academic Press, New York.

Bowman, R. I. 1961. Morphological differentiation and adaptation in the Galápagos finches. Univ. Calif. Publ. Zool. 58:1–302.

Bryant, E. H., S. A. McCommas, and L. M. Coombs. 1986. The effect of an experimental bottleneck upon quantitative genetic variation in the housefly. Genetics 114:1191–2011.

Bryant, E. H., and L. Meffert. 1995. An analysis of selectional response in relation to a population bottleneck. Evolution 49:626–634.

Bryant, E. H., and L. Meffert. 1996. Nonadditive genetic structuring of morphometric variation in relation to a population bottleneck. Heredity 77:168–176.

Bulmer, M. G. 1971a. The effect of selection on genetic variability. Am. Nat. 105:201–211.

Bulmer, M. G. 1971b. Stable equilibria under the migration matrix model. Heredity 27:419–430.

Bulmer, M. G. 1972. The genetic variability of polygenic characters under optimizing selection, mutation and drift. Genet. Res., Camb. 18:17–25.

Bulmer, M. G. 1989. Maintenance of genetic variability by mutation-selection balance: a child's guide through the jungle. Genome 31:761–767.

Caballero, A., and P. D. Keightley. 1994. A pleiotropic nonadditive model of variation in quantitative traits. Genetics 138:883–900.

Caballero, A., P. D. Keightley, and W. G. Hill. 1995. Accumulation of mutations affecting body weight in inbred mouse lines. Genet. Res., Camb. 65:145–149.

Cain, A. J., and P. M. Sheppard. 1954. Natural selection in *Cepaea*. Genetics 39:89–116.

Cheverud, J. M., and E. J. Routman. 1996. Epistasis as a source of increased additive genetic variance at population bottlenecks. Evolution 50:1042–1051.

Dhondt, A. A. 1982. Heritability of blue tit tarsus length from normal and cross-fostered broods. Evolution 36:418–419.

Dhondt, A. A. 1991. How unreliable are cuckoldry estimates using heritability analyses? Ibis 133:91–93.

Diaz, M. 1994. Variability in seed size selection by granivorous passerines: effects of bird size, bird size variability, and ecological plasticity. Oecologia 99:1–6.

Dixon, A., D. Ross, S. L. C. O'Malley, and T. Burke. 1994. Paternal investment inversely related to degree of extra-pair paternity in the reed bunting. Nature 371:698–700.

Doebeli, M. 1996. An explicit genetic model for ecological character displacement. Ecology 77:510–520.

Ellner, S., and N. G. Hairston, Jr. 1994. Role of overlapping generations in maintaining genetic variation in a fluctuating environment. Am. Nat. 143:403–417.

Endler, J. A. 1986. Natural Selection in the Wild. Princeton University Press, Princeton, N.J.

Falconer, D. 1989. Introduction to Quantitative Genetics, 3rd ed. Longman, New York.

Feinsinger, P., and L. A. Swarm. 1982. "Ecological release," seasonal variation in food supply, and the hummingbird *Amazilia tobaci* on Trinidad and Tobago. Ecology 63:1574–1587.

Fitzpatrick, S. 1997. Patterns of morphometric variation in birds' tails: length, shape and variability. Biol. J. Linn. Soc. 62:145–162.

Gavrilets, S., and A. Hastings. 1995. Dynamics of polygenic variability under stabilizing selection, recombination, and drift. Genet. Res., Camb. 65:63–74.

Gebhardt-Henrich, S. G., and H. L. Marks. 1993. Heritabilities of growth curve parameters and age-specific expression of genetic variation under two different feeding regimes in Japanese quail (*Coturnix coturnix japonica*). Genet. Res., Camb. 62:45–56.

Gebhardt-Henrich, S. G., and R. G. Nager. 1991. Can extra-pair fertilizations be detected by differences in maternal and paternal heritability estimates? Ibis 133:93–94.

Gebhardt-Henrich, S. G., and A. J. van Noordwijk. 1991. The genetical ecology of nestling growth in the great tit. I. Heritability estimates under different environmental conditions. J. Evol. Biol. 3:341–362.

Gibbs, H. L., and P. R. Grant. 1987. Oscillating selection on Darwin's finches. Nature 327: 511–513.

Gillespie, J. H., and M. Turelli. 1989. Genotype-environment interactions and the maintenance of polygenic variation. Genetics 121:129–138.

Gimmelfarb, A., and J. H. Willis. 1994. Linearity versus nonlinearity of offspring-parent regressions: an experimental study of *Drosophila melanogaster*. Genetics 138:-343–352.
Goodnight, C. J. 1987. On the effect of founder events on epistatic genetic variance. Evolution 41:80–91.
Goodnight, C. J. 1988. Epistasis and the effect of founder events on the additive genetic variance. Evolution 42:441–454.
Grant, B. R., and Grant, P. R. 1989a. Evolutionary Dynamics of a Natural Population: The Large Cactus Finch of the Galápagos. University of Chicago Press, Chicago.
Grant, B. R., and Grant, P. R. 1989b. Natural selection in a population of Darwin's finches. Am. Nat. 133:377–393.
Grant, B. R., and Grant, P. R. 1993. Evolution of Darwin's finches caused by a rare climatic event. Proc. R. Soc. Lond. B 251:111–117.
Grant, B. R., and P. R. Grant. 1996a. High survival of Darwin's finch hybrids: effects of beak morphology and diets. Ecology 77:500–509.
Grant, B. R., and P. R. Grant. 1996b. Cultural inheritance of song and its role in the evolution of Darwin's finches. Evolution 50:2471–2487.
Grant, B. R., and P. R. Grant. 1998. Hybridization and speciation in Darwin's finches: the role of sexual imprinting on a culturally transmitted trait. Pp. 404–422 *in* D. J. Howard and S. H. Berlocher, eds., Endless Forms: Species and Speciation. Oxford University Press, New York.
Grant, P. R. 1979a. Ecological and morphological variation of Canary Island blue tits, *Parus caeruleus* (Aves:Paridae). Biol. J. Linn. Soc. 11:103–129.
Grant, P. R. 1979b. Evolution of the chaffinch, *Fringilla coelebs*, on the Atlantic Islands. Biol. J. Linn. Soc. 11:301–332.
Grant, P. R. 1981a. The feeding of Darwin's finches on *Tribulus cistoides* (L.) seeds. Anim. Behav. 29:785–793.
Grant, P. R. 1981b. Patterns of growth in Darwin's finches. Proc. R. Soc. Lond. B 212: 403–432.
Grant, P. R. 1983. Inheritance of size and shape in a population of Darwin's finches, *Geospiza conirostris*. Proc. R. Soc. Lond. B 220:219–236.
Grant, P. R. 1986. Ecology and Evolution of Darwin's Finches. Princeton University Press, Princeton, N.J.
Grant, P. R. 1993. Hybridization of Darwin's finches on Isla Daphne Major, Galápagos. Phil. Trans. R. Soc. Lond. B 340:127–139.
Grant, P. R. 1994. Population variation and hybridization: comparison of finches from two archipelagos. Evol. Ecol. 8:598–617.
Grant, P. R. 1998. Speciation. Pp. 83–101 *in* P. R. Grant, ed., Evolution on Islands. Oxford University Press, Oxford.
Grant, P. R., I. Abbott, D. Schluter, R. L. Curry, and L. K. Abbott. 1985. Variation in the size and shape of Darwin's finches. Biol. J. Linn. Soc. 25:1–39.
Grant, P. R., and B. R. Grant. 1992a. Demography and the genetically effective sizes of two populations of Darwin's finches. Ecology 73:766–784.
Grant, P. R., and B. R. Grant. 1992b. Hybridization of bird species. Science 256:193–197.
Grant, P. R., and B. R. Grant. 1994. Phenotypic and genetic effects of hybridization in Darwin's finches. Evolution 48:297–316.
Grant, P. R., and B. R. Grant. 1995a. The founding of a new population of Darwin's finches. Evolution 49:229–240.
Grant, P. R., and B. R. Grant. 1995b. Predicting microevolutionary responses to directional selection on heritable variation. Evolution 49:241–251.

Grant, P. R., and B. R. Grant. 1996a. Finch communities in a climatically fluctuating environment. Pp. 343–390 *in* M. L. Cody and J. A. Smallwood, eds., Long-term Studies of Vertebrate Communities. Academic Press, New York.

Grant, P. R., and Grant, B. R. 1996b. Speciation and hybridization of island birds. Phil. Trans. R. Soc. Lond. B 351:765–772.

Grant, P. R., and B. R. Grant. 1997a. Hybridization, sexual imprinting and mate choice. Am. Nat. 149:1–28.

Grant, P. R., and B. R. Grant. 1997b. Mating patterns of Darwin's finch hybrids determined by song and morphology. Biol. J. Linn. Soc. 60:317–343.

Grant, P. R., and Grant, B. R. 1997c. Genetics and the origin of bird species. Proc. Natl. Acad. Sci. USA 94:7768–7775.

Grant, P. R., B. R. Grant, J. N. M. Smith, I. J. Abbott, and L. K. Abbott. 1976. Darwin's finches: population variation and natural selection. Proc. Natl. Acad. Sci. USA 73:257–261.

Grant, P. R., and T. D. Price. 1981. Population variation in continuously varying traits as an ecological genetics problem. Am. Zool. 21:795–811.

Gustafsson, L., and J. Merilä. 1994. Foster parent experiment reveals no genotype-environment correlation in the external morphology of *Ficedula albicollis*, the collared flycatcher. Heredity 73:124–129.

Hartley, I. R., M. Shepherd, T. Robson, and T. Burke. 1993. Reproductive success of polygynous male corn buntings (*Miliaria calandra*) as confirmed by DNA fingerprinting. Behav. Ecol. 4:310–317.

Hasselquist, D., S. Bensch, and T. von Schantz. 1995. Estimating cuckoldry in birds: the heritability method and DNA fingerprinting give different results. Oikos 72:173–178.

Hõrak, P., and T. Tammaru. 1996. Between-year variation in breeding conditions biases heritability estimates for body size in birds. Ardea 84:127–135.

Houle, D. 1989. The maintenance of polygenic variation in finite populations. Evolution 43:1767–1780.

Houle, D. 1992. Comparing evolvability and variability of quantitative traits. Genetics 130:195–204.

Johannesson, K., E. Rolán-Alvarez, and J. Erlandsson. 1997. Growth rate differences between upper and lower shore ecotypes of the marine snail *Littorina saxatalis* (Olivi) (Gastropda). Biol. J. Linn. Soc. 61:267–279.

Keightley, P. D., T. F. C. Mackay, and A. Caballero. 1993. Accounting for bias in the estimates of the rate of polygenic mutation. Proc. R. Soc. Lond. B 253:291–296.

Keightley, P. D., and W. G. Hill. 1990. Variation maintained in quantitative traits with mutation-selection balance: pleiotropic side-effects on fitness traits. Proc. R. Soc. Lond. B 242:95–100.

Kondrashov, A. S., and M. Turelli. 1992. Deleterious mutations, apparent stabilizing selection and the maintenance of quantitative variation. Genetics 132:603–618.

Kimura, M. 1965. A stochastic model concerning the maintenance of genetic variability in quantitative characters. Proc. Natl. Acad. Sci. USA 54:731–736.

Lack, D. 1945. The Galápagos finches (Gepospizinae): a study in variation. Occas. Pap. Cal. Acad. Sci. 21:1–159.

Lack, D. 1947. Darwin's Finches. Cambridge University Press, Cambridge.

Lande, R. 1976. The maintenance of genetic variability by mutation in a polygenic character with linked loci. Genet. Res., Camb. 26:221–235.

Lande, R. 1979. Quantitative genetic analysis of multivariate evolution, applied to brain:body size allometry. Evolution 33:402–416.

Lande, R., and S. J. Arnold. 1983. The measurement of selection on correlated characters. Evolution 37:1210–1226.

Larsson, K. 1993. Inheritance of body size in the barnacle goose under different environmental conditions. J. Evol. Biol. 6:195–208.

Levene, H. 1953. Genetic equilibrium when more than one ecological niche is available. Am. Nat. 87:331–333.

Levin, D. A., and H. W. Kerster. 1967. Natural selection for reproductive isolation in *Phlox*. Evolution 21:679–687.

Ludwig, W. 1950. Zur theorie der konkurrenz. Neue Ergebn. Prob. Zool., Klatt-Festschrift: 516–537.

Lynch, M., and W. G. Hill. 1986. Phenotypic evolution by neutral mutation. Evolution 40:915–935.

Mackay, T. F. C. 1981. Genetic variation in varying environments, Genet. Res., Camb. 37:79–93.

Mackay, T. F. C., R. F. Lyman, and W. G. Hill. 1995. Polygenic mutation in *Drosophila melanogaster*: non-linear divergence among unselected lines. Genetics 139:849–859.

Merilä, J. 1997. Expression of genetic variation in body size of the collared flycatcher under different environmental conditions. Evolution 51:526–536.

Merilä, J., M. Björklund, and L. Gustafsson. 1994. Evolution of morphological differences with moderate genetic correlations among traits as exemplified by two flycatcher species (*Ficedula*: Muscicapidae). Biol. J. Linn. Soc. 52:19–30.

Merilä, J., and L. Gustafsson. 1996. Temporal stability and micro-geographic homogeneity of heritability estimates in a natural bird population. J. Hered. 87:199–204.

Millington, S. J., and P. R. Grant. 1983. Feeding ecology and territoriality of the cactus finch *Geospiza scandens* on Isla Daphne Major, Galápagos. Oecologia 58:76–83.

Møller, A. P. 1987. Behavioral aspects of sperm competition in swallows (*Hirundo rustica*). Behaviour 100:92–104.

Møller, A. P., and T. R. Birkhead. 1992. Validation of the heritability method to estimate extra-pair paternity in birds. Oikos 64:485–488.

Mousseau, T. A., and D. A. Roff. 1987. Natural selection and the heritability of fitness components. Heredity 59:181–197.

Neigel, J. E. 1997. A comparison of alternative strategies for estimating gene flow from genetic markers. Annu. Rev. Ecol. Syst. 28:105–128.

Payne, R. B., and L. L. Payne. 1989. Heritability estimates and behaviour observations: extra-pair matings in indigo buntings. Anim. Behav. 38:457–467.

Petren, K, B. R. Grant, and P. R. Grant. 1999a. Low extrapair paternity in the cactus finch (*Geospiza scandens*). Auk 116: 252–256.

Petren, K, B. R. Grant, and P. R. Grant. 1999b. A phylogeny of Darwin's finches based on microsatellite DNA length variation. Proc. Roy. Soc. Lond. B. 266: 321–329.

Phillips, P. C. 1996. Maintenance of polygenic variation via a migration-selection balance under uniform selection. Evolution 50:1334–1339.

Polis, G. A. 1984. Age structure component of niche width and intraspecific resource partitioning: can age groups function as ecological species? Am. Nat. 123:541–564.

Pomiankowski, A., and A. P. Møller. 1995. A resolution of the lek paradox. Proc. R. Soc. Lond. B 260:21–29.

Price, T. D. 1985. Reproductive responses to varying food supply in a population of Darwin's finches: clutch size, growth rates and hatching asynchrony. Oecologia 66:411–416.

Price, T. D. 1987. Diet variation in a population of Darwin's finches. Ecology 68: 1015–1028.

Price, T. D. 1991. Environmental and genotype-by-environment influences on chick size in the Yellow-browed leaf warbler *Phylloscopus inornatus*. Oecologia 86:535–541.

Price, T. D. 1998. Maternal and paternal effects in birds: effects on offspring fitness. Pp.

202–226 *in* T. A. Mousseau and C. W. Fox, eds., Maternal Effects as Adaptations. Oxford University Press, New York.

Price, T. D., and P. T. Boag. 1987. Selection in natural populations of birds. Pp. 257–287 *in* F. Cooke and P. A. Buckley, eds., Avian Genetics. Academic Press, New York.

Price, T. D., and P. R. Grant. 1984. Life history traits and natural selection for small body size in a population of Darwin's finches. Evolution 38:483–494.

Price, T. D., P. R. Grant, and P. T. Boag. 1984a. Genetic changes in the morphological differentiation of Darwin's ground finches. Pp. 49–66 *in* K. Wöhrmann and V. Loeschcke, eds., Population Biology and Evolution. Springer, New York.

Price, T. D., P. R. Grant, H. L. Gibbs, and P. T. Boag. 1984b. Recurrent patterns of natural selection in a population of Darwin's finches. Nature 309:787–789.

Richner, H., P. Schneiter, and H. Stirnimann. 1989. Life-history consequences of growth rate depression: an experimental study on carrion crow (*Corvus corone corone* L.). Funct. Ecol. 3:617–624.

Ricklefs, R. E., and Peters, S. 1981. Parental components of variance in growth rate and body size of nestling European starlings (*Sturnus vulgaris*). Auk 98:39–48.

Roff, D. A. 1997. Evolutionary Quantitative Genetics. Chapman and Hall, New York.

Rothstein, S. 1973a. The niche-variation model — is it valid? Am. Nat. 107:598–620.

Rothstein, S. 1973b. Relative variation of avian morphological features: relation to the niche. Am. Nat. 107:796–798.

Roughgarden, J. 1972. Evolution of niche width. Am. Nat. 106:683–718.

Routman, E. J., and J. M. Cheverud. 1997. Gene effects on a quantitative trait: two-locus epistatic effects measured at microsatellite markers and at estimated QTL. Evolution 51:1654–1662.

Rowe, L., and D. Houle. 1996. The lek paradox and the capture of genetic variance by condition dependent traits. Proc. R. Soc. Lond. B 263:1415–1421.

Sasaki, A., and S. Ellner. 1997. Quantitative genetic variance maintained by fluctuating selection with overlapping generations: variance components and covariances. Evolution 51:682–696.

Schluter, D. 1984. Morphological and phylogenetic relations among the Darwin's finches. Evolution 38:921–930.

Schluter, D. 1988. Estimating the form of natural selection on a quantitative trait. Evolution 42:849–861.

Schluter, D. 1996. Adaptive radiation along genetic lines of least resistance. Evolution 50:1766–1774.

Schluter, D., and D. Nychka. 1994. Exploring fitness surfaces. Am. Nat. 143:597–616.

Schluter, D., T. D. Price, and P. R. Grant. 1985. Ecological character displacement in Darwin's finches. Science 227:1056–1059.

Schluter, D., and Smith, J. N. M. 1986. Natural selection on beak and body size in the song sparrow. Evolution 40:221–231.

Schnol, E. E., and A. S. Kondrashov. 1993. The effect of selection on the phenotypic variance. Genetics 134:995–996.

Schoener, T. W. 1986. Resource partitioning. Pp. 91–126 *in* J. Kikkawa and D. J. Anderson, eds., Community Ecology. Patterns and Process. Blackwell Scientific, Oxford.

Selander, R. K. 1966. Sexual dimorphism and differential niche utilization in birds. Condor 68:113–151.

Shaw, F. H., R. G. Shaw, G. S. Wilkinson, and M. Turelli. 1995. Changes in genetic variances and covariances: G whiz! Evolution 49:1260–1267.

Slatkin, M. 1978. Selection and polygenic characters. J. Theor. Biol. 70:213–238.

Slatkin, M. 1987. Gene flow and the geographic structure of natural populations. Science 236:787–792.

Smith, H. G. 1993. Heritability of tarsus length in cross-fostered broods of the European starling (*Sturnus vulgaris*). Heredity 71:318–322.

Smith, H. G., and K. J. Wettermark. 1995. Heritability of nestling growth in cross-fostered European Starlings *Sturnus vulgaris*. Genetics 141:657–665.

Smith, J. N. M., and A. A. Dhondt. 1980. Experimental confirmation of heritable morphological variation in a natural population of song sparrows. Evolution 34:1155–1160.

Smith, J. N. M., and R. Zach. 1979. Heritability of some morphological characters in a song sparrow population. Evolution 33:460–467.

Smith, T. B. 1987. Bill size polymorphism and intraspecific niche utilization in an African finch. Nature 329:717–729.

Smith, T. B. 1993. Disruptive selection and the genetic basis of bill size polymorphism in the African finch *Pyrenestes*. Nature 363:618–620.

Sokal, R. R., and F. J. Rohlf. 1981. Biometry. W. H. Freeman, San Francisco.

Soulé, M. E., and B. R. Stewart. 1970. The "niche-variation" hypothesis: a test and alternatives. Am. Nat. 104:85–97.

Stern, D. L., and P. R. Grant. 1996. A phylogenetic reanalysis of allozyme variation in the Galápagos finches. Zool. J. Linn. Soc. 118:119–134.

Taper, M. L., and T. J. Case. 1985. Quantitative genetic models for the coevolution of character displacement. Ecology 66:355–371.

Taper, M. L., and T. J. Case. 1992. Coevolution among competitors. Oxf. Surveys Evol. Biol. 8:63–109.

Thessing, A., and J. Ekman. 1994. Selection on the genetical and environmental components of tarsal growth in juvenile willow tits (*Parus montanus*). J. Evol. Biol. 7:713–726.

Turelli, M., and N. H. Barton. 1994. Genetical and statistical analyses of strong selection on polygenic traits: what, me normal? Genetics 138:913–941.

van Noordwijk, A. J. 1984. Quantitative genetics in natural populations of birds illustrated with examples from the great tit *Parus major*: Pp. 67–79 *in* K. Wöhrmann and V. Loeschcke, eds., Population Biology and Evolution. Springer, New York.

van Noordwijk, A. J. 1987. Quantitative ecological genetics of great tits. Pp. 363–380 *in* F. Cooke and P. A. Buckley, eds., Avian Genetics. Academic Press, New York.

Van Valen, L. 1965. Morphological variation and the width of the ecological niche. Am. Nat. 99:377–390.

Van Valen, L., and Grant, P. R. 1970. Variation and niche width reexamined. Am. Nat. 104:589–590.

Via, S., and R. Lande. 1985. Genotype-environment interaction and the evolution of phenotypic plasticity. Evolution 39:505–523.

Via, S., and R. Lande. 1987. Evolution of genetic variability in a spatially heterogeneous environment: effects of genotype-environment interaction. Genet. Res., Camb. 49:147–156.

Vrijenhoek, R. C. 1984. Ecological differentiation among clones: the frozen niche variation model. Pp. 217–231 *in* K. Wöhrmann and V. Loeschcke, eds., Population Biology and Evolution. Springer, New York.

Ward, P. 1994. Parent-offspring regression and extreme environments. Heredity 72:574–581.

Wayne, M. L., J. B. Hackett, and T. F. C. Mackay. 1997. Quantitative genetics of ovariole number in *Drosophila melanogaster*. I. Segregating variation and fitness. Evolution 51:1156–1163.

Werner, E. E., and J. F. Gilliam. 1984. The ontogenetic niche and species interactions in size-structured populations. Annu. Rev. Ecol. Syst. 15:393–425.

Werner, T. K., and T. W. Sherry. 1987. Behavioral feeding specialization in *Pinaroloxias inornatus*, the "Darwin's finch" of Cocos Island, Costa Rica. Proc. Natl. Acad. Sci. USA 84:5506–5510.

Westneat, D. F., and M. S. Webster. 1994. Molecular analysis of kinship in birds: interesting questions and useful techniques. Pp. 91–126 *in* B. Schierwater, B. Streit, G. P. Wagner, and R. DeSalle, eds., Molecular Ecology and Evolution: Approaches and Applications. Birkhäuser, Basel.

Wiggins, D. A. 1989. Heritability of body size in cross-fostered tree swallow broods. Evolution 43:1808–1811.

Wiggins, D. A. 1991. Natural selection on body size and laying date in the tree swallow. Evolution 45:1169–1174.

Willis, J. H., and H. A. Orr. 1993. Increased heritable variation following population bottlenecks: the role of dominance. Evolution 47:949–956.

Yablakov, A. V. 1974. Variability of Mammals (Izmenchivost' mlekopitaynshchikh). Amerind Publishing Co., New Delhi.

Yang, S.-Y., and J. L. Patton. 1981. Genetic variability and differentiation in the Galápagos finches. Auk 98:230–242.

Yauk, C. L., and J. S. Quinn. 1996. Multilocus DNA fingerprinting reveals high rate of heritable genetic mutation in herring gulls nesting in an industrialized urban site. Proc. Natl. Acad. Sci. USA 93:12137–12141.

2

Adaptation, Natural Selection, and Optimal Life-History Allocation in the Face of Genetically based Trade-offs

BARRY SINERVO

Optimality arguments from life-history theory predict that females investing an intermediate level of energy in reproduction will have higher fitness compared to females with the smallest or largest investment. Adaptational concepts of optimal life-history allocation were tested in the side-blotched lizard, *Uta stansburiana*, by measuring whether natural selection on reproduction is stabilizing in the long run. Five years (generations) of selection data were analyzed for female side-blotched lizards. Clutch size and egg size of free-ranging lizards were significantly heritable, and a negative genetic correlation was also found between the two traits. I was specifically interested in how natural selection acts on the genetically based trade-off between clutch size and egg-mass. Accordingly, I used a path analytic model to decompose variation in reproductive allocation about the trade-off between clutch size and egg-mass. Clutch-size residuals were computed from the regression of clutch size on body size and year-to-year factors. Egg-mass residuals were computed from the regression of egg-mass on clutch size, body size, and year-to-year factors. Egg-mass residuals represent deviations from the trade-off between clutch size and egg-mass. Natural selection on clutch-size and egg-mass residuals was decomposed into two life-history episodes: (1) survival selection of adult females from the first to later clutches as a function of current allocation and (2) number of progeny that survived to maturity from first and later clutches. A striking pattern of multivariate selection acted on the clutch size and egg-mass trade-off. Variation in clutch size was subjected to disruptive selection, while variation in egg-mass about the trade-off was subjected to stabilizing selection. Stabilizing selection on egg-mass residuals can be visualized as a ridge of high fitness that runs along the length of the egg-mass and clutch size trade-off. Disruptive selection would tend to pull the population distribution for clutch size to extreme values. This form of multivariate selection would serve to refine the mechanistic trade-off between clutch size and egg size while at the same time leading to divergence in clutch size in the population. I discuss functional and selective constraints that limit optimal allocation to offspring size in terms of Lack's (1954) and Williams's (1966) principle.

2.1 Optimizing Selection: Fact or Fiction?

A central tenet of adaptationism predicts that natural selection on traits should be optimizing if the population is at evolutionary equilibrium (Lande and Arnold 1983, Orzack and Sober 1994). Individuals that express intermediate values of the trait should have high fitness compared to individuals that express either large or small values of the trait. Despite the pervasiveness of the optimality assumption in life-history theory (Stearns 1976, 1977, Charlesworth 1990), few studies have demonstrated optimizing natural selection on suites of life-history traits. Lack (1947, 1954) formalized notions of life-history optimality and adaptation in terms of a trade-off between quantity and quality of offspring. An organism should produce a clutch size that maximizes the number of reproductive offspring in a given reproductive period (Lack 1954). Fecundity selection favors production of large clutches, whereas survival selection favors production of large offspring. The trade-off between clutch size and egg size leads to a selective constraint on evolution of the life-history. Fecundity and offspring survival selection oppose each other and combine to produce stabilizing selection. Williams (1966) refined Lack's (1947, 1954) ideas to include selective costs of reproduction that act on allocation by parents. Investment in current reproduction is expected to be somewhat costly in terms of fecundity or survival to future reproductive episodes. Costs of reproduction are generally thought to result in an optimal level of investment on a given clutch that maximizes reproductive and survival schedules (Gadgil and Bossert 1970).

Demonstrations of Lack's (Sinervo et al. 1992, Vander Werf 1992) or Williams's principles (Nur 1984, 1988, Gustafsson and Sutherland 1988, Landwer 1994, Nilsson and Svensson 1996, Sinervo and DeNardo 1996) have typically relied on experimentally induced variation to assess the role of selection in shaping life-history trade-offs. It is rare that selective trade-offs are demonstrated in natural variation (for an exception, see Sinervo et al. 1992, Sinervo and DeNardo 1996). Experiments are necessary to demonstrate causation (Mitchell-Olds and Shaw 1987, Schluter 1988, Wade and Kalisz 1990); however, studies on natural variation must be made to validate the strength of natural phenotypic selection. Finally, field studies must also include analysis of heritability to verify a key aspect of Darwin's (1859) theory of natural selection as it relates to adaptation: differential survival or reproductive success must act on heritable variation (Endler 1986) to have an evolutionary impact.

The side-blotched lizard, *Uta stansburiana*, is ideally suited to addressing the role of natural selection in shaping patterns of life-history variation. Side-blotched lizards have no parental care, and the offspring quantity and quality trade-off is described by the relations between egg size and clutch size (Sinervo and Licht 1991a,b) and by how egg size influences survival of offspring to maturity (Sinervo et al. 1992). Maturation occurs in 1 year in side-blotched lizards, and we can collect information on long-term effects of selection that would not be possible in organisms with overlapping generations, a long time to maturity, or a long life span (e.g., many birds and mammals). Previous experimental studies of natural selection on side-blotched lizards have validated Lack's original hypothesis concerning organisms without parental care (Sinervo et al. 1992, Sinervo and Doughty 1996). Egg size is under stabilizing selection owing to the multiplicative effects of two opposing directional selection surfaces. Females that produce small eggs produce larger clutches; however, offspring from small eggs generally have lower survival. Observed egg size produced by females is close to optimal egg size as assessed by cohort survival to maturity.

Given that egg size is optimal from the perspective of fecundity and offspring survival selection, it remains to be determined whether the allocation to reproduction on the first clutch yields the optimal pattern of reproduction and survival as adults. Previous experimental studies of postlaying adult survival have demonstrated significant natural selection on total clutch mass; however, the strength and magnitude of directional selection changes among years. This work has not included an explicit test of the long-term optimality predictions generated from life-history theory. The optimality of egg-mass and clutch size can be addressed in a straightforward manner by asking how investment on the first clutch affects survival to the second clutch (Sinervo and DeNardo 1996). Most reproductive output from females occurs on the first and the second clutches of the reproductive season. Even though some females survive to produce a third and fourth clutch during their first season, high adult mortality leaves few egg-laying females past the third clutch. Moreover, the contribution of these later clutches to growth rate of the population is small because survival of late-season progeny, particularly after the third clutch, is greatly reduced. Thus, to a close approximation, reproductive success estimated from the first and second clutch yields a reasonable estimate of lifetime reproductive success. Lifetime reproductive success provided by long-term longitudinal studies of cohort fitness used in this study (Clutton-Brock 1988) are regarded as superior to cross-sectional analysis of fitness (Arnold and Wade 1984).

In this chapter, I present a new analysis of long-term selection on lifetime reproductive success (1990–1994, five generations). In previous work I sought to document year-to-year variation in selective regimes and how such temporal variation affects selection on life-history (Sinervo and DeNardo 1996). This study focuses on the average effects of selection across many generations. I also estimated the heritability and genetic correlation for clutch size and egg size. The hypothesis that natural selection on reproductive allocation is optimizing was tested by multivariate fitness regression (Lande and Arnold 1983, Schluter and Nychka 1994) of survival to the second clutch as a function of the clutch size and egg size produced on the first clutch. I also tested whether the observed pattern of reproductive allocation maximizes recruitment of progeny to the next generation and thereby maximizes the estimate of lifetime reproductive success. Long-term multivariate phenotypic selection on clutch size and egg size should also shape the negative genetic correlation between these traits. Accordingly, my analysis focused on how fitness varies relative to the trade-off between clutch size and egg size. I computed female fitness as a function of a female's deviation from the clutch size and egg-mass trade-off. I discuss the trade-off in terms of physiological, functional, and selective constraints that limit optimal allocation to offspring size defined by Lack's (1954) and Williams's (1966) principles.

2.2 Materials and Methods

Side-blotched lizards (*Uta stansburiana*) were studied in a population located on a 250-m long outcropping of sandstone adjacent to Billy Wright Road, Merced County, California, USA (~2 km east of Los Baños Creek). Data are from previous studies of natural selection on offspring size (Sinervo et al. 1992, Sinervo and Doughty 1996), natal dispersal (Doughty and Sinervo 1994, Doughty et al. 1994), heritability (Sinervo and Doughty 1996), sexual selection (Sinervo and Lively 1996), proximate physiological control of clutch size and egg size (Sinervo and Licht 1991a,b), and costs of reproduction (Sinervo and DeNardo 1996).

2.2.1 Heritable variation in egg size, manipulations of yolk volume, and control of maternal effects

My co-workers and I individually mark all progeny obtained from known mothers to estimate heritable variation in life-history traits of free-ranging females. We rely on dam–daughter correlations in assessing genetic variation in the wild. However, nongenetic maternal effects remain confounded with correlations of parents and offspring (Falconer and Mackay 1996). We controlled incubation environment of eggs to eliminate maternal effects that arise from among-dam variation in oviposition sites. In addition, we released siblings on the outcrop randomly with respect to the dam's territory to remove confounding influences that would arise from a correlated rearing environment between dams and daughters. Finally, two key manipulations of egg size, which is a likely maternal effect, were performed. Upon discovering freshly laid eggs, we removed a portion of the yolk from half of the eggs in a clutch. Yolk was aspirated using a sterile syringe (25 gauge). Hatchlings were miniaturized in proportion to the amount of yolk removed (Sinervo 1990). The remaining control eggs in the clutch were sham-manipulated by inserting a syringe, but no yolk was withdrawn. Approximately 20% of the yolk by mass was removed from eggs. Although induced maternal effects of yolk volume were considered in these experiments, maternal effects of yolk composition may still confound estimates of heritability.

Eggs from all females were individually incubated in moist vermiculite (-200 kPa) at 28°C, and the vermiculite was changed weekly. Incubation conditions in the laboratory control for maternal effects arising from oviposition site choice that would otherwise arise in natural populations. Newly emerged hatchlings were permanently marked with a unique toe-clip and released at nest sites based on natural density of female territories (Doughty and Sinervo 1994). We released 558 hatchlings in the 1990 cohort (partitioning of hatchlings by treatment can be found in Sinervo et al. 1992). All progeny matured during March and April of the following year. Reproductive condition of female offspring was carefully monitored by abdominal palpation when they matured in 1991. When eggs were detected in the oviduct, progeny were brought into the laboratory to lay eggs as per methods described for female parents. We returned female progeny to their territory after they laid eggs and recaptured them on second and third clutches.

2.2.2 Assessing survival costs of reproduction

During 1990 through 1994, we captured all females in the population in early March. Females received a unique toe-clip which individually identifies them for life. Periodic abdominal palpation of free-ranging animals allows ovulation to be predicted to within 4 days (Sinervo and Licht 1991a). Females were brought into the laboratory in early April to lay their eggs as per methods described above. The mass of individual eggs in the clutch was weighed to the nearest 0.01 g within 12 h of laying. Clutch size was also measured and total clutch mass (g) was calculated by summing weights of individual eggs. Postlaying mass was used as an index of female size. We returned females to territories after oviposition, and they invariably resumed residence on the territory used before to the brief laboratory stay (<10 days).

Postlaying survival of females was monitored to estimate survival costs of reproduction that might arise from allocation to reproduction on the first clutch. During 1990 through 1994, we studied natural selection on cohorts of adult female side-blotched lizards for which we had previously measured clutch size, egg size, and postlaying mass on the first clutch (n = 32, 62, 62, 54, and 55 for 1990–94 respectively). Females are sedentary. Survival of individually marked females was measured to the production of a second clutch by walking daily transects of the study site and recapturing individually paint-marked females with known reproductive history. Follicle development was monitored by abdominal palpation, and I censused females that survived to produce a second clutch when I detected ovulated eggs (~1 month after the first clutch). Daily censuses continued for at least 1 month to ensure that surviving females were not missed.

2.2.3 Path analysis of life-history trade-offs and selection

Path analysis (Li 1975) was used to decompose proximate sources of natural variation in clutch size and egg size variation within and among years. The path analytic model of reproductive allocation reflected a causal chain of reproductive events that were analyzed by computing successive multiple regressions and residuals (Li 1975, Sinervo and DeNardo 1996). Clutch size is determined by body size (e.g., postlaying body mass), and clutch size regulation occurs earlier than egg size regulation. The former assumption is a pervasive pattern among reptiles, while the later assumption has been explicitly validated by experimental manipulation of hormones in side-blotched lizards (Sinervo and Licht 1991a). I hypothesized that clutch size development was influenced by a general year effect. Changes in clutch size among years could be due to either microevolutionary forces (e.g., response to selection, see Sinervo and DeNardo 1996), or changes in the abiotic environment that have a plastic effect on clutch size such as rainfall (James and Whitford 1994). In a cascading fashion, egg size variation should be influenced by body-size–free residuals for clutch size, body size, and a year effect. I hypothesized that female survival to the second clutch was in turn influenced by year effects, body size, clutch-size residuals (with effects of yearly variation and body size variation removed by regression), and egg-mass residuals (with effects of yearly variation, body size variation, and clutch size residuals removed by regression). I used a linear and quadratic term for clutch-size and egg-mass residuals in the analysis of survival to the second clutch. The linear effect of clutch-size and egg-mass residuals on survival, when pooled across years, specifically tests for long-term directional selection on the mean value of the trait (e.g., clutch-size or egg-mass residuals). This is because gross among-year differences in the mean value of the traits were removed by regression analysis. A significant quadratic term would be indicative of stabilizing selection on the clutch-size and egg-mass residuals if the sign of the parameter is negative. I also fitted a nonparametric multivariate fitness surface of postlaying female survival as a function of clutch-size and egg-mass residuals.

2.2.4 Progeny survival to maturity

I measured the number of progeny that survived to maturity for each adult female that produced eggs. Beginning in early in March, I censused the entire study site where progeny were released the previous year. In addition, I censused a large 600- to 1000-m buffer zone. It is highly unlikely that a substantial number of progeny were missed during our intensive 4-month censuses that run from March to June (Doughty and Sinervo, 1994). Estimates of progeny survival when combined with selection analysis of adult female survival (see above) yield an estimate of fitness from egg production to egg production or one complete life cycle. As was the case for selection on adult female survival, results from analysis of the 1990–94 cohorts were pooled to estimate the long-term effects of natural selection.

2.3 Results

2.3.1 Heritability of egg size and clutch size

Clutch size of the female parent was positively correlated with the clutch size of her daughter at maturity (fig. 2.1A). Similarly, egg-mass of the female parent was positively correlated with egg-mass of the daughter (fig. 2.1C). Experimental removal of yolk from freshly laid eggs had no effect on the clutch size or egg size of daughters (fig. 2.1B, D). Thus, maternal effects of yolk volume did not confound heritable covariation measured for egg size and clutch size of dams and daughters. A genetic correlation between traits was measured from covariation between mother's egg size and daughter's clutch size. Egg size of dams was negatively correlated with clutch size of daughters, which is indicative of a genetically based trade-off between egg size and clutch size (fig. 2.1E). The covariation between dam and daughters for egg size (Cov_{yy}), clutch size (Cov_{xx}), and the covariation between daughter's clutch size and the dam's egg size (Cov_{xy}) allows us to compute the genetic correlation (Falconer and Mackay 1996).:

$$r_{xy} = 2\, Cov_{xy} \Big/ \sqrt{Cov_{xx}\, Cov_{yy}}$$

The genetic correlation between clutch size and egg size of side-blotched lizards was equal to -0.92. The mechanistic source of this genetic correlation has been previously elucidated in experiments (fig. 2.1F; Sinervo and Licht, 1991a,b). Ablation of follicles during early vitellogenesis causes females to produce a smaller clutch of enlarged eggs. Conversely, administration of exogenous follicle-stimulating hormone during early vitellogenesis causes females to produce an enlarged clutch of smaller eggs. The experimentally derived trade-off corresponds quite closely to the negative genetic correlation between clutch size and egg-mass.

2.3.2 The clutch size and egg-mass trade-off and survival costs of reproduction

Egg size, clutch size, and total clutch mass were significantly variable among the years of this study (1990–94; fig. 2.2). Similarly, survival varied significantly among the years of this study (table 2.1). The source of between-year changes has been treated in detail elsewhere and is summarized in the Discussion (Sinervo and DeNardo 1996, Sinervo 1998a,b). Clutch size was positively affected by postlaying body mass of the female (fig. 2.3A). Paths linking effects of among-year variation (fig. 2.2) and postlaying mass (fig. 2.3B) to clutch size (fig. 2.4) were significant. My path model assumed that year and body size influenced ovarian

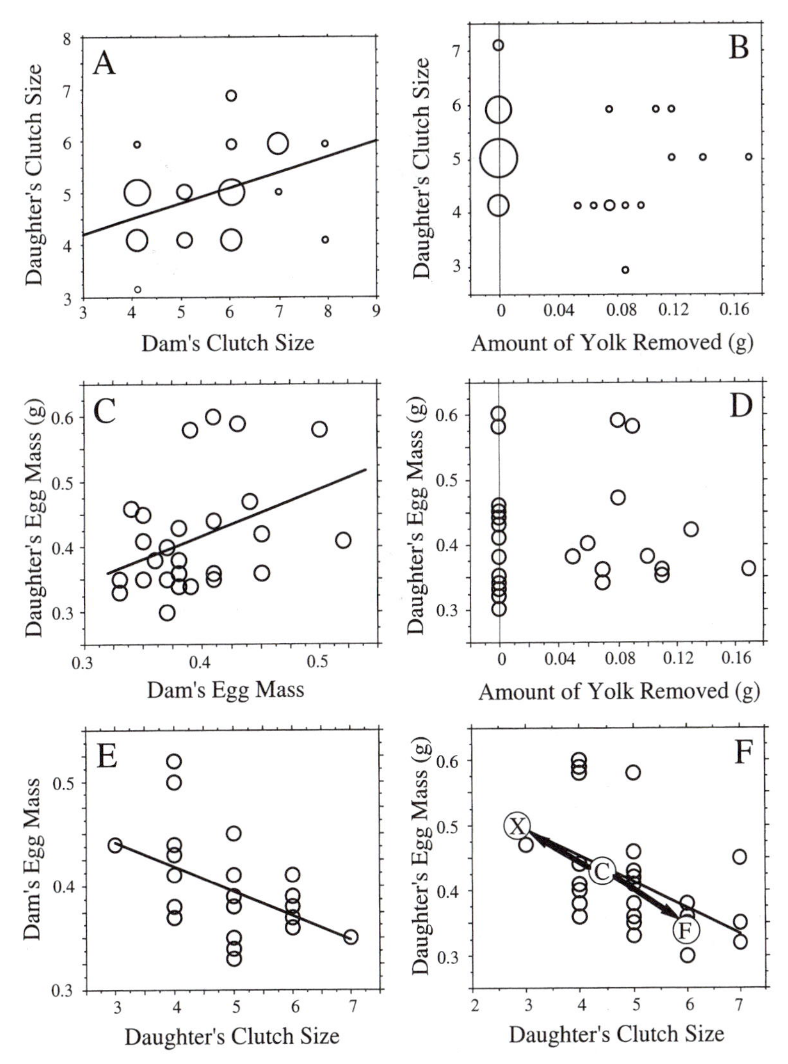

Figure 2.1 (A) Heritability for clutch size was significant ($F_{1,30} = 5.04, P < .05$). (B) Effect of yolk removal during embryogenesis on the clutch size produced by daughters when they matured the following year was not significant ($P > .87$). (C) Heritability for egg mass was also significant ($F_{1,30} = 4.23, P < .05$). (D) Effect of yolk removal during embryogenesis on the egg mass produced by daughters at maturity was not significant ($P > .87$). (E) Negative correlation between a daughter's clutch size and the dam's egg mass was significant ($F_{1,30} = 9.67, P < .01$), which is indicative of an underlying genetic correlation (Falconer and Mackay 1996). (F) The mechanistic source of this genetic correlation has been elucidated in experiments (data from Sinervo and Licht, 1991a,b). Ablation of follicles (X) during early vitellogenesis causes females to produce a significantly smaller clutch of enlarged eggs relative to the average controls (C). Conversely, administration of exogenous follicle-stimulating hormone (F) during early vitellogenesis causes females to produce an enlarged clutch of smaller eggs relative to controls. The phenotypic trade-off between clutch size and egg size of the 1990 cohort of offspring are presented for comparison (circles) with the experimentally induced covaration. Thus, the phenotypic trade-off, genotypic trade-off, and experimentally induced trade-off are all concordant.

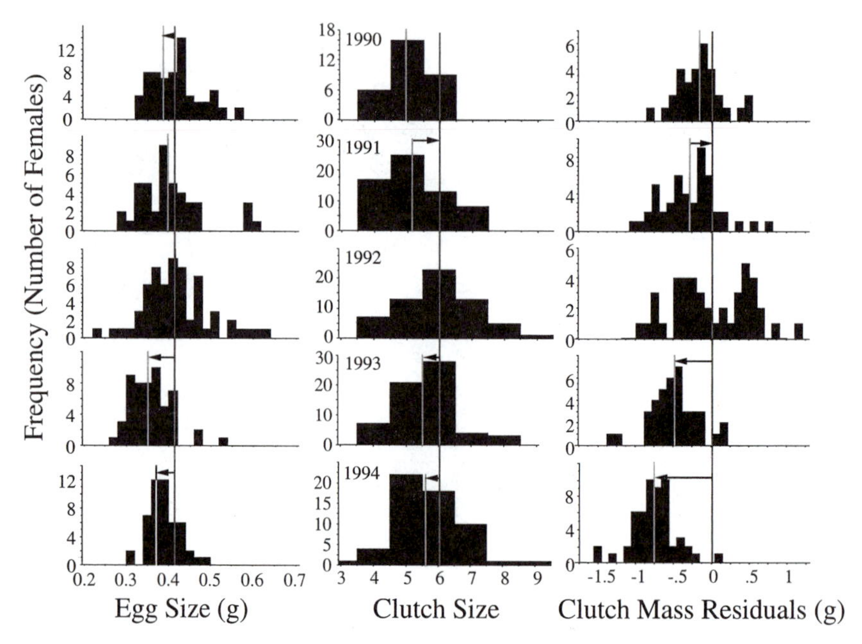

Figure 2.2 Significant among-year variation was detected for (a) egg size ($F_{4,261} = 7.59$, $P < .0001$), (B) clutch size ($F_{4,261} = 7.51$, $P < .0001$), and (C) clutch mass residuals [effects of body size on clutch mass (= egg mass x clutch size) were removed by regression analysis]. Significant changes in reproductive traits among years (1990–1994) are denoted by arrows.

development or clutch size and that this allocation "decision" has a cascading effect on egg-mass. Accordingly, I computed clutch-size residuals (CSRs) which have phenotypic effects of body size or environmental effects of year removed by regression analysis (figs. 2.3B, 2.4). Egg-mass of females was inversely related to clutch-size residuals, reflecting the trade-off between egg-mass and clutch size (fig. 2.3C). In addition, egg-mass was also influenced significantly by year effects (figs. 2.2, 2.4), but only weakly by postlaying body mass (figs. 2.3B, 2.4). I also computed egg-mass residuals in my path model. The clutch-size and egg-mass residuals are independent owing to the form of the path model. The egg-mass residuals

Table 2.1 Variation in probability of adult female survival among years.

Year	Count	Survival probability	SE
1990	32	0.750	0.078
1991	62	0.742	0.056
1992	63	0.810	0.050
1993	54	0.556	0.068
1994	55	0.691	0.063

Survival was indexed from egg laying on the first clutch to production of a second clutch of eggs. Difference in survival among years is significant ($F_{4,261} = 2.536$, $P < .05$).

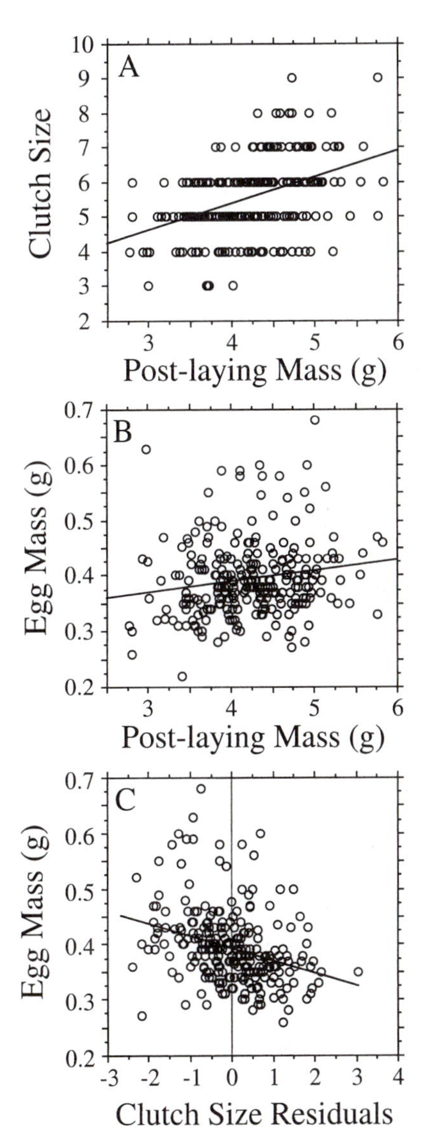

Figure 2.3 Univariate plots describing significant effects of (A) postlaying mass on clutch size, (B) postlaying mass on egg size, and (c) clutch size residuals on egg mass. Variables were used in the multivariate path model of figure 2.4.

(EMRs) have a simple geometric intrepretation. Egg-mass residuals reflect the deviation of individual females from the trade-off between clutch size and egg-mass (fig. 2.3C).

Clutch size residuals and (EMRs) derived from the path analysis described above were used in an analysis of variance model testing for directional and stabilizing selection on each trait. Linear (directional) and quadratic (stabilizing) terms (Lande and Arnold 1983) were estimated for CSR and EMR in two separate analyses (table 2.2). Fitness consequences of the

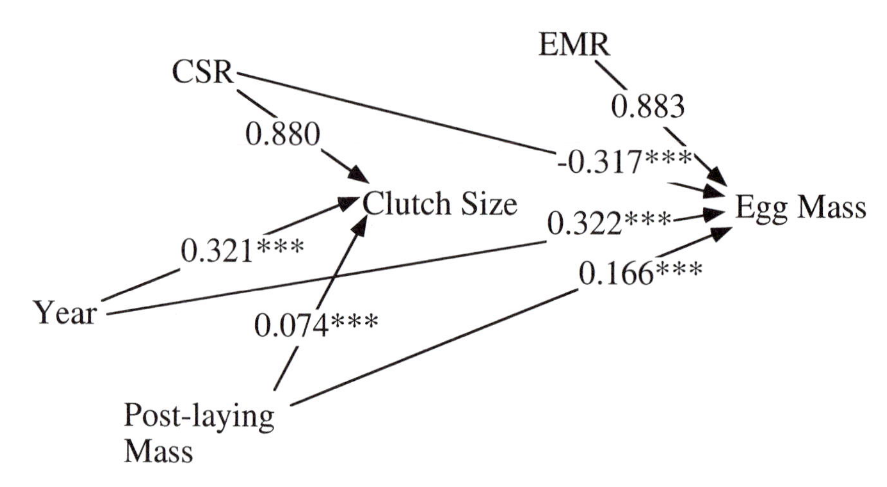

Figure 2.4 Path analytic model describing the hypothesized causal relationships between factors for year (fig. 2.2), postlaying mass (fig. 2.3A, C), clutch size, and egg mass. The effect of clutch size on egg mass is hypothesized to act through clutch-size residuals (e.g., CSRs in which effects of postlaying mass and year on clutch size are removed by multivariate regression analysis). Likewise, egg-mass residuals (EMRs) are derived from multivariate regression in effects of postlaying mass, year, and CSR. The EMR and CSR were used in analysis of long-term natural selection.

EMR and CSR were estimated from postlaying survival censused from the first clutch to ovulation of their second clutch ($n = 265$). Clutch-size residuals were not under significant stabilizing or directional selection (table 2.2), but a significant effect of year on general survival was detected ($P = .03$). Significant stabilizing selection was detected for the egg-mass residuals (table 2.2). While stabilizing selection on the EMR was relatively strong and highly significant ($P = .002$; fig. 2.5), directional selection on the EMR was not significant ($P > .81$). Yearly variation led to significant differences in level of survival ($P = .004$). Given that directional selection on EMR was not significant, the optimum for the EMR is centered on the long-term average value for EMR in the population (e.g., year effects removed by regression).

2.3.3 Multivariate fitness surface and reproductive allocation

I used Schluter and Nychka's (1994) mutlivariate cubic spline algorithm to visualize the shape of the multivariate fitness surface that acted on survival of the female parent as a function of variation around the trade-off between clutch size and egg-mass (EMR and CSR). Because variation in the EMR and CSR reflect residuals computed as deviations of the phenotypically based trade-off, the surface fitted to the observed data would reflect multivariate phenotypic selection that shapes the genetically based trade-off between clutch size and egg size (fig. 2.2). If the fitted surface were at an angle to the trade-off, then selection would act to alter the genetic correlation between clutch and egg size. Alternatively, if the fitted surface lined up along one or more axes of the CSR versus EMR plot, selection would stabilize the trade-off. The EMR and CSR were standardized by the standard deviation in each trait

Table 2.2 Analysis of variance for probability of adult female survival as a function of year and post-laying mass (PLM)

Source	df	Sum of squares	Mean square	*F*-value	*P*-value
CSR	1	0.039	0.039	0.195	.66
CSR^2	1	0.126	0.126	0.620	.43
Year	1	0.238	0.238	1.173	.28
PLM	4	2.170	0.542	2.667	.03
Residual	257	52.28	0.203		
EMR	1	0.010	0.010	0.055	.81
EMR^2	1	1.871	1.871	9.586	.002**
PLM	1	0.133	0.133	0.682	.41
Year	4	3.143	0.785	4.025	.004**
Residual	257	50.17	0.195		

Selection on clutch-size residuals (CSR) or egg-mass residuals (EMR) were included as linear (i.e., directional) or quadratic (i.e., stabilizing) terms.

Effects significant at $^{**}P < .01$ and at $^{**}P < .05$.

before analysis. As per methods described in Schluter and Nychka (1994), a grid search of $\ln(\lambda)$ was performed for 40 values $[-20 < \ln(\lambda) < 20]$. The minimum generalized cross validation score (GCV) was obtained for $\ln(\lambda) = -10$. The effective number of parameters for the best fit surface was 12.

A visual inspection of the surface indicates that a ridge of high fitness runs orthogonal to the EMR axis (fig. 2.6A,B). Given that the EMR represents deviations in egg mass from the clutch size-versus-egg size trade-off, the observed stabilizing selection would refine variation about the clutch size and egg-mass trade-off. Although the shape of the surface along the CSR appears disruptive, parametric analysis (above) indicates it is not significant. The shape of the multivariate fitness surface confirms interpretations derived from parametric fitness regression analysis. A ridge of high fitness runs along the length of the trade-off between clutch size and egg-mass. Females that had large positive or negative EMRs from the clutch size and egg-mass trade-off had low fitness. Females with residuals close to the trade-off had high fitness.

2.3.4 The clutch size and egg-mass trade-off and offspring survival to maturity

A comprehensive estimate of natural selection entails measuring progeny survival to maturity. I pooled data on female progeny from first and later clutches into a single analysis. Previous analysis which partitioned hatchlings into first and later clutches indicated that seasonal effects of lay date have negligible effects on progeny reproduction at maturity (see Sinervo and Doughty, 1996). Similarly, survival of male progeny from first and later clutches was pooled into a single analysis. Fitness regression of survival of female progeny to maturity as a function the female parent's EMR and CSR indicated that clutch size variation was under significant disruptive selection ($P = .01$; fig. 2.7 and table 2.3), whereas egg-mass variation was under weak stabilizing selection ($P = .08$). Selection on male progeny was not signficant (see table 2.3; data not shown).

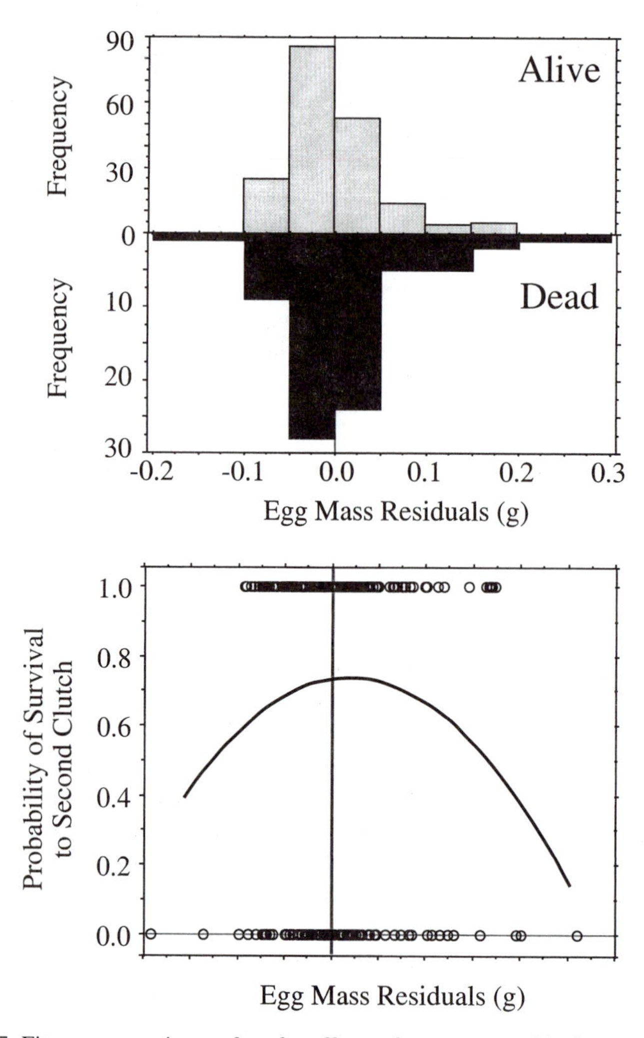

Figure 2.5 Fitness regression surface for effects of egg-mass residuals on survival probability of adult females to the second clutch of the reproductive season for the pooled 1990–1994 data set. Stabilizing selection is significant, but directional selection is not significant (table 2.2). Thus, the long-term optima are centered on average egg mass.

I used Schluter and Nychka's (1994) mutlivariate cubic spline algorithm to visualize the shape of the multivariate fitness selection that acted on progeny survival as a function of variation around the trade-off between clutch size and egg-mass (EMR and CSR). The minimum GCV score was obtained for $\ln(\lambda) = -8$. The effective number of parameters for the best fit surface was 11.6. Multivariate selection on progeny survival seems to amplify the selection observed on survival of the female parent, particularly for clutch size. Selection can be visualized as a saddle that is oriented along the length of the trade-off between clutch size and egg-mass (fig. 2.8). Stabilizing selection acts on egg mass, while at the same time disruptive selection eliminates females with intermediate clutch size and favors females with either

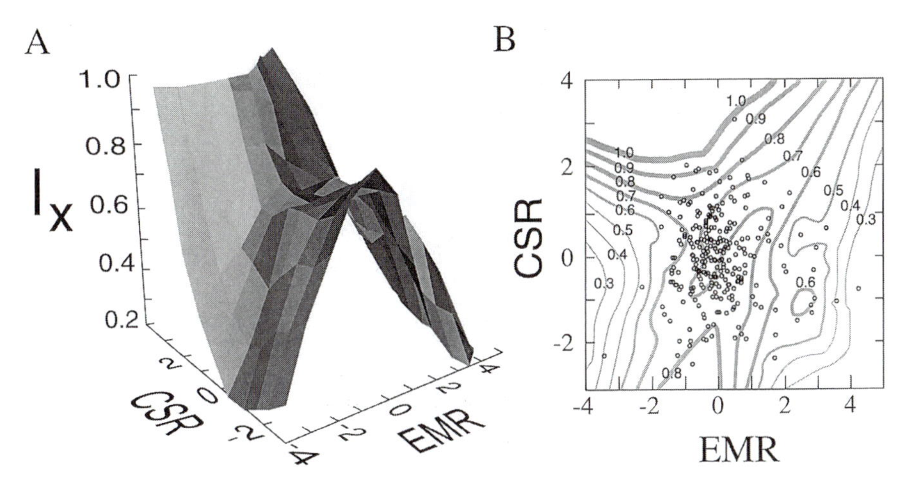

Figure 2.6 Nonparametric fitness surface for effects of clutch-size residual (CSR) and egg-mass residual (EMR) on postlaying female survival (l_x) resembles a fitness ridge. (A) Stabilizing selection is visualized as a ridge that runs along the CSR axis and perpendicular to the EMR axis. (B) Contour plot of fitness corresponding to the three-dimensional fitness surface. Contours describe probability of survival as a function of the individual variation in CSR and EMR.

large or small clutch size. Finally, I remapped the fitness surface for progeny survival back onto the observed trade-off between clutch size and egg-mass (fig. 2.9). The fitness saddle is oriented along the axis of variation describing the trade-off, and two optima are found at the extremes of clutch size.

2.4 Discussion

A key component of natural selection analysis of adaptation concerns the heritable basis of traits (Darwin 1859, Endler 1986). Both egg size and clutch size are heritably transmitted in side-blotched lizards (fig. 2.1A, C), and the traits are coupled by a strong negative genetic correlation (fig. 2.1F). A second issue concerns potential confounding maternal effects on the correlation between dam and daughters traits (Falconer 1965, Schluter and Gustafsson 1993, Sinervo and Doughty 1996). In the case of lizards, many maternal influences such as those that arise from oviposition site choice (e.g., temperature and hydric conditions) were controlled by incubating eggs in a common laboratory environment and then releasing hatchlings into the wild to undergo the process of growth and maturation. In addition, hatchlings were randomly released with respect to the dam's territory to avoid common environmental factors between dams and daughters. Finally, specific maternal effects associated with yolk volume were experimentally manipulated by removing yolk from freshly laid eggs and studying cascading effects on offspring when they mature. Yolk volume manipulations did not have a significant effect on the egg size or clutch size produced by offspring at maturity.

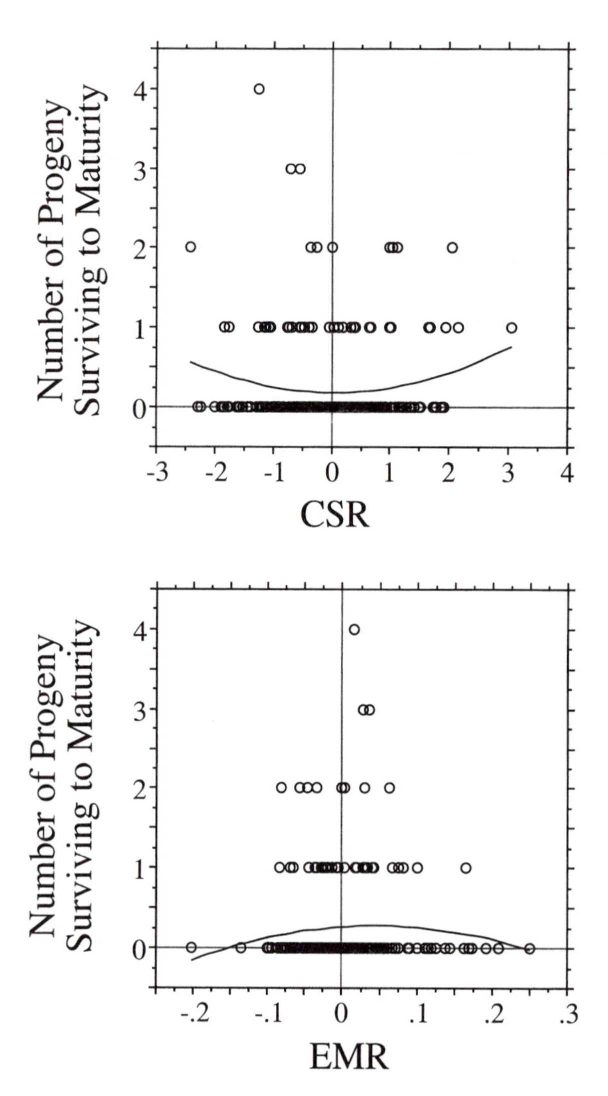

Figure 2.7 Fitness regression surface for effects of egg-mass residuals (EMRs) on survival probability of female progeny to maturity for the pooled 1990–1994 data set (survival of male progeny not shown; see table 2.3). Disruptive selection on progeny survival as a function of the female parent's clutch-size residual (CSR) is significant (table 2.3). Stabilizing selection on progeny survival as a function of the female parent's EMR is marginally significant (table 2.2). Thus, the long-term optima are centered on average egg mass, but two optima were found for the clutch size extremes.

Table 2.3 Analysis of variance for probability of progeny survival to maturity as a function of year and postlaying mass (PLM)

Source	df	Sum of squares	Mean square	F-value	P-value
EMR	1	0.519	0.519	1.579	.21
EMR^2	1	1.010	1.010	3.074	.08
CSR	1	0.069	0.069	0.211	.64
CSR^2	1	2.078	2.078	6.320	.01**
PLM	1	0.001	0.001	0.003	.95
Year	4	1.829	0.457	1.391	.23
Residual	255	83.852	0.328		
EMR	1	0.034	0.034	0.189	.66
EMR^2	1	0.203	0.203	1.113	.29
CSR	1	0.222	0.222	1.220	.27
CSR^2	1	0.053	0.053	0.293	.58
PLM	1	0.190	0.190	1.044	.30
Year	4	0.801	0.200	1.097	.35
Residual	255	46.565	0.182		

Selection on clutch-size residuals (CSR) or egg-mass residuals (EMR) were included as linear (i.e., directional) or quadratic (i.e., stabilizing) terms.

Effects significant at $^{**}P < .01$ and at $^{**}P < .05$.

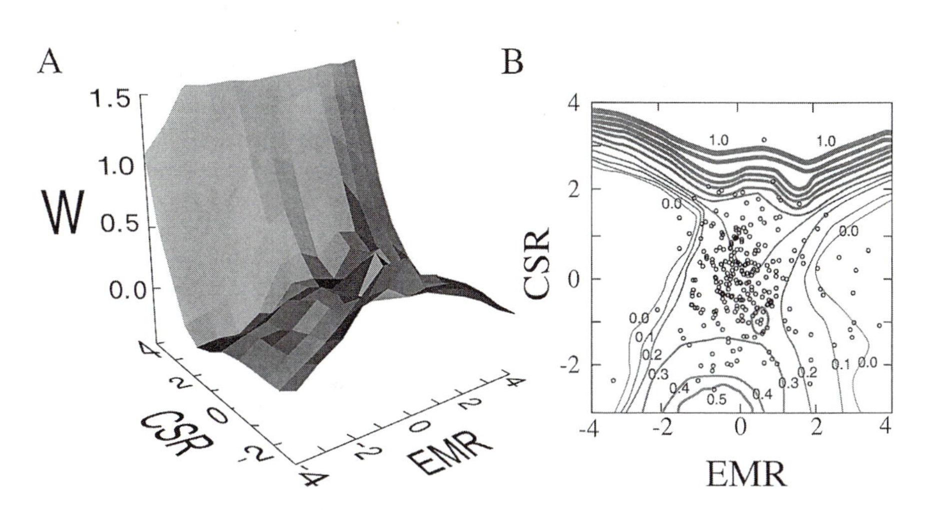

Figure 2.8 Nonparametric fitness surface for effects of clutch-size residual (CSR) and egg-mass residual (EMR) on number of female progeny (W) that recruit into the population resembles a saddle. (A) Stabilizing selection on the EMR is visualized as a ridge that runs along the CSR axis, while disruptive selection on the CSR favors extreme values of clutch size in the population. (B) Contour plot of fitness corresponding to the three-dimensional fitness surface. Contours describe the number of female progeny that survive to maturity as a function of the individual variation in the female parent's CSR and EMR.

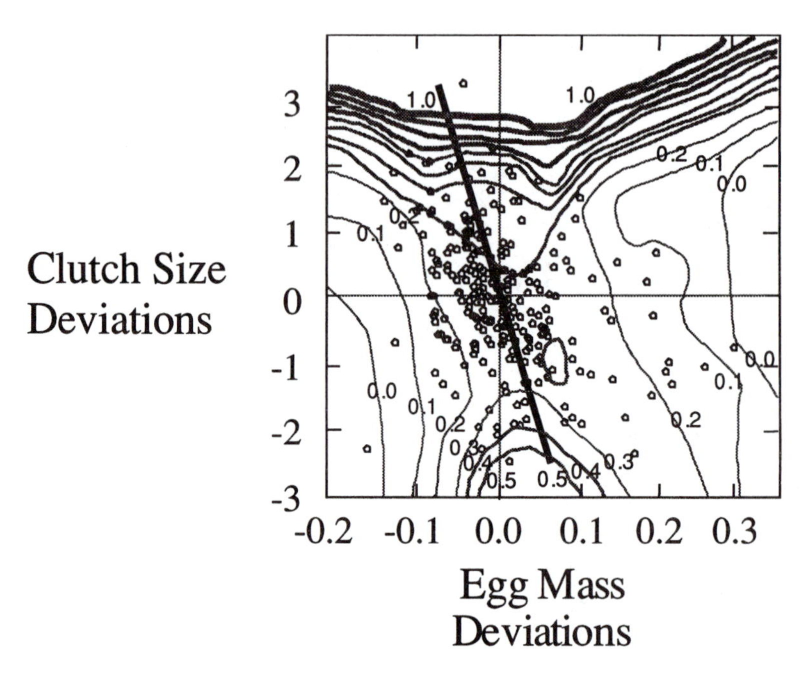

Figure 2.9 Summary of parametric and nonparametric analysis of multivariate selection remapped back on the phenotypic trade-off between clutch size and egg size (solid line). Fitness contours (gray lines) illustrate the number of female progeny that survived as a function of the clutch size and egg mass of the female parent on the first clutch of the season (deviations in clutch size and egg mass from the mean in the population are displayed). Selection can be visualized as a ridge that runs along the length of the clutch size and egg mass trade-off. Weak stabilizing selection serves to refine additive genetic covariation between clutch size and egg mass because the negative genetic correlation (fig. 2.1E) coincides with the phenotypic and experimentally induced trade-offs (fig. 2.1F). Disruptive selection on clutch size results in two optima located at the extremes for clutch size in the population.

It is generally thought that long-term, stable natural selection should deplete additive genetic variation in natural populations (Falconer and Mackay 1996). Studies on levels of heritable variation for reproductive traits in breeding birds have yielded mixed results, with some studies reporting high heritability (Ojanen et al. 1979, van Noordwijk et al. 1981, Lessels et al. 1989). Recent reviews of levels of additive genetic variation in natural populations have indicated a surprising level of additive genetic variation (Mousseau and Roff 1987). Moreover, heritability estimated from the laboratory provides a reliable index of heritability in the wild under some conditions (Weigensberg and Roff 1996). Nevertheless, field estimates of heritable variation as found in this study are superior to laboratory estimates. The maturational environment of the progeny may be quite different between laboratory and field, and this might affect magnitude of heritability.

The theoretical expectation that additive genetic variation is depleted in natural populations is not necessarily the case if correlations among competing life-history functions are coupled by strong negative genetic correlations (Falconer and Mackay 1996). The negative genetic correlation between clutch size and egg size is large and negative in side-blotched lizards (r = -.92). In addition, the effects of selection can be balanced by input of additive genetic variation by mutation (Charlesworth 1987).

2.4.1 Origins of the physiologically based life-history trade-off between clutch size and egg size

A third issue related to adaptive genetic variation concerns the mechanistic basis of life-history trade-offs and how selection operates on the trade-off (Pease and Bull 1988, Schluter et al. 1991). A fundamental prediction of Lack's hypothesis is a negative genetic correlation between clutch size and offspring size. The prediction of a negative genetic correlation between brood size and egg size may not hold in organisms with parental care (Lessels et al. 1989). However, negative phenotypic correlations are pervasive in organisms without parental care. The prediction of a negative genetic correlation between egg mass and clutch size is validated for the side-blotched lizards in this study (fig. 2.1E). A female that produces small clutches and large offspring also produces offspring that lay small clutches and large offspring when they mature.

Moreover, manipulations of the endocrine system of lizards verify the mechanistic source of the negative genetic correlation (Sinervo and Licht 1991a,b, Sinervo and DeNardo 1996). Phenotypic manipulations of clutch and egg size regulation (i.e., manipulation of the endocrine system) should perturb the fecundity trade-off by the same mechanisms as those initiated by genetic control (Sinervo 1998a, b, Sinervo and Svensson 1998). Follicle-stimulating hormone is a likely candidate for such control. Surgical ablation of ovarian follicles reduce clutch size and increase egg size. Inducing more follicles via follicle-stimulating hormone increases clutch size and reduces egg size. These two complementary manipulations indicate that clutch size regulation is linked to egg size regulation in a fashion concordant with the genetically based trade-off (fig. 2.1F).

Reznick (1985, 1992) and others (Chippindale et al. 1993, Rose et al. 1996) have argued that manipulations of phenotype cannot be used to study life-history trade-offs and that life-history trade-offs can only be studied by reference to assessing the underlying negative genetic correlations that underlie life-history trade-offs. Results presented in this paper indicate that life-history trade-offs induced by endocrine manipulations of phenotype are remarkably similar to trade-offs measured via a negative genetic correlation (fig. 2.1F). We have argued that genetic approaches afforded by breeding studies and artificial selection experiments are complementary to manipulations of the mechanisms that underlie life-history trade-offs (Sinervo 1998a, Sinervo and Svensson 1998). Couplings demonstrated by endocrine manipulations should in principle arise from the same mechanisms as the pleiotropic effect of a single endocrine regulation gene.

2.4.2 Negative genetic correlation and the clutch size and egg-mass trade-off

Considerable interest in the stability of the genetic variance–covariance matrix has been generated in recent years, in a large measure due to renewed interest in the methodology and measurement of natural selection (Bock 1980, Arnold 1983, Lande and Arnold 1983, Arnold and Wade 1984, Mitchell-Olds and Shaw 1987, Schluter 1988, Wade and Kalisz 1990). The genetic variance–covariance matrix plays a central role governing microevolutionary change in quantitative genetic models (Lande 1983). Genetic correlations arise from either the proximate effects of linkage between two or more genes or from the pleiotropic effects of one gene on two or more traits (Lande 1983). Pleiotropy is generally thought to give rise to a more stable form of genetic correlation. Pleiotropy only changes if the gene responsible for the pleiotropic effect evolves such that the effect on one trait becomes more or less uncoupled from the effect on another trait. The different sources for genetic correlations, linkage versus pleiotropy, are differentially stable. While pleiotropy is thought to be relatively stable, linkage disequilibrium arising from natural or sexual selection can decay if selection changes or if selection is relaxed.

Disentangling the relative roles of linkage versus pleiotropy is essential to understanding long-term constraints on evolution of the life-history. Endocrine manipulations (fig. 2.1E) provide a plausible explanation for the negative genetic correlation between clutch size and egg size in lizards (fig. 2.1F). Given that the negative genetic correlation between clutch size and egg size has an identifiable cause, it is illuminating to map the fitness surface onto the negative genetic correlation as expressed in phenotypic trade-off between clutch size and egg-mass. This mapping is possible because the genetic correlation, phenotypic correlation, and experimentally induced covariation between clutch size and egg size are all concordant (fig. 2.1).

2.4.3 Multivariate stabilizing selection acts on the trade-off between clutch size and egg size

The observation that postlaying selection on EMRs is purely stabilizing (table 2.2 and fig. 2.5) has a simple geometrical interpretation. Selection can be visualized as a ridge of high fitness that runs along the length of the egg mass and clutch-size trade-off (figs. 2.6, 2.7). However, additional selection on survival of progeny from different-sized clutches further refines the shape of multivariate selection (figs. 2.8, 2.9). Selection on the trade-off also acts through disruptive selection on clutch size, which would increase phenotypic variation in clutch size over the long term. This form of multivariate selection would serve to refine the mechanistic trade-off between clutch size and egg size. Any phenotype that departs from the phenotypic and genotypic trade-off are selected against in the long term. Females that produce above average or below average egg-mass for their body size and clutch size are selectively eliminated early in the spring and do not contribute to later reproductive episodes in the summer. Multivariate stabilizing selection promotes the stability of the negative genetic correlation between clutch size and egg-mass, while at the same time leading to divergence in clutch size in the population.

The genetic correlation is likely to be mechanistically determined by plasma levels of gonadotropin that serve to simultaneously regulate clutch size and egg size (fig. 2.1F). However, egg-mass is undoubtedly determined by levels of vitelloprotein production by the liver in relation to the number of growing follicles. Vitelloprotein production is governed by levels of estrogen (Gerstle and Callard 1972). Estrogens are produced by growing follicles and estrogens are potent negative regulators of follicle-stimulating hormone production by the pituitary. In addition, estrogens can bind directly to promoters on DNA via the action of carrier proteins. I hypothesize that females are selected to produce the optimal level of follicle-stimulating hormone, estrogen, and vitelloprotein because long-term optimizing phenotypic survival selection would refine underlying mechanisms of egg size regulation. Endocrine systems are based on negative and positive regulatory feedback loops that entail the interaction between two or more hormones (e.g., follicle-stimulating hormone and estrogen). It is likely that other hormone regulatory networks (e.g., those underlying growth and maturation) generate many classic multitrait trade-offs of life-history biology.

Given the important role of the endocrine system in shaping life-history variation, we might also expect changes in hormone levels to lead to selective consequences that are independent of their selective consequence on clutch size and egg size. For example, hormones such as estrogen affect behavior. Behavioral effects of plasma hormone levels might have a direct effect on survival. The paradoxical low survival of females that lay relatively small eggs is contrary to life-history theories of a cost of reproduction. However, small egg size may be correlated with low plasma estrogen or with some other socially important hormone, and low plasma estrogen may reduce survival owing to socially mediated selective costs. Life-history theory has typically ignored the importance of mechanism in generating trade-offs. An understanding of mechanism is crucial to interpret natural selection. Further work is needed to address these issues.

2.4.4 Costs of reproduction and stabilizing selection on postlaying survival

In this study, the postlaying survival costs of reproduction in side-blotched lizards were found to arise from stabilizing selection on egg size. Females that laid the largest clutches had postlaying survival to the second clutch that was similar to survival of females laying small clutches. Moreover, females laying either large clutches or small clutches had significantly higher fitness than females that laid intermediate clutches when considering number of progeny that recruited to maturity. The latter observations on fitness run contrary to an extrapolation of Williams's premise of selective trade-off between current and future reproductive success. The former observations on fitness also contradict Lack's premise of optimizing selection on clutch size and offspring size. Despite the contradiction between our observations and these classic premises, it is intuitively appealing that long-term selection on adult female survival stabilizes the clutch size and egg-mass trade-off. Multivariate stabilizing selection serves to refine the genetic architecture of a species, as reflected in the genetic correlation structure among traits.

Elucidating the morphologically, physiologically, and ecologically mediated causes of reproductive costs is a central goal of life-history analysis (Partridge and Harvey 1985). The agents of selection on the costs of reproduction are likely to be multifactorial. In this regard, previously published experimental manipulations of reproductive effort and effects on adult survival are useful to dissect the underlying selective causes of long-term phenotypic stabilizing selection on natural variation in reproductive effort observed in this population. Maximum egg size is subject to proximate limitations such as the pelvic girdle of vertebrates (Luetenegger 1976, 1979, Congdon and Gibbons 1987, Sinervo and Licht 1991b). In side-blotched lizards, females become egg-bound, and eggs burst when egg size is increased beyond 0.5 g. Although burst eggs only affect offspring survival, a female that becomes egg-bound (dystocia) may have an elevated risk of mortality from egg necrosis. Many females lay eggs that are 0.2 g heavier in the second compared to the first clutch (mean = 0.38–0.41 g; Sinervo et al. 1992). Thus, females laying 0.5 g eggs on the first clutch would be laying 0.7-g eggs on the second clutch, which could potentially result in a 65% drop in a survival (e.g., see fig. 2.6). Finding evidence of such dystocia is likely to be difficult in nature, but we have found many cases of dystocia in females experimentally induced to lay a small clutch of large eggs in both laboratory-maintained (Sinervo and Licht, 1991b) and free-ranging females (Sinervo and DeNardo, unpublished data).

Selection on maximum egg size is also subject to the burdening effect that the yolking eggs have on females. Reproductive burden decrements female sprint speed (Bauwens and Thoen 1981, Sinervo et al. 1991) and concomitantly elevates risk of mortality to predatory snakes (Shine 1980, Sinervo and DeNardo 1996). Such ecologically based costs of reproduction would reinforce selection arising from constraints of the pelvic girdle. Indeed, Sinervo and DeNardo (1996) have found that experimentally inducing large per offspring investment with corticosterone implants reduces survival in years with abundant snake predators, suggesting that the burdening effect of large eggs can lower survival. In addition, inducing large clutch size with follicle-stimulating hormone implants also reduces survival in years with abundant snake predators, and reducing clutch size by follicle ablation elevates survival.

Agents of selection acting against the production of excessively large eggs are readily identified, but environmental agents that selectively eliminate females laying small eggs for a given clutch size are more difficult to rationalize. These females are not subject to the burdening effect of eggs or constraints imposed by the pelvic girdle. Why should females with the smallest investment in offspring have low survival? In a year with few snake predators, corticosterone implants enhanced survival relative to control females. Corticosterone implants also cause females to produce enlarged eggs (Sinervo and DeNardo, 1996). Corticosterone is well known in its role of modulating the stresses of the social environment in vertebrates (Sapolsky 1982, 1992, Johnson et al. 1992). The pattern of enhanced survival for females laying large eggs is quite robust, as it was observed in both natural and experimental variation. Females that produce small eggs may be at a selective disadvantage in some years because production of small eggs may be correlated with production of low levels of corticosterone or some other correlated hormone that affects female survival (e.g., estrogen; see above). Causes of natural variation in corticosterone levels in females remains obscure, and only additional experimental manipulations will resolve these issues.

2.4.5 Microevolutionary change and the clutch size and egg-mass trade-off

The long-term stabilizing selection observed on the egg mass and clutch-size trade-off in side-blotched lizards may arise from a balance between selective factors arising from social environment, predation regime, and functional constraints. Manipulations of the putative selective environment such as predator exclusions or manipulations of female density would elucidate the multifarious causes and environmental agents of selection (Wade and Kalisz 1990). Regardless of the actual agents of selection, optimizing selection of the form observed in this study would have a stabilizing effect on genetic correlations in the long term.

The stabilizing effect of postlaying survival selection on life-history traits does not impose a dramatic constraint on short-term microevolutionary change among years; rather, it channels change in certain directions but not in others. Egg-mass and clutch size are both significantly heritable (fig. 2.1A, C) and can readily respond to selection. For example, reproduction is thought to be costly, and increased current effort should lower survival or future reproductive success (Williams 1966). However, costs can arise from cryptic natural selection that eliminates individuals during maturation. For example, the large drop observed in average clutch size, egg-mass, and total clutch mass observed between 1992 and 1993 (see fig. 2.2) arises from cryptic natural selection on clutch size that acts on females during maturation (Sinervo, unpublished data). Likewise, the long-term pattern of disruptive selection on clutch size observed in this study also arises from cryptic natural selection that acts on females at maturity (see Sinervo, unpublished data). Survival selection that selectively eliminates females as a function of clutch size would rapidly change the average egg-mass and clutch size within a single generation (fig. 2.2) because the two traits have high heritability and they are genetically correlated (fig. 2.1E). However, the mean of each trait should slowly return to equilibrium, presumably under the long-term background force of stabilizing selection that acts on survival of adult females as a function of egg-mass and disruptive selection that favors clutch size extremes.

We probably have a comprehensive picture of long-term patterns of natural selection, particularly in light of the diverse abiotic and biotic selection regimes that the lizards have experienced during the course of our studies (see Sinervo and DeNardo, 1996). If we had collected data from only 1 or 2 years, the picture of natural selection would have been far different. This highlights the importance of multigenerational studies of selection to understand the process of adaptation. It is interesting that comparable long-term studies of natural selection on clutch size in birds have found that components of fitness are likewise under stabilizing selection (Gustafsson and Sutherland 1988). Our study provides partial support for the optimality concept in life-history analysis, particularly with reference to long-term patterns of microevolutionary forces that act on adaptive genetic variation in the wild. However, the pattern of disruptive selection on clutch size begs an explanation outside the scope of life-history analysis. Female side-blotched lizards are found in two heritable throat-color morphs that differ in clutch size. The pattern of phenotypic selection found in this study would be consistent with a stable life-history polymorphism. However, the selective mechanisms promoting two fitness peaks on the clutch size axis are likely to be more complicated than the selective trade-offs of life-history theory. Frequency-dependent selection is known to be responsible for maintaining variation in three throat-color morphs of male side-blotched lizards (Sinervo and Lively, 1996). Frequency-dependent selection may likewise contribute to the pattern of disruptive selection found in female lizards.

Acknowledgments Thanks to Ryan Calsbeek, Tosha Comendant, Bruce Lyon, Erik Svensson, and Michelle Wainstein for discussions that inspired this chapter and to Kelly Zamudio, Dale DeNardo, Karla Allred, and Jen Graff for assistance with field work. Data were collected with support from National Science Foundation.

References

Arnold, S. J. 1983. Morphology, performance and fitness. Am. Zool. 23:347–361.

Arnold, S. J., and M. J. Wade. 1984. On the measurement of natural and sexual selection: theory. Evolution 38:709–719.

Bauwens, D., and C. Thoen. 1981. Escape tactics and vulnerability to predation associated with reproduction in the lizard *Lacerta vivipara*. J. Anim. Ecol. 50:733–743.

Bock, W. J. 1980. The definition and recognition of biological adaptation. Am. Zool. 20:217–227.

Charlesworth, B. 1990. Optimization models, quantitative genetics and mutation. Evolution 44:520–538.

Charlesworth, B. 1987. The heritability of fitness. Pp. 21–40 *in* J. W. Bradbury and M. B. Andersson, eds., Sexual Selection: Testing the Alternatives. New York: John Wiley & Sons.

Chippindale, A. K., A. M. Leroi, S. B. Kim, and M. R. Rose. 1993. Phenotypic plasticity and selection in *Drosophila* life-history evolution. I. Nutrition and costs of reproduction. J. Evol. Biol. 6:171–193.

Clutton-Brock, T. H. 1988. Reproductive Success: Studies of Individual Variation in Contrasting Breeding Systems. Chicago: University of Chicago Press.

Congdon, J. D., and J. W. Gibbons. 1987. Morphological constraint on egg size: a challenge to optimal egg size theory? Proc. Natl. Acad. Sci. USA 84:4145–4147.

Darwin, C. 1964 [1859]. The Origin of Species by Means of Natural Selection. Facsimile of the first edition Cambridge, MA, Harvard University Press. Originally published London: John Murray.

Doughty, P., G. Burghardt, and B. Sinervo. 1994. Dispersal in the side-blotched lizard: a test of the mate-defense dispersal theory. Anim. Behav. 47:227–229.

Doughty, P., and B. Sinervo. 1994. The effects of habitat, time of hatching, and body size on dispersal in *Uta stansburiana*. J. Herp et al. 28:485–490.

Endler, J. A. 1986. Natural selection in the wild. Princeton, N.J.: Princeton University Press.

Falconer, D. S. 1965. Maternal effects and selection response. Pp. 763–774 *in* S. J. Geerts, ed., Genetics Today, Proceedings of the XI International Congress of Genetics. Oxford: Pergamon Press.

Falconer, D. S., and T. F. C. Mackay. 1996. Introduction to Quantitative Genetics. Essex, UK: Longman.

Gadgil, M., and W. H. Bossert. 1970. Life historical consequences of natural selection. Am. Nat. 104:1–24.

Gerstle, J. F., and I. P. Callard. 1972. Reproduction and estrogen-induced vitellogenesis in *Dipsosaurus dorsalis*. Comp. Biochem. Physiol. A42:791–801.

Gustafsson, L., and W. J. Sutherland. 1988. The costs of reproduction in collared flycatcher *Ficedula albicollis*. Nature 335:813–815.

James, C. D., and W. G. Whitford. 1994. An experimental study of phenotypic plasticity in the clutch size of a lizard. Oikos 70:49–56.

Johnson, E. O., T. C. Kamilaris, G. P. Chrousos, and P. W. Gold. 1992. Mechanisms of stress: a dynamic overview of hormonal and behavioral homeostasis. Neurosci. Biobehav. Rev. 16:115–130.

Lack, D. 1947. The significance of clutch size. Ibis 89:302–352.

Lack, D. 1954. The Natural Regulation of Animal Numbers. Oxford: Clarendon Press.

Lande, R. 1983. A quantitative genetic theory of life-history evolution. Ecology 63:607–615.

Lande, R., and S. J. Arnold. 1983. The measurement of selection on correlated characters. Evolution 37:1210–1226.

Landwer, A. J. 1994. Manipulation of egg production reveals costs of reproduction in the tree lizard (*Urosaurus ornatus*). Oecocologia 100:243–249.

Lessels, C. M., F. Cooke, and R. F. Rockwell. 1989. Is there a trade-off between egg weight and clutch size in wild lesser snow geese (*Anser c. caerulescens*)? J. Evol. Biol. 2:457–472.

Li, C. C. 1975. Path Analysis — A Primer. Pacific Grove, CA: Boxwood Press.
Luetenegger, W. 1976. Allometry of neonatal size in eutherian mammals. Nature 263:229–230.
Luetenegger, W. 1979. Evolution of litter size in primates. Am. Nat. 114:525–531.
Mitchell-Olds, T., and R. G. Shaw. 1987. Regression analysis of natural selection: statistical and biological interpretation. Evolution 41:1149–1161.
Mousseau, T. A., and D. A. Roff. 1987. Natural selection and the heritability of fitness components. Heredity 59:181–197.
Nilsson, J.-A., and E. Svensson. 1996. The cost of reproduction: a new link between current reproductive effort and future reproductive success. Proc. R. Soc. Lond. B 263:711–714.
Nur, N. 1984. The consequences of brood size for breeding blue tits II. Nestling weight, offspring survival and optimal brood size. J. Anim. Ecol. 53:497–517.
Nur, N. 1988. The costs of reproduction in birds: an examination of the evidence. Ardea 76:155–168).
Ojanen, M., M. Orell, and R. A. Vaisanen. 1979. Role of heredity in egg size variation in the great tit *Parus major* and the pied flycatcher *Ficedula hypoleuca*. Ornis. Scand. 10:22–28.
Orzack, S. H., and E. Sober. 1994. Optimality models and the test of adaptationism. Am. Nat. 143:361–380.
Partridge, L., and P. H. Harvey. 1985. Costs of reproduction. Nature 316:20–21.
Pease, C. M., and J. J. Bull. 1988. A critique of methods for measuring life-history trade-offs. J. Evol. Biol. 1:293–303.
Reznick, D. 1985. Costs of reproduction: an evaluation of the empirical evidence. Oikos 44:257–267.
Reznick, D. 1992. Measuring costs of reproduction. Trends Ecol. Evol. 7:42–45.
Rose, M. R., T. J. Nusbaum, and A. K. Chippindale. 1996. Laboratory evolution: the experimental wonderland and the Cheshire Cat Syndrome. Pp. 221–243 *in* M. R. Rose and G. V. Lauder, ed., Adaptation. San Diego, CA: Academic Press.
Sapolsky, R. M. 1982. The stress-response and social status in the wild baboon. Horm. Behav. 16:279–292.
Sapolsky, R. M. 1992. Neuroendocrinology of the stress-response. Pp. 00–00 *in* J. B. Becker, S. M. Breedlove, and D. Crew, eds., Behavioral Endocrinology. Cambridge, MA: MIT Press.
Schluter, D. 1988. Estimating the form of natural selection on a quantitative trait. Evolution 42:849–861.
Schluter, D., and L. Gustafsson. 1993. Maternal inheritance of condition and clutch size in the collared flycatcher. Evolution 47:658–667.
Schluter, D., and D. Nychka. 1994. Exploring fitness surfaces. Am. Nat. 143: 597–616.
Schluter, D., T. D. Price, and L. Rowe. 1991. Conflicting selection pressures and life-history trade-offs. Proc. R. Soc. Lond. B 246:11–17.
Shine, R. 1980. 'Costs' of reproduction in reptiles. Oecologia 46:92–100.
Sinervo, B. 1990. The evolution of maternal investment in lizards: an experimental and comparative analysis of egg size and its effects on offspring performance. Evolution 44:279–294.
Sinervo, B. 1998a. Adaptation of maternal effects in the wild: path analysis of natural variation and experimental tests of causation. Pp. 288–306 *in* T. Mousseau and C. Fox, eds., Maternal Effects as Adaptations. Oxford: Oxford University Press.
Sinervo, B. 1998b. Mechanistic analysis of natural selection and a refinement of Lack's and William's principles. Am. Nat.
Sinervo, B., and D. F. DeNardo. 1996. Costs of reproduction in the wild: path analysis of natural selection and experimental tests of causation. Evolution 50:1299–1313.

Sinervo, B., and P. Doughty. 1996. Interactive effects of offspring size and timing of reproduction on offspring reproduction: experimental, maternal, and quantitative genetic aspects. Evolution 50:1314–1327.

Sinervo, B., P. Doughty, R. B. Huey, and K. Zamudio. 1992. Allometric engineering: a causal analysis of natural selection on offspring size. Science 258:1927–1930.

Sinervo, B., R. Hedges, and S. C. Adolph. 1991. Decreased sprint speed as a cost of reproduction in the lizard *Sceloporus occidentalis*: variation among populations. J. Exp. Biol. 155:323–336.

Sinervo, B., and P. Licht. 1991a. The physiological and hormonal control of clutch size, egg size, and egg shape in *Uta stansburiana*: constraints on the evolution of lizard life histories. J. Exp. Zool. 257:252–264.

Sinervo, B., and P. Licht. 1991b. Proximate constraints on the evolution of egg size, egg number, and total clutch mass in lizards. Science 252:1300–1302.

Sinervo, B., and C. M. Lively. 1996. The rock-paper-scissors game and the evolution of alternative male reproductive strategies. Nature 380:240–243.

Sinervo, B., and E. Svensson. 1998. Mechanistic and selective causes of life-history tradeoffs and plasticity. Oikos 83:432–442.

Stearns, S. C. 1976. Life-history tactics: a review of the ideas. Q. Rev. Biol. 51:3–47.

Stearns, S. C. 1977. The evolution of life-history traits: a critique of the theory and a review of the data. Annu. Rev. Ecol. Syst. 8:145–171.

Tinkle, D. W. 1967. The life and demography of the side-blotched lizard, *Uta stansburiana*. Misc. Publ. Mus. Zool. Univ. Mich. 132.

Vander Werf, E. 1992. Ecology 73:1699–1705.

van Noordwijk, A. J., J. H. Balen, and W. Scharloo. 1981. Genetic and environmental variation in clutch size of the great tit. Neth. J. Zool. 31:342–372.

Wade, M. J., and S. Kalisz. 1990. The causes of natural selection. Evolution 44:1947–1955.

Weigensberg, I., and D. A. Roff. 1996. Natural heritabilities: can they be reliably estimated in the laboratory? Evolution 50:2149–2157.

Williams, G. C. 1966. Natural selection, the costs of reproduction, and a refinement of Lack's principle. Am. Nat. 100:687–690.

3

Natural Selection and the Evolution of Adaptive Genetic Variation in Northern Freshwater Fishes

BEREN W. ROBINSON AND DOLPH SCHLUTER

The fishes inhabiting depauperate northern postglacial lakes and rivers have lately experienced high rates of divergence and species formation. The study of these fishes provides insight into how genetic variation in populations is molded by divergent environmentally based selection and how genetic variation and selection interact to cause rapid diversification. We summarize patterns of variation and divergence in the trophic traits of these fishes and show that (1) populations in depauperate environments often have elevated levels of niche-based phenotypic variation, which regularly takes the form of trophic or resource polymorphisms, and which appear to represent cases of character release; (2) much of this variation is genetically based and is the product of disruptive selection; (3) parallel patterns of diversification occur within and among populations; (4) in some instances this process has led to stable species coexistence despite gene flow and incipient to full-blown sympatric speciation; and (5) a reserve of genetic variation persists in most postglacial fishes that may allow a response to certain kinds of environmental change. Contemporary patterns of variation reflect a recent history of natural selection that has operated since the last glaciation some 15,000 years ago, but they may also reflect species-level selection that has occurred over the much longer time scale of millions of years representing many glacial periods.

3.1 Genetic Variation in Freshwater Fishes

The origins of diversity cannot be identified without understanding the ecological processes that drive divergence and speciation. The goal of this chapter is to examine the origins and patterns of phenotypic diversity at low taxonomic levels in northern freshwater fishes, in order to assess whether phenotypic diversity is generally adaptive. We define "adaptive genetic variation" as heritable phenotypic variation that is sorted by natural selection into different environmental niches. In practice, adaptive genetic variation is identified by an association between heritable phenotypic variation and the exploitation of different niches. We focus on the fishes of recently deglaciated northern freshwater systems because they are in the process of rapid diversification.

Northern fish faunas are not as species rich as the cichlid species flocks of African rift lakes, but they are simpler and perhaps better suited for study of the ecological causes of phenotypic diversification and adaptive radiation. They are also characterized by a remarkable amount of variation at low taxonomic levels (Svardson 1970, Behnke 1972, Lindsey 1981, Nordeng 1983, Klemetsen 1984, Noakes et al. 1989, Bell and Foster 1994, Robinson and Wilson 1994, Foster et al. 1998), which has produced confusion over their systematics (see Lindsey 1981, Nordeng 1983, Bell and Foster 1994). The relative youth of many taxa in postglacial freshwater systems is supported by the extremely low levels of genetic divergence among many species (Schluter 1996, Bernatchez and Wilson 1998). All of this suggests that the high rate of origination and phenotypic divergence may be a response to abundant ecological opportunities in novel environments created since the last major ice sheets receded. In this chapter, we present evidence that this abundance of variation at low taxonomic levels is in large measure adaptive and is maintained by divergent natural selection.

The most extraordinary feature of variation in postglacial fishes is the apparent process of its conversion from within species to between species. This process is suggested by the observation that niche-based variation is packaged in similar ways across different biological levels, even within single lakes. For example, the shallow littoral margins and the deeper open-waters of many lakes are distinct environments that are inhabited in some cases by slightly different phenotypes within a single population, by distinct morphs in other cases, and by different species in still other cases. All intermediate stages of this continuum are found and are linked in the sense that natural selection is believed to be the mechanism that sorts variation between the littoral and open-water niches in most cases (Robinson and Wilson 1994, Robinson et al. 1996, Schluter 1995, 1996). Note that the occupation of these two niches by a continuum of divergent biological units does not imply that the conversion from the lowest to the highest levels is inevitable, only that differences at the species level may have accumulated through this sequence (Wilson 1989, Skulason and Smith 1995, Smith and Skulason 1996). In other words, adaptive diversification within species is sometimes a precursor to speciation. We consider this possibility further at the end of the chapter.

Our goal is to review the emerging evidence that divergent selection is the primary mechanism that sorts heritable variation into ecological niches at all of these levels. Support for natural selection's dominant role is suggested by six attributes of northern freshwater taxa, which we will discuss in turn. (1) Similar patterns of niche differentiation and associated patterns of body shape differentiation occur at all taxonomic levels up to and including species (in lakes, usually one type is planktivorous, and the other is benthivorous or littoral). (2) Excessive variation is common among ecotypes in the form of trophic or resource polymorphisms and is most dramatic in depauperate environments where competitors and predators are rare. These polymorphisms likely represent cases of character release. (3) Ecotypes persist in single populations (extreme sympatry) in the face of gene flow, as do sibling species. (4) Parallel patterns of divergence have independently evolved on numerous occasions. (5) More direct evidence for natural selection's preeminent role comes from the preliminary confirmation that ecologically intermediate forms exhibit reduced fitness compared to more specialized ecotypes. Divergent selection acts against hybrids between species and against ecologically intermediate phenotypes within species. (6) Niche-based variation can sometimes be converted from within to between species.

3.1.1 Evolutionary context of northern lakes

From the perspective of colonizing fish taxa, northern lakes resemble oceanic islands, and this has influenced contemporary patterns of variation and diversification. Northern lakes were formed following the most recent retreat, starting 15,000 years ago, of massive glaciers that covered the northern regions of North America and Eurasia (Svardson 1961, Pielou 1991). Fish taxa colonized these lakes either from a few widely dispersed freshwater refugia or from the sea. Colonization was impeded by the limited number and duration of passage routes among lakes and rivers (McPhail and Lindsey 1986). Marine access to coastal rivers was limited to salt-tolerant species, and changes in sea and land levels meant that many lakes were accessible for only short periods. Freshwater species that could not spread through coastal marine waters dispersed inland via infrequent changes in the drainage patterns of rivers during deglaciation. Thus, lakes and sometimes whole watersheds of previously glaciated regions share many islandlike features: movement between them was sporadic, their faunas are depauperate and heterogeneous, and separate drainages constitute evolutionarily independent units (Schluter 1996). An additional feature of lakes is that they are environmentally heterogeneous and frequently include multiple niches accessible to fishes. The contrast between the open-water and shallow inshore habitats, in particular, can impose strong divergent selection on a variety of traits related to resource use, defense against predation, and life-history (Smith and Todd 1984, Robinson and Wilson 1994, Schluter 1995, 1996, Robinson et al. 1996). It is in this ecological and historical context that we must search for the factors that have influenced modern patterns of variation.

3.2 Trophic Polymorphism in Depauperate Environments

To say that phenotypic variation runs riot in numerous northern fish taxa is not an overstatement. In most cases, the trophic or resource polymorphism represents the most dramatic form of niche-based variation (Robinson et al. 1993, Robinson and Wilson 1994, Wimberger 1994, Skulason and Smith 1995, Smith and Skulason 1996). We define a trophic polymorphism as excessive niche-based phenotypic variation within a single population. By excessive, we mean that the variation is greater than seems be to the norm in fishes, as judged by the authors that have described them. Variation need not be discrete; it can be continuous as long as it is unusually high. Robinson and Wilson (1994) catalogued cases of coexisting littoral and open-water phenotypes in 37 species of freshwater fishes. Some cases involved discrete trophic phenotypes that rival the differences observed among species, while other cases were more continuously distributed. Four attributes of trophic polymorphisms in fish are consistent with the idea that they represent adaptive genetic variation: (1) parallel patterns of phenotypic variation that accompanies trophic polymorphism in lakes has evolved multiple times among diverse taxa using similar environments; (2) phenotypic variation within taxa parallels adaptive divergence between taxa; (3) some of the phenotypic variation is heritable;

and (4) preliminary evidence suggests that trophic polymorphisms represent instances of character release. We discuss each of these attributes next, and an alternative possibility: that trophic polymorphisms arise through the hybridization of already differentiated populations from different environments.

3.2.1 Patterns of trophic polymorphism in northern fishes

Similarity of patterns of phenotypic variation among distantly related taxa is exemplified by the trophic polymorphisms in the cold water arctic char, and the warmwater pumpkinseed sunfish. Arctic char (*Salvelinus alpinus*) are polymorphic through much of their range, with two forms regularly coexisting in a single lake (reviewed in Griffiths 1994). Usually, a large form is found in the open-water habitat of lakes where it is either planktivorous or piscivorous, and a dwarf form is found feeding on benthic macroinvertebrates inshore. In extreme cases, four sympatric forms can coexist in some Icelandic populations: piscivores, planktivores, large benthivores, and small benthivores (fig. 3.1). The adult morphs are distinguished by an array of trophically related traits. Open-water forms have terminal mouths, larger eyes, more and longer gill rakers (comblike projections used to process planktonic prey in the buccal cavity), and a more slender body shape compared to benthic forms, which display subterminal mouths, smaller eyes, and shorter rakers that are more widely spaced (Sandlund et al. 1987, Snorrason et al. 1989). The pumpkinseed sunfish (*Lepomis gibbosus*) is also trophically polymorphic in some lakes of the mountainous Adirondack region of New York (Robinson et al. 1993, 1996, Robinson and Wilson 1996). Sunfish sampled from the open-water forage extensively on zooplankton and exhibit a more slender body shape and smaller gaps among gill rakers than fish sampled from littoral habitats that do not include plankton in their diet (fig. 3.2). More cases of trophic polymorphisms are listed in table 3.1 (from Robinson and Wilson 1994, Wimberger 1994, Skulason and Smith 1995, Smith and Skulason 1996). A functional interpretation of such body shape variation is possible because it repeatedly arises among independent populations and species inhabiting these lake habitats and because tests of swimming ability and foraging efficiency have been made at the species level (references in Robinson and Wilson 1994).

3.2.2 Heritable basis of trophic polymorphism

It is clear that trophic polymorphisms in fishes reflect elevated genetic variation in addition to phenotypic plasticity associated with ontogenetic changes in diet and habitat use. For example, in the Icelandic char system described above, genetic variation in jaw width, jaw position, and possibly foraging behavior between the piscivore and large benthivore is indicated by its early onset in ontogeny despite rearing juveniles in a common laboratory environment (fig. 3.3; Skulason et al. 1989, 1993). Svedang (1990) has also found genetic variation for life-history and developmental traits among northern European char forms. In sunfish, Robinson and Wilson (1996) demonstrated how phenotypic plasticity and genetic variation both contribute to morphological variation within single populations (fig. 3.4). Svardson (1970) has demonstrated that heritable phenotypic variation for gill raker number in a natural whitefish population can rival the variation frequently observed among sibling species. Extreme parental phenotypes were mated with each other (high × high with 41 gill rakers each and

Figure 3.1 The four sympatric morphs of arctic char (*Salvelinus alpinus*) from Thangvallavatn in Iceland. All individuals are adults. From the top down: Large benthic (33 cm long), small benthic (8 cm.; both found in the inshore littoral habitat), piscivore (35 cm), and planktivore (19 cm.; both present in the open water habitat). (Redrawn from Skulason and Smith 1995.)

low × low with only 32 gill rakers) to demonstrate that their respective progenies would maintain the widely divergent gill raker numbers (high progeny = 38.3, low progeny = 33.9).

Heritable phenotypic variation is not limited to sexual fish species. Naturally occurring asexual populations of topminnows (*Poeciliopsis* sp.) found in depauperate desert streams in Mexico also exhibit considerable amounts of variation in the form of phenotypically differentiated clones that occupy alternate trophic niches (Vrijenhoek 1978, 1994). Heritable variation exists for many ecologically important life-history, physiological, and morphological traits in salmon (table 3.2). Phenotypic plasticity also plays a significant role in other well known cases of trophic polymorphism (e.g., Todd et al. 1981, Meyer 1987, 1990, Wimberger 1991, 1992, 1994). In these cases, however, genetic variation in morphology or behavior cannot be ruled out because appropriate experiments combining a common garden approach in a reciprocal transplant design have not been performed (e.g., fig. 3.4). The relative importance of, and

Table 3.1 Geographic locations of trophic or resource polymorphisms in North American freshwater fishes (used in fig. 3.5), and also found elsewhere

Common name	Genus	Species	Family	Spp. range center (lat.)	TP (lat.)	Var	Env	Reference
North America								
Brook char	*Salvelinus*	*fontinalis*	Salmonidae	49	43		S	McLaughlin and Grant (1994)
Artic char	*Salvelinus*	*alpinus*		61	69[a]		L	D. Griffiths (1994, personal communication)
Sockeye	*Oncorhynchus*	*nerka*		55	59		L	Blair et al. (1993)
Cisco	*Coregonus*	*artedi*		57	42		L	Behnke (1972)
				57	45		L	Behnke (1972)
				57	44		L	Behnke (1972)
				57	66		L	Behnke (1972)
Least cisco	*Coregonus*	*sardinella*		71	69		L	Mann and McCart (1981)
				71	71		L	Lindsey and Kratt (1982)
				71	68		L	Lindsey and Kratt (1982)
				71	69		L	Lindsey and Kratt (1982)
Lake whitefish	*Coregonus*	*clupeaformis*		53	61		L	Bernatchez et al. (1996)
				53	48		L	Bodaly et al. (1991)
				53	46		L	Bodaly et al. (1991)
Pygmy whitefish	*Prosopium*	*coulteri*		56	59		L	McCart (1970)
Pumpkinseed sunfish	*Lepomis*	*gibbosus*	Centrarchidae	40	44[b]	G, E	L	Robinson (1994)
Bluegill sunfish	*Lepomis*	*macrochirus*		37	43		L	Ehlinger and Wilson (1988)
				37	35[c]		L	Layzer and Clady (1987)
Three-spine stickleback	*Gasterosteus*	*aculeatus*	Gasterosteidae	49	62		L	Cresko and Baker (1996)
				49	50		L	Schluter and McPhail (1992)
Rainbow smelt	*Osmerus*	*mordax*	Osmeridae	56	45[d]	G	L	Taylor and Bentzen (1993b)
				56	46		L	Taylor and Bentzen (1993a)
				56	48		L	Taylor and Bentzen (1993a)
Johny darter	*Etheostoma*	*nigrum*	Percidae	43	43	G	L	Hubbs (1961)
Yellow perch	*Perca*	*flavescens*		46	49		L	Hubbs (1961)
Cichlid	*Cichlasoma*	*minckleyi*	Cichlidae		27[e]	G	L	Sage and Selander (1975)
Tui chub	*Gila*	*bicolor*	Cyprinidae	40	40		L	Galat and Vucinich (1983)
Gila topminnow	*Poeciliopsis*	*occidentalis*	Poeciliidae	29	29[f]	G	S	Vrijenhoek (1978, 1994)
Western brook lamprey	*Lampreta*	*richardsoni*	Petromyzontidae	50	50	G	S	Beamish (1987)

Central and South America (including central and southern Mexico)								
Midas cichlid	*Cichlasoma*	*citrinellum*	Cichlidae	12	12[g]	E	L	Meyer (1989, 1990)
Cichlid	*Cichlasoma*	*managuense*		10	12[g]	E	L	Meyer (1987)
	Cichlasoma	*labiatum*		12		E	L	Meyer (1993)
	Cichlasoma	*haitensis*		19	19		L	Meyer (1993)
Goodeid	*Ilyodon*		Goodeidae	19	19	E	S	Grudzien and Turner (1984)
Saccodon	*Saccodon*		Paredontidae		8		S	Roberts (1974)
Percichthys	*Percichthys*	*trucha*	Percichthyidae		40°S			Ruzzante et al. (in review)
Europe								
Artic char	*Salvelinus*	*alpinus*	Salmonidae	61	60[h]		L	D. Griffiths (1994, personal communication)
Other								
Ayu fish (Japan)	*Plecoglossus*	*altivelus*	Osmeridae		35		L	Azuma (1987)
Cichlid (Israel)	*Serotherodon*	*gallaeus*	Cichlidae		33		L	Shapiro and Chervinski (1990)
Cichlid (Africa)	*Astatoreochromis*	*alluaudi*			1°S	E	L	Greenwood (1959, 1965)
Cichlid (Africa)	*Labidochromis*	*caeruleus*			12°S		L	Meyer (1993)
Cichlid (Africa)	*Hemichromis*	*letourneauxi*			12°S		L	Meyer (1993)
Cichlid (Africa)	*Perissodus*	*microlepis*			4°S	G	L	Hori (1993)

Shown are the latitudes (degrees north) of each species' range center (Spp. range center), and each known case of a trophic polymorphism (TP). Polymorphisms with heritable genetic variation (G) or environmentally induced variation (E) are noted in the "Var" column. The local environment (Env) of each polymorphism is given as stream (S), lake (L), or marine (M).

[a] This is the mean latitude for 10 polymorphic arctic char populations (latitudes: 50, 57, 62, 67, 2x68, 74, 3x80).
[b] This is the mean latitude for 17 polymorphic pumpkinseed sunfish populations in the Adirondacks, NY (43, 15x44, 45)
[c] Wichitaw Mountains lake, southwest Oklahoma.
[d] Possible sympatric speciation from trophic polymorphism.
[e] Quatro Cienegas basin, Chihuahua desert, Mexico.
[f] Hybrid clones, Sonoran desert streams, Mexico.
[g] Crater lake, Nicaragua.
[h] This is the mean latitude for 59 cases of polymorphic arctic char populations in Europe.

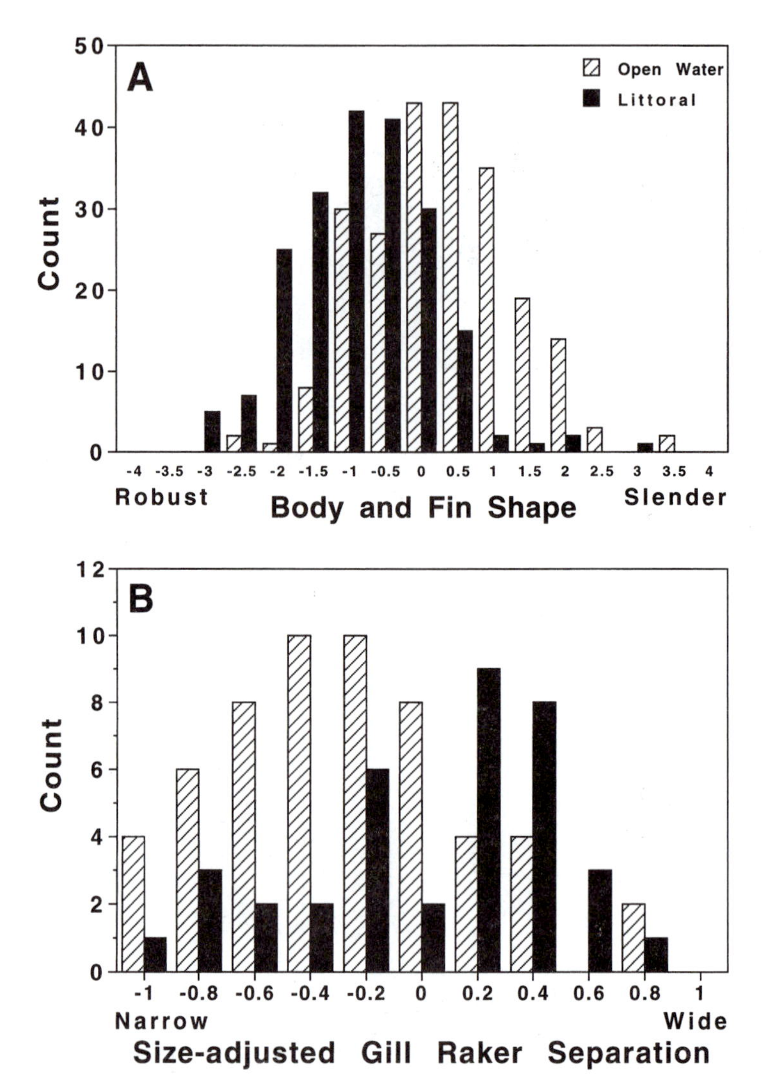

Figure 3.2 (A) Body-shape distributions of pumpkinseed sunfish phenotypes (*Lepomis gibbosus*) sampled from the shallow littoral zone and deep open water habitat of Paradox Lake (Adirondack region of New York) between 1990 and 1993 (Robinson 1994). The horizontal axis represents a complex of morphological traits obtained from a discriminant function analysis of 21 body-shape characters. Open water fish exhibit on average a more slender body shape and less robust paired fins than their littoral counterparts. Seventy percent of the 433 fish were correctly classified to habitat (Wilks' $\lambda = 0.746$, $F = 6.99$, df = 15,415, $P < .0001$). (B) Distances between gill rakers in a subset of open water and littoral sunfish in (A). The size of the gap between rakers was up to 38% smaller in the open water compared to littoral fish (ANOVA of size-adjusted average gap: $F = 7.9$, df = 1, 63, $P = .007$). The width of three gaps between the first four gill rakers below the apical gill raker on the descending gill arch were averaged together in this analysis (Robinson et al. 1993.)

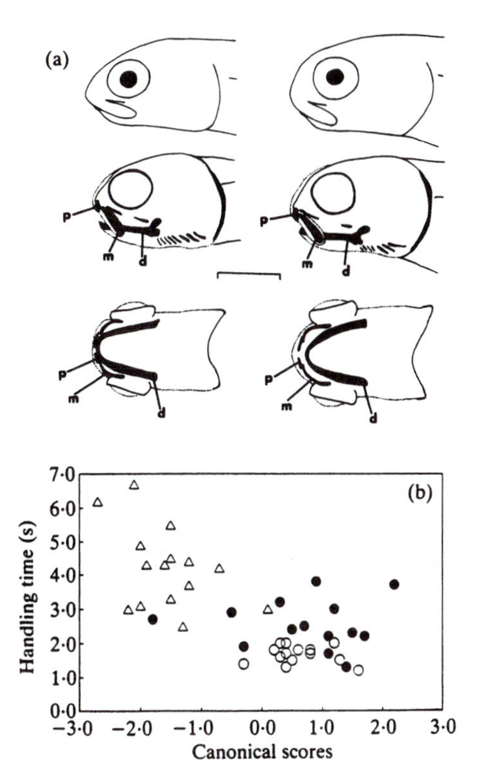

Figure 3.3 Heritable variation in (a) head morphology, and (b) foraging behavior of arctic char from Thingvallavatn, Iceland. In each case, progeny were reared in a common laboratory environment. (a) The progeny of piscivores (left column) have mouths that are narrower and more terminal than their large benthic counterparts (right column) at 225 days of age. The diagram at the top shows the preserved head; in the middle is the same head after clearing and staining (only a few bones are shown in black), and the bottom shows a ventral view (d, dentary; m, maxilla; p, premaxilla); scale bar = 3 mm. (Reprinted from Skulason et al. 1989). (b) Relationship between body shape and handling time of Daphnia (a zooplankton prey common in the open water habitat of many northern lakes) in foraging tests with 15-month-old laboratory-reared juvenile arctic char (r = -0.63; P << .01). Small benthivores (triangles) have higher handling times than do planktivores (solid circles) or piscivores (open circles). (Reprinted from Skulason et al. 1993.)

interaction between, genetic variation and phenotypic plasticity in producing phenotypic variation associated with trophic polymorphisms and even closely related species (Day et al. 1994) awaits a wider comparative analysis among taxa (Robinson and Wilson 1996).

3.2.3 Character release

Trophic polymorphisms in fishes are believed to be examples of character release (Robinson et al. 1993, Robinson and Wilson 1994, Skulason and Smith 1995). Character release is an increase in the variation of certain characters associated with the enlargement of a species'

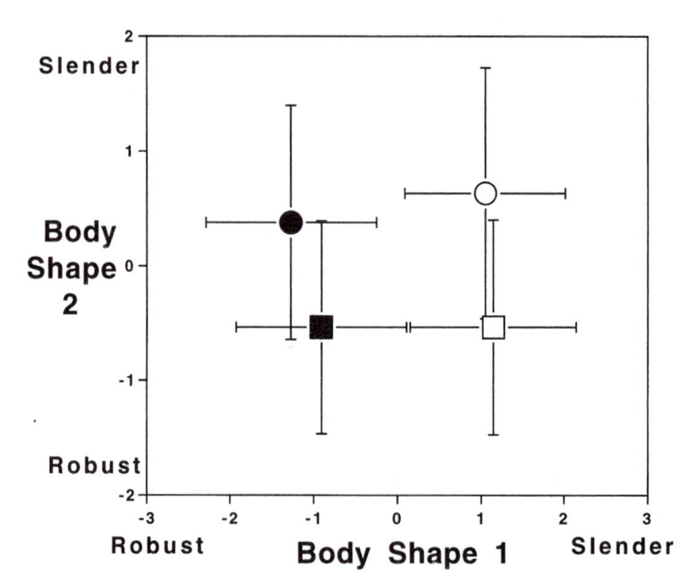

Figure 3.4 Mean body shape (± 1 SD) of pumpkinseed sunfish progeny whose parents were collected from shallow littoral (squares) or open water (circles) habitats in Paradox Lake (Adirondack, NY). Each axis represents a complex of morphological traits obtained from a discriminant function analysis of 21 body-shape characters measured on the progeny. Both progeny groups were split and reared for 113 days in cages placed in the open water (white) and littoral (black) habitat of an artificial pond. Sunfish reared in open water cages (white symbols) were on average more slender than fish reared in shallow water cages (black symbols) regardless of parental type, as expected under an adaptive plastic response to rearing environment. Heritable variation in morphology also exists because the progeny of open water fish (circles) were on average more slender than the littoral progeny (squares) regardless of rearing environment. Environmentally induced variation (primarily along the horizontal axis) and heritable variation (vertical axis), respectively, accounted for 53% and 14% of the total morphological variation in this experiment. The four treatment groups were distinct in multivariate morphological space using a discriminant function analysis (Wilks' λ = 0.322, F = 11.75, df = 60, 1525, P < .0001), or a two-factor MANOVA of three sets of factor scores (factors: environment Wilks' λ = 0.02, F = 164, df = 3, 10, P < .0001; offspring type Wilks' λ = 0.22, F = 11.84, df = 3, 10, P = .001; interaction Wilks' λ = 0.425, F = 4.51, df = 3, 10, P = .03). (Redrawn from Robinson and Wilson 1996: fig. 2.)

niche due to the absence of other restricting species, such as competitors or predators (Van-Valen 1965). We may then predict that instances of trophic polymorphisms should occur disproportionately often in depauperate and/or novel environments (Robinson and Schluter, unpublished data). Figure 3.5 compares the latitudinal distribution of trophic polymorphisms with two estimates of fish species diversity in North America. The first is the distribution of all species' geographic range centers. The second is an independent estimate of fish species diversity based on location data for 501 species of fish on a 1° grid across North America between 25° and 50° north (McAllister et al. 1986). Our range center data are remarkably

Table 3.2 Phenotypic differences in morphology and life history between the anadromous (searun form) of sockeye salmon (*Oncorhynchus nerka*) and its purely freshwater form

Trait	Searun sockeye	Freshwater kokanee
Early development		
Growth rate	High	Low
Growth rate variation	Low	High
Smolt size	Large	Small
Physiology		
Swimming performance	Stronger	Weaker
Seawater adaptability	Early	Late
(hybrids: intermediate)		
Maturation		
Age at maturation	Older	Younger
Size at maturation	Large	Small
Morphology		
Gill raker number	Fewer	Many

These differences appear to be adaptations for migration to and maturation in the marine environment versus for living exclusively in freshwater habitats. Heritable variation underlies of all of these traits, as they are maintained despite the rearing of progeny in common laboratory environments. From Foote et al. (1992) and Wood and Foote (1996).

congruent with McAllister et al's. (1986) estimates of species diversity at middle latitudes, indicating that our range center distribution correctly reflects a northward decline in fish species diversity. Two features of the trophic polymorphism distributions stand out. First, most cases occur at northern latitudes where species diversity is declining or low: 94% are above 39° north. Second, this latitude represents the maximum southerly glacial advance of the Wisconsin ice sheet in North America approximately 15,000 years ago (Crossman and McAllister 1986). In the Americas several trophic polymorphisms are found below 39°, but these all arguably occur in depauperate environments, such as crater lakes, alpine lakes, desert streams, and desert springs (see table 3.1). In other words, trophic polymorphisms overwhelmingly occur in relatively depauperate and young environments, as predicted. This is consistent with a scenario of excessive phenotypic variation being expressed as a consequence of ecological release from competitors and predators. Character release may also have occurred in the North American cyprinid minnows (Eastman and Underhill 1973) and in lampreys (e.g., Beamish 1987; reviewed in Bell and Andrews 1998).

A second prediction of character release is that it should be most common in heterogeneous environments. A prerequisite of this prediction is that the niches are unoccupied by other species, and so an association between cases of character release and niche diversity will only be found in depauperate environments (as discussed above). Niche-based variation is frequently related to environmental heterogeneity in freshwater fishes. Fish species diversity is positively related to lake area which parallels species-area relationships observed in other taxa (e.g., Barbour and Brown 1974, Griffiths 1997). Diversity increases because lake area is thought to be positively correlated with environmental heterogeneity. This in turn permits greater niche-based separation of coexisting forms.

Trophic polymorphisms in fish also appear to be positively associated with lake size in

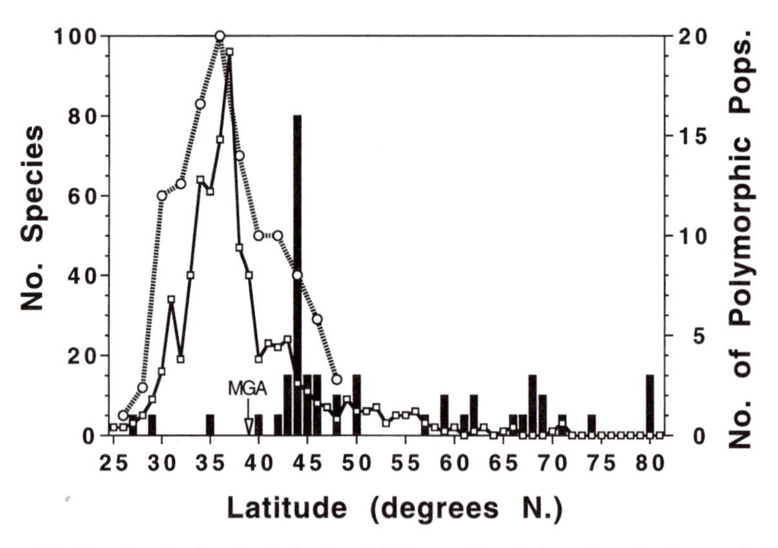

Figure 3.5 The distributions by latitude of all known North American freshwater fish populations exhibiting trophic polymorphisms (histogram scaled to right axis), superimposed under two estimates of fish species diversity (lines scaled to left axis). The first is based on the number of species range centers (solid line with squares), and the second is an independent estimate of species diversity of 501 species (hatched line with circles, also scaled to left axis, but should be multiplied by a factor of five). The patterns of species diversity are congruent, reflecting a peak diversity at 36.5° and a rapid reduction in diversity northward. Ninety-four percent of trophic polymorphisms occur in the relatively depauperate and young aquatic environments above 39° north, which was the maximum glacial advance (MGA) southward during the last (Wisconsin) glaciation. Cases below this latitude occur in older but still depauperate environments (see text and table 3.1). We calculated each species range center as half the distance between the maximum northward and southward extensions of its geographic range to the nearest 1° latitude using range maps from Page and Burr (1991) and Lee et al. (1980). The fish species diversity curve was redrawn from McAllister et al. 1986: fig. 2.5). Details of the 53 trophically polymorphic populations are given in table 3.1.

two ways. First, cases of trophic polymorphisms in both arctic char and pumpkinseed sunfish are significantly more common in larger than in smaller lakes (Hindar and Jonsson 1982, Robinson 1994, unpublished data). Griffiths (1994) analyzed 173 published reports in an attempt to test factors that influence arctic char polymorphism. He found that body size (and trophic) polymorphism was positively related to lake area and depth and negatively related to the number of resident fish species as expected. Second, the morphological separation of open and shallow water forms appears to increase with lake size (Griffiths 1994, Robinson unpublished data). In other words, the minimum requirements for trophic polymorphisms are distinct and productive open-water and littoral environments. This preliminary evidence is consistent with the prediction that niche-based phenotypic variation is both more common and greater in larger and more environmentally heterogeneous lakes. In the future, it should be possible to test if trophic polymorphisms are rare below some minimum lake size (or some other metric incorporating lake area and depth).

3.2.4 Hybridization and the generation of variation

We must consider the possibility that trophic polymorphisms result from the mixing of already differentiated populations that inhabited different environments. If so, then trophic polymorphism may be an epiphenomenon arising out of hybridization and not an evolutionary phenomenon resulting from ecological release. Hybridization is extremely common among north temperate fishes (Hubbs 1955), and Svardson (1970) and Nordeng (1983) consider hybridization to be a significant variance-generating mechanism in European arctic char. Grant (this volume) is also studying how hybridization increases variation in species of Darwin's finches (*Geospiza*) on the Galapagos Islands. However, although hybridization may explain elevated levels of genetic variation in some instances in northern freshwater fishes, it is not expected to increase levels of phenotypic plasticity, which is known to be a significant factor in numerous cases of trophic polymorphism (Meyer 1987, Wimberger 1991, 1992, Robinson and Wilson 1996). As noted above, we know little about the interaction between phenotypic plasticity and genetic divergence in the evolution of trophic polymorphisms. Some trophic polymorphisms also occur in such species-poor regions that recent hybridization seems unlikely. This may be the case in Iceland, which is home to the most extreme example of trophic polymorphism known in arctic char (Sandlund et al. 1987, Skulason et al. 1989, Snorrason et al. 1989). The role that hybridization plays in other cases of trophic polymorphism remains to be studied in a systematic fashion. Molecular genetic approaches may yield some insights into this question if markers can be found that distinguish between sympatric origins versus the hybridization of divergent allopatric populations.

3.3 Stable Divergence Despite Gene Flow

Gene flow among demes or species is a common feature of northern freshwater fishes and counteracts any buildup of an association between heritable phenotypic variation and habitats. Natural selection is the only process that can maintain such an association over time because no random mechanism is expected to consistently partition variation in the face of the homogenizing effects of gene flow.

Hybridization may be more common among northern freshwater species than among any other group of fishes in the world (Hubbs 1955), and it is likely more common at lower taxonomic levels. For example, hybridization is low but persistent among many closely related species (e.g., Svardson 1970, Hammar et al. 1989, 1991, Bodaly et al. 1992, McPhail 1994), moderate among populations or races (e.g., Svardson 1970, McPhail 1994), and probably high among morphs within polymorphic populations (e.g., Vuorinen et al. 1993). Whether the persistence of coexisting trophic morphs in the face of gene flow is evidence of selection depends on the genetic basis of the traits. For example, although there is gene flow (no assortative mating) between the beak morphs in the trophically polymorphic African finch, *Pyrenestes ostrinus* (Smith 1987, 1990a,b), this will not cause the collapse of this trophic polymorphism even in the absence of selection. This is because the beak polymorphism is coded for by a single gene with two alleles (one dominant), and hybridization creates no intermediate forms. In contrast, phenotypic variation associated with trophic polymorphisms in northern fish taxa involve many traits for body size and shape, mouth size and position, and gill raker architecture, and it is unlikely that this variation is coded by a simple genetic system such as that found in the African finches. In the absence of selection, hybridization is expected to produce an intermediate hybrid "swarm." For example, Hatfield (1997) demonstrated that some of the important morphological differences between sympatric sibling species of three-spined stickleback (*Gasterosteus aculeatus*) are polygenic, since F_1 and F_2 hybrids exhibit an array of phenotypes intermediate to the original species.

Although empirical verification is required, trophically polymorphic fishes presumably exhibit a high degree of gene flow among forms, particularly in those populations that do not exhibit marked bimodal variation. Yet habitat-specific variation is found with respect to a variety of morphological and behavioral traits (e.g., Robinson et al. 1993, 1996, Robinson and Wilson 1994, Skulason and Smith 1995, Smith and Skulason 1996). An exemplary case of diversification despite moderate gene flow occurs among parapatric demes of kokanee, a completely freshwater form of sockeye salmon (*Oncorhynchus nerka*). Kokanee are born in the tributary streams of large lakes. As juveniles, they must migrate into the resource-rich lake in order to grow to maturity and then return usually to their natal stream to reproduce. Average gene flow of up to 5% occurs among tributaries indicating that individuals sometimes reproduce in streams other than their natal stream (Wood and Foote 1996). Raleigh (1967) tested the association between migration behavior of recently hatched wild kokanee fry and the direction of current found in their natal streams (i.e., inlet versus outlet streams). If migration behavior is adaptive, then fry placed in an artificial stream should migrate in opposite directions (e.g., inlet fish moving downstream and outlet fish moving upstream). Ninety-eight percent of inlet fry migrated downstream over 24 h as expected, almost always at night (96% moving in the dark). In contrast, 30% of the fry from the outlet stream moved upstream, with an equal frequency moving during the day and night. Although it is curious that 70% of the outlet fry also moved downstream, other lakes occur below this system, and so downward migration by outlet fry may reflect movement into these lakes. This pattern requires verification. The timing of migration may also be under natural selection in this case because predatory lake fishes congregate at the mouths of inlet streams to forage on fry, flushed into the lake (predators do not congregate similarly at outlet streams because prey are not so concentrated). Piscivores are visual predators and presumably can be avoided by migrating into the lake at night, although this needs to be tested. Differences in migration behavior between inlet and outlet kokanee fry are likely maintained by natural selection imposed both by resource demands and by predators despite moderate amounts of gene flow among streams. Many instances of divergence in migration behavior, life history, and morphology have also recently and independently evolved between stream- and lake shore-spawning kokanee (Carl and Healey 1984, Swain and Holtby 1989, Blair et al. 1993, Burger et al. 1995,Taylor et al. 1997). All of these are candidate systems in which to test for divergent patterns of natural selection.

Marked phenotypic divergence is also common among closely related species despite historic and current low but constant levels of gene flow. For example, the marine three-spined stickleback has colonized a vast array of northern freshwater environments (Bell and Foster 1994). Some coastal lakes of British Columbia contain sympatric stickleback species pairs (McPhail 1984, 1992, 1993, 1994, Schluter and McPhail 1992). A "limnetic" species forages on zooplankton prey in the open-water habitat of lakes and is smaller, more slender, and distinct in other ways from a "benthic" species that forages in the shallower inshore habitat (fig. 3.6). Most of the trophically related traits that distinguish benthics from lim-

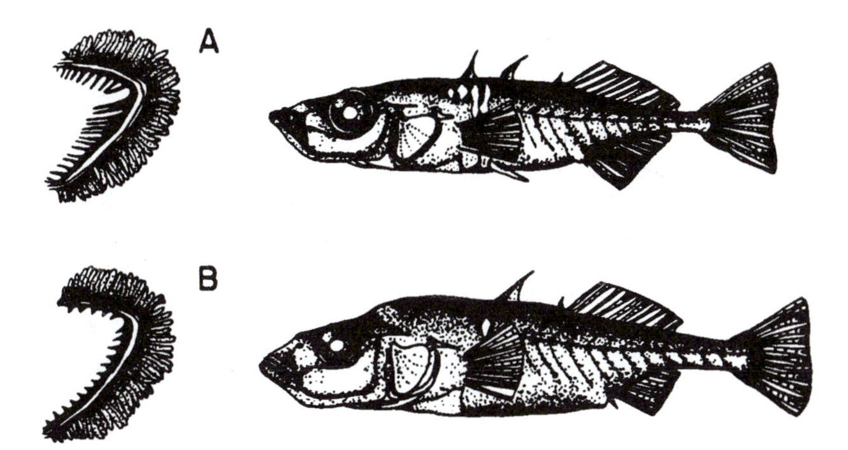

Figure 3.6 Adult males from sympatric sibling species of three-spined stickleback (*Gasterosteus aculeatus*) found in Paxton Lake, British Columbia. (A) The open water "limnetic" species, and (B) the shallow inshore "benthic" species. The limnetic male is more slender and less robust than the benthic male. Limnetics also exhibit on average larger eyes, smaller mouths, more and longer gill rakers (the comblike projections shown on the gill arches to the right) for their size compared to benthics. (Reprinted from Schluter 1993.)

netics are heritable (McPhail 1984), but it is estimated that hybrids persistently make up 1–2% of the adult individuals in a lake (McPhail 1984, 1992). Evidence of past or present gene flow is also indicated by molecular genetic studies (Taylor et al. 1997). Nonetheless, these species pairs are maintained despite current and likely higher historic levels of hybridization. These and other examples of marked phenotypic variation in the presence of gene flow are strong evidence for the important role played by natural selection in sorting variation among lake habitats.

3.4 Parallel Patterns of Phenotypic Divergence

Natural selection is the only process expected to produce parallel patterns of diversification among independent groups experiencing similar environmental conditions. Niche-based variation that is similarly expressed by evolutionarily independent taxa is therefore additional evidence that natural selection is sorting variation among different niches (Schluter and Nagel 1995). Such parallel patterns of phenotypic variation are seen among independent populations and, most remarkably, among distantly related families that share common environments. Parallel patterns of divergence are found across a number of pairs of ecological niches in aquatic systems.

One of the best examples of a parallel diversification is between sea run and freshwater resident sockeye salmon (*O. nerka*), which has a large range in freshwater habitats draining into the north Pacific. The sockeye is the classic anadromous salmon born in freshwater streams. After 1 or 2 years in freshwater, it migrates as a juvenile to the ocean, where it grows into an adult before migrating 2–4 years later, sometimes 1000 miles inland, to its natal creek to spawn and die. As identified earlier, a purely freshwater form of sockeye salmon, the kokanee, also exists. The anadromous and freshwater forms of sockeye are different in traits related to early development, physiology, maturation, and morphology that appear functionally useful for their respective life-history strategies (table 3.2). Remarkably, kokanee have been independently derived from anadromous sockeye on numerous occasions. Taylor et al. (1996) have identified at least two phenotypically similar kokanee stocks that are more closely related to local sockeye than they are to each other (fig. 3.7; see also Kurenkov 1978, Wilmot and Burger 1985, Wood and Foote 1990, 1996, Foote et al. 1992,). Kokanee have also spontaneously arisen in sockeye populations transplanted to rivers outside of their natural range (Taylor et al. 1996). Parallel shifts in life-history and phenotype occur with the loss of anadromy in many salmonid species (Hindar et al. 1986, Stahl 1987, Skaala and Naevdal 1989, Birt et al. 1991, reviewed in Behnke 1972), sticklebacks (Bell and Foster 1994, Foster et al. 1998), lampreys (reviewed in Bell and Andrews 1998), the Japanese ayu fish (Azuma 1987), and rainbow smelt (Taylor and Bentzen 1993a, 1993b), indicating that divergent selection operates between sea run and freshwater forms in most of these species (reviewed in Bell and Andrews 1998).

An even more remarkable pattern of parallel divergence is exhibited by distantly related taxa that colonize the open (limnetic) versus shallow (benthic) habitats of lakes. Across diverse taxa such as whitefish, salmonids, sticklebacks, smelt, and, to a lesser extent, sunfish, limnetic forms all possess terminal mouths, slender bodies, and either more or longer gill rakers compared to benthic forms (see, e.g., Lindsey 1981, Schluter and McPhail 1992, Robinson and Wilson 1994, Bell and Andrews 1998). This remarkable parallelism in habitat-specific phenotypes can only be accounted for if natural selection plays a dominant role in molding patterns of variation in taxa using these two lake habitats. Such a consistent pattern also suggests that considerable heritable variation exists to exploit these different habitats.

Developmental constraints inherited from a common ancestor may facilitate the origin of parallel patterns of diversification. This is most conceivable within a lineage of closely related forms, such as in the diverging sockeye/kokanee system described above. However, it is much less likely to be a factor promoting parallel evolution among more distantly related families of fish (e.g., whitefish, salmonids, sticklebacks, smelt, and sunfish). In the future, it will also be possible to test for parallel patterns of divergence between the distantly related postglacial fishes of North America and South America in similar environments (e.g., Ruzzante et al., in review). Among closely related species or divergent populations, we can determine whether parallel divergence is the result of natural selection or developmental constraints by a simple test: is divergence the same along other environmental gradients? If so, then developmental constraints are supported because the phenotypic response to the environment is limited. Alternatively, if the pattern of divergence changes along different environmental gradients, then natural selection is supported.

3.5 Divergent Selection and Niche-based Variation

Divergent natural selection on traits that enhance the efficient use of resources is thought to be a major mechanism in the evolution of trophic polymorphisms and adaptive radiation in postglacial fishes. The repeated evolution of limnetic and benthic forms in many lake-

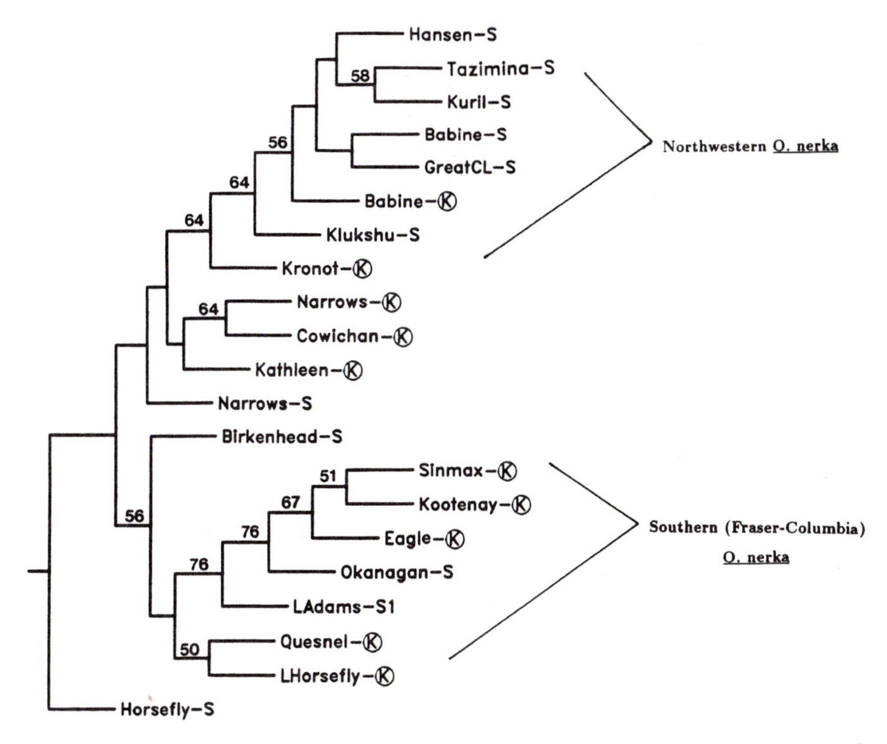

Figure 3.7 Evolutionary relationships among sockeye salmon (S) and kokanee (K) populations in northwestern North America. Both are members of the species *Oncorhynchus nerka*. The relationships are represented by the consensus tree from 100 bootstrap maximum-likelihood analyses of the allele frequencies at 3 loci (2 minisatellite and 1 mtDNA). Nodes receiving at least 50% support from 100 bootstrap replicates are indicated by numbers at branch points. (Reprinted from Taylor et al. 1996.)

dwelling taxa suggests that natural selection is often divergent between the open and shallow water habitats. The presence of divergent selection can tell us two things. First, natural selection is an important factor in sorting variation among niches. Second, when divergent selection occurs on different phenotypes within a species, we can infer that elevated amounts of heritable variation probably exist (unlike stabilizing selection, which constantly reduces variation). Below, we outline preliminary evidence for divergent selection acting between open-water and littoral habitats at two taxonomic levels.

Using the stickleback species pairs, Schluter (1995) performed reciprocal transplant experiments that demonstrated superior growth (a fitness-related trait) by adults of each species in their respective preferred lake habitats (fig. 3.8). Hybrids, which possessed a phenotype intermediate between the two species, were on average inferior to each species in both environments. This suggests an adaptive landscape with two fitness peaks corresponding to each habitat, separated by a valley of lower fitness occupied by the hybrids. This may be due to how well each species forages on the habitat-specific prey. The divergent trait combinations of the benthic and limnetic sticklebacks result in significant foraging efficiency differ-

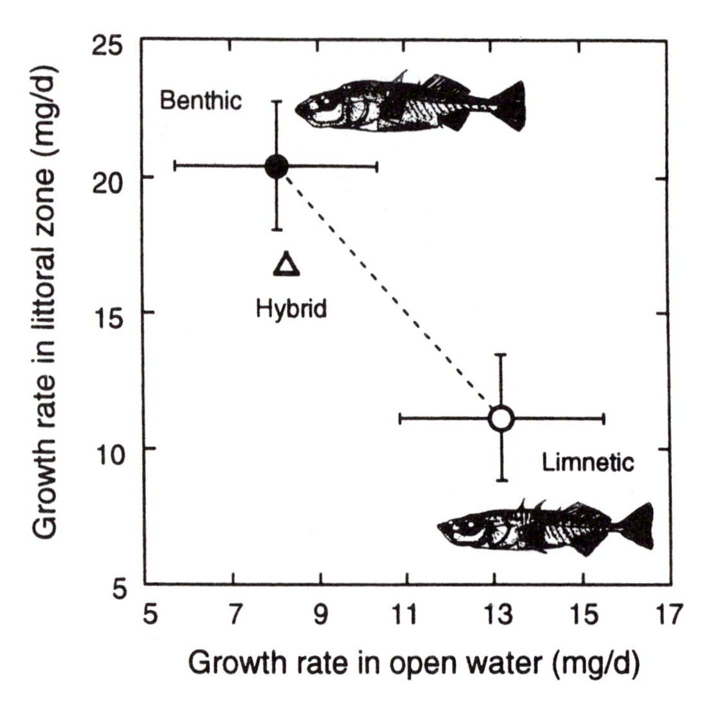

Figure 3.8 Divergent selection favors benthic and limnetic stickleback species respectively in the littoral and open water habitats of Paxton Lake, British Columbia, because of the severe trade-off in growth rate experienced by each species between lake habitats and because of the relative inferiority of intermediate hybrid phenotypes. Symbols indicate species means ± 1 SE. (Reprinted from Schluter 1995.)

ences in laboratory trials (Schluter 1993). It is also unlikely that these hybrids experience reduced fitness due to nonecological effects of their genotype (Futuyma and Moreno 1988) because they exhibit strong viability in the laboratory and in the field (Hatfield 1995).

Divergent selection also appears to favor different pumpkinseed sunfish forms in the open and shallow water habitats in some Adirondack lakes. Robinson and Wilson (1996) demonstrated that extreme phenotypes resident in each lake habitat generally performed better than did individuals of intermediate phenotype (fig. 3.9). This pattern was significant in the open-water and weak in the littoral environment. Performance was estimated using two fitness-related traits: condition factor, measured as lipid content in skeletal muscle, and relative growth rates, estimated using scale measurements. Thus, it seems reasonable that divergent selection affecting traits involved in resource use can be a major mechanism resulting in trophic polymorphism and adaptive radiation in northern fishes.

Divergent selection may act in other cases (e.g., Ehlinger 1990, McPhail 1994) and along other environmental gradients, such as between marine versus freshwater habitats (e.g., see table 3.2; Wood and Foote 1996; reviewed in Bell and Andrews 1998), lakes versus streams (e.g., Reimchen et al. 1985, Lavin and McPhail 1993, Thompson et al. in press), inlet versus outlet streams of lakes (Raleigh 1967), riffles versus pools in streams (McLaughlin et al. 1994, McLaughlin and Grant 1994), and perhaps even between different spawning times

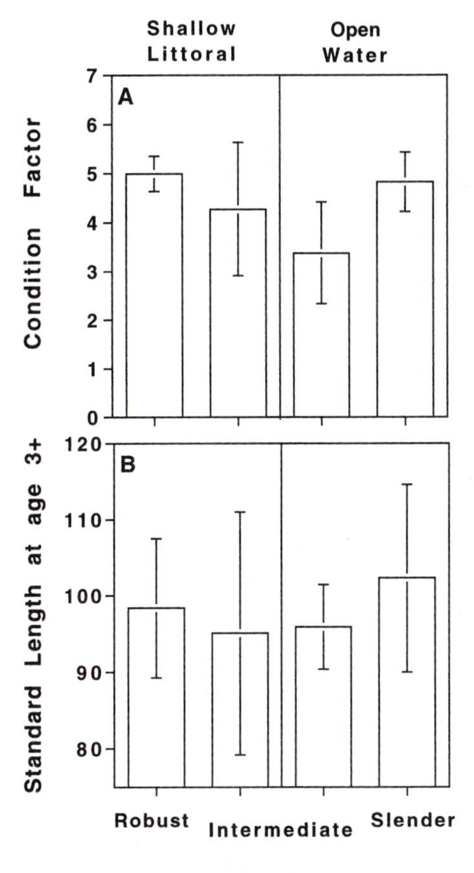

Figure 3.9 Divergent selection acting on pumpkinseed sunfish phenotypes in Paradox Lake, NY, based on two fitness related traits: (A) condition factor (lipid content of muscle tissue), and (B) growth rate (standard length at age 3+ calculated from scale measurements). The average performance of extreme open water and littoral phenotypes in their respective habitats was compared to individuals with intermediate phenotypes present in each habitat. The fitness of the extreme morphs resident in each habitat is significantly greater than intermediate morphs in the open water ($P < .05$) but not in the littoral habitat. Means ± 1 SE are shown for the 10% most extreme individuals in each habitat. (Modified from Robinson et al. 1996: figs. 2, 3.)

(e.g., spring versus fall), because of the cascading effects of temperature on many life-history traits in northern fishes (Smith and Todd 1984). Although these cases are intriguing, most provide only weak evidence for divergent selection because the fitness (or at least performance) of intermediate phenotypes are rarely compared to habitat-specific forms (but see, e.g., Wood and Foote 1990). Future studies addressing the importance of divergent selection must measure selection on intermediate forms relative to extreme phenotypes.

3.6 Conversion of Niche-based Variation into Species

Northern fishes exhibit strikingly similar patterns of niche-based variation at a variety of biological levels, and the sympatric occurrence of forms is common. We have reviewed the evidence that much of this variation is adaptive with respect to specific environmental niches in postglacial lakes. These observations challenge us to consider the origins of closely related species. Two alternative speciation models can account for sibling taxa found in sympatry, and both require niche-based variation. The models are distinguished by whether the variation is the cause or the consequence of speciation.

The ecological speciation model (Schluter 1996) assumes that niche-based variation is under strong divergent selection and that speciation results (even with gene flow) either as a direct byproduct of natural selection acting against intermediate phenotypes (Rice and Hostert 1993) or by reinforcement of sexual selection against hybrids (Liou and Price 1994). Ecological speciation can occur in sympatry, parapatry, allopatry, or any sequential combination of these. In the alternative nonadaptive speciation model, speciation occurs for reasons other than divergent natural selection, such as genetic drift or founder effects. As sibling species became sympatric, slight differences resulted in further divergence, perhaps through a process of character displacement (Lack 1947). This model requires a significant period of allopatry to permit speciation. We will not review theories of speciation here, but instead suggest that many attributes of northern freshwater taxa are consistent with the first "ecological speciation" model.

The view that divergent ecologically based selection is the primary force driving speciation is supported by five attributes of northern fishes found in postglacial lakes (Schluter 1996). (1) Both phenotypic differentiation and assortative mating evolve rapidly (often in less than 15,000 years). In many cases, premating isolation is linked to the morphological traits (such as body size) that have diverged between species (Ridgeway and McPhail 1984, McKinnon 1995, Rundle and Schluter 1998). In other cases temporal or spatial variation in mating occurs (Svardson 1970, Behnke 1972, Lindsey 1981, Smith and Todd 1984). (2) Species persist in sympatry despite gene flow and do not collapse into hybrid swarms. (3) Phenotypic diversification is associated with a high degree of niche differentiation. (4) Assortative mating and niche-based phenotypic divergence have independently evolved in similar environments on numerous occasions (Schluter and Nagel 1995). (5) Hybrids are viable and fertile, but exhibit reduced fitness in specific natural environments. Hybrid disadvantage results from the ecological consequences of having an intermediate phenotype and is not caused by purely genetic factors. Random drift and nonecological processes that cause sterility and inviability of hybrids among isolated populations require much longer periods of time than are characteristic of recently deglaciated northern freshwater systems. Instead, northern lineages exhibit recent postglacial bursts of speciation (Bernatchez and Wilson 1998). Divergent selection can cause reproductive isolation much sooner than drift and may allow stable sympatry even after relatively brief periods of geographical isolation where drift could effect little change. In other words, ecological environments most be considered to understand the origin of species and adaptive radiation in northern freshwater systems.

Divergence, and ultimately speciation, without natural selection seems more of an exception than the rule in northern fishes. There is evidence that some sympatric kokanee populations may be reproductively isolated, with little niche-based morphological divergence (e.g., Taylor et al. 1997), but this is rare. It is also clear that isolation in glacial refugia has influenced patterns of divergence in this system (Bernatchez and Wilson 1998), although this does not by itself deny an important diversifying role played by natural selection. We also do not expect every trophically polymorphic population to continue to diverge, ultimately forming separate species (even in the absence of future glaciations). Trophic polymorphisms are, however, candidate systems for ecological speciation along the lines suggested by byproduct and/or reinforcement models (Rice and Hostert 1993, Liou and Price 1994). Finally, it is noteworthy that most of the best cases of sympatric speciation occur in the freshwater fishes of lakes (table 3.3). Clearly, heritable, niche-based variation within a species can sometimes be converted into variation between species. Unfortunately, we cannot yet say how frequently this has occurred in sympatry because spurious molecular genetic signals of divergence are easily created by the high degree of gene flow among taxa, and because diagnostic alleles are still rare.

A rich array of diversifying mechanisms occur in this system, but this does not dilute the fact that the fish of postglacial lakes are providing many opportunities to understand the role played by divergent selection in speciation. We require further field studies of selection, on traits determining reproductive isolation. We need to clarify the mechanisms of selection and the precise basis of pre- and postmating isolation. Are the genes underlying hybrid inferiority generally those that cause hybrids to be phenotypically intermediate? Are the traits involved in premating behavioral isolation associated with resource use? Postglacial fishes are ideal for studying these questions because of their tractability for ecological experiments, the simplicity of their relatively depauperate environments, and the remarkable amounts of niche-based phenotypic variation exhibited at numerous biological levels.

3.7 Responses to Human-induced Environmental Change

We now briefly turn our attention to how northern taxa may respond to human-induced changes to the global environment. Although reserves of genetic variation are likely available for further diversification, they may be inadequate for certain types of environmental change or may not provide protection from too rapid an environmental change (see reviews by Tonn 1990, Carpenter et al. 1992). For example, gradual long-term warming at lower elevations and latitudes will mean warmer water and reduced oxygen concentrations, a problem for most cold-adapted species such as salmonids. Although such a change will adversely affect their eggs and fry (Meisner et al. 1988), it may be offset by increased overwinter survival due to warmer winters. At higher latitudes and elevations, warmer temperatures are expected to lengthen productive summer seasons, which could benefit all life-history stages. This would permit the colonization of even more northerly lakes (Shuter and Post 1990, Magnuson et al. 1990). Too rapid or extreme an environmental change, however, may cause extinction through habitat loss and genetic homogenization due to hybridization in refugia, rather than further northerly range shifts and diversification (Smith and Todd 1984, Miller et al. 1989). Further work in this regard is required.

The consequences of introduction of non-native species on endemic northern taxa may also be severe if driven by environmental warming because adaptive genetic variation may not persist in the face of species interactions. Most diversification in northern taxa has occurred in depauperate environments (fig. 3.5). The distribution of southern species and their impact on northern taxa seems to be currently limited by thermal barriers, which, if removed by global warming, could have devastating effects on northern fish taxa. Stickle-

Table 3.3 Suggested instances of "intralacustrine" speciation in freshwater fishes

Species	Location	Replication	Time (years)	Unique genotype	Reference
Sockeye kokanee	Pacific NW America	>2	15,000		Taylor et al. (1996)
	British Columbia	3	15,000		Foote et al. (1989) Wood and Foote (1996)
Sockeye races	Alaska	1	15,000	Yes	Wilmot and Burger (1985)
	Kamchatka	1	12,500		Kurenkov (1978)
Rainbow smelt	New Brunswick	1	15,000	Yes	Taylor and Bentzen (1993a, 1993b)
Whitefish	Yukon	1	15,000		Bernatchez et al. (1996)
Cichlids	Cameroon[a]	2		Yes	Schliewen et al. (1994)

[a] Not a northern lake, although this endemic radiation occurred in a species-poor volcanic crater lake.

backs, salmonids, and other cold-water species may not be generally resistant to invasion by non-native fishes (e.g., Crowder 1986, Mandrake 1989, reviewed in Carpenter et al. 1992), perhaps because they have evolved in response to selection imposed by habitat-specific resources in depauperate environments and not to interactions with many other species (e.g., McPhail 1994). Southerly warm-water fishes are known to be more omnivorous and have rapid generation times compared to northern taxa, making them successful invaders of disturbed habitats (Carpenter et al. 1992). In short, northern fish taxa may have little adaptive variation to resist southerly taxa invading because of global warming.

3.8 Niche-based Divergent Selection versus Species Selection

At the beginning of this chapter, we defined adaptive genetic variation as heritable phenotypic variation sorted by natural selection into different environmental niches. We then reviewed evidence for the importance of environmentally based divergent selection in contemporary northern fish taxa. Because of the youth of postglacial northern environments, contemporary patterns of adaptive genetic variation are primarily a consequence of relatively recent natural selection. However, this short-term view ignores a potentially much longer history of species-level selection that may be equally important.

Over the past 2 million years of our current glacial age, there have been some 20 glacial cycles lasting approximately 100,000 years, with glaciations of between 60,000 and 90,000 years separated by shorter warmer interglacials (Pielou 1991). Many northern fish taxa are believed to have survived these glaciations in a few refugia, from which they colonized the ice-scoured northern regions as the planet warmed (see Bernatchez and Wilson 1998). Consequently, the northern fish fauna has been exposed to repeated cycles of range expansion as taxa colonized new freshwater habitats, followed by extinction over much of their range. The small remnant populations surviving glaciation in isolated refugia then provided the colonists for the next period of range expansion during the warmer interglacials. To what extent has selection, imposed by such long-term environmental fluctuations, influenced contemporary patterns of genetic variation? Perhaps the capacity of many northern fish taxa to survive gradual global warming reflects species-level selection that resulted from nonrandom patterns of extinction. Robinson and Wilson (1994) proposed that this long-term cycle may have influenced the evolution of phenotypic plasticity within some northern fish species, but it may also have favored the survival and evolution of northern species that exhibit a certain "glacial genotype." In particular, species that were good at dispersing into newly created freshwater habitats during the warmer interglacials and that survived in a few glacial refugia during the cooler glacial periods may have been favored under this cycle.

Species that exhibit enhanced dispersal into newly created northern rivers and lakes may be at a selective advantage over slower colonizing species for two reasons. First, the ability to colonize newly created lakes may be important if possession of a particular environmental niche in a newly created and potentially resource-poor lake confers a competitive advantage over following taxa. This would require verification. Second, selection may have favored those species that when faced with competition could shift into other vacant local niches within a lake. Consider the outcomes of a dispersing species on arrival in a new lake. If it is the first to colonize the lake, then it may have a moderate chance of survival. For a variety of reasons, however, even taxa that are good dispersers will sometimes arrive in a lake after the littoral or open-water niche is already inhabited. If they cannot invade the inhabited niche, then the capacity to shift into an available vacant niche would be advantageous. Traits involved with long-range dispersal may have a positive influence on the ability to switch niches within a lake (short-range dispersal), a hypothesis that could be tested if taxa having the largest northerly geographic range also use the largest array of environmental niches.

In short, the favored glacial genotype would have the capacity to produce phenotypes that are advantageous in a variety of freshwater environmental niches during the dispersal phase of the cycle. Such a genotype may be phenotypically plastic (e.g., Via et al. 1995), or perhaps it has the capacity to maintain hidden genetic variation that can be mobilized or expressed in new environments (e.g., Schmalhausen 1949), or maybe it can infrequently hybridize. At the moment, we have little quantitative information on the dispersal abilities of northern fish taxa. However, a comparative phylogenetic analysis of dispersal ability among northern (and perhaps distantly related postglacial Southern Hemisphere) fish taxa may be possible in the future that could test if excellent dispersal ability is a feature that has evolved in parallel. At this point, the most reasonable expectation about genetic variation in northern fishes is that it reflects a recent history of natural selection imposed by different freshwater environmental niches since the last glaciation, and possibly species selection over at least 1 million years.

Acknowledgments We thank J. D. McPhail, E. Taylor, A. Mooers, S. Vamosi, J. Pritchard, S. Otto, M. Whitlock, A. Margosian, and two anonymous reviewers for helpful discussions and comments. M. Bell, L. Bernatchez, R. McLaughlin, and D. Ruzzante all kindly provided unpublished manuscripts. This research was supported by a National Science and Engineering Research Council (NSERC) of Canada postdoctoral fellowship to B.W.R. and by an NSERC operating grant to D.S.

References

Azuma, M. 1987. On the origin of Koayu, a landlocked form of amphidromous Ayu-fish, *Plecoglossus altivelis*. Verhandlungon, International Association of Theoretical and Applied Limnology 21: 1291–1296.

Barbour, C. D., and J. H. Brown. 1974. Fish diversity in lakes. American Naturalist 108: 473–489.

Beamish, R. J. 1987. Evidence that parasitic and nonparasitic life-history types are produced by one population of lamprey. Canadian Journal of Fisheries and Aquatic Sciences 44: 1779–1782.

Behnke, R. J. 1972. The systematics of Salmonid fishes of recently glaciated lakes. Journal of the Fisheries Research Board of Canada 29: 639–671.

Bell, M. A., and C. A. Andrews 1998. Evolutionary consequences of postglacial colonization of fresh water by primitively anadromous fishes. In B. Streit, T. Stadler, and C. M. Lively, eds., Evolutionary Ecology of Freshwater Animals. Birkhauser Verlag, Basel.

Bell, M. A., and S. A. Foster (eds). 1994. The Evolutionary Biology of the Threespine Stickleback. Oxford University Press, Oxford.

Bernatchez, L., J. A. Vuorinen, R. A. Bodaly, and J. J. Dodson. 1996. Genetic evidence for reproductive isolation and multiple origins of sympatric trophic ecotypes of whitefish (*Coregonus*). Evolution 50: 624–635.

Bernatchez, L., and C. C. Wilson. 1998. Comparative phylogeography of Nearctic and Palearctic fishes. Molecular Ecology, 7:431–452.

Birt, T. P., J. M. Green, et al. 1991. Mitochondrial DNA variation reveals genetically distinct sympatric populations of anadromous and nonanadromous Atlantic salmon, *Salmo salar*. Canadian Journal of Fisheries and Aquatic Sciences 48: 577–582.

Blair, G. R., D. E. Rogers, and T. P. Quinn. 1993. Variation in life-history characteristics and morphology of sockeye salmon in Kvichak River system, Bristol Bay, Alaska. Transactions of the American Fisheries Society 122: 550–559.

Bodaly, R. A., J. Vuorinen, and V. Macins. 1991. Sympatric presence of dwarf and normal forms of the lake whitefish, *Coregonus clupeaformis*, in Como Lake, Ontario. Canadian Field Naturalist 105: 87–90.

Bodaly, R. A., J. W. Clayton, C. C. Lindsey. and J. Vourinen. 1992. Evolution of lake whitefish (*Coregonus clupeaformis*) in North America during the Pleistocene: genetic differentiation between sympatric populations. Canadian Journal of Fisheries and Aquatic Sciences 49: 769–779.

Burger, C. V., J. E. Finn, and L. Holland-Bartels. 1995. Pattern of shoreline spawning by sockeye salmon in a glacially turbid lake: evidence for subpopulation differentiation. Transactions of the American Fisheries Society 124: 1–15.

Carl, L. M., and M. C. Healey. 1984. Differences in enzyme frequency and body morphology among three juvenile life-history types of Chinook salmon (*Oncorhynchus tshawytscha*) in the Nanaimo River, British Columbia. Canadian Journal of Fisheries and Aquatic Sciences 41: 1070–1077.

Carpenter, S. R., S. G. Fisher, N. B. Grimm, and J. F. Kitchell. 1992. Global change and freshwater ecosystems. Annual Review of Ecology and Systematics 23: 119–139.

Cresko, W. W., and J. A. Baker, 1996. Two morphotypes of lacustrine threespine stickleback, *Gasterosteus aculeatus*, in Benka Lake, Alaska. Environmental Biology of Fishes 45: 343–350.

Crossman, E. J., and D. E. McAllister. 1986. Zoogeography of freshwater fishes of the Hudson Bay Drainage, Ungava Bay and the Arctic archipelago. Pp. 53–104 *in* C. H. Hocutt and E. O. Wiley, eds., The Zoogeography of North American Freshwater Fishes. John Wiley and Sons, New York.

Crowder, L. B. 1986. Ecological and morphological shifts in Lake Michigan fishes: glimpses of the ghost of competition past. Environmental Biology of Fishes, 16: 147–157.

Day, T., Pritchard, J., and D. Schluter. 1994. Ecology and genetics of phenotypic plasticity: a comparison of two sticklebacks. Evolution 48: 1723–1734.

Eastman, J. T., and J. C. Underhill. 1973. Intraspecific variation in the pharyngeal tooth formulae of some cyprinid fishes. Copeia 1: 45–53.

Ehlinger, T. J. 1990. Habitat choice and phenotype-limited feeding efficiency in bluegill: individual differences and trophic polymorphisms. Ecology 71: 886–896.

Ehlinger, T. J., and D. S. Wilson. 1998. Complex foraging polymorphism in bluegill sunfish. Proceedings of the National Academy of Sciences USA 85: 1878–1882.

Foote, C. J., C. C. Wood, and R. E. Withler, 1989. Biochemical genetic comparison of sockeye salmon and kokanee, the anadromous and nonanadromous forms of *Oncorhynchus nerka*. Canadian Journal of Fisheries and Aquatic Sciences 46: 149–158.

Foote, C. J., C. C. Wood, and R. E. Withler. 1992. Biochemical genetic comparison of sockeye salmon and kokanee, the anadromous and nonanadromous forms of *Oncorhynchus nerka*. Canadian Journal of Fisheries and Aquatic Sciences 46: 149–158.

Foster, S. A., Scott, R. J., and W. A. Cresko. 1998. Nested biological variation and speciation. Philosophical Transactions of the Royal Society of London B 353: 207–218.

Futuyma, D. J. and G. Moreno. 1988. The evolution of specialization. Annual Review of Ecology and Systematics 19: 207–233.

Galat, D. L., and N. Vucinich. 1983. Food partitioning between young of the year of two sympatric Tui chub morphs. Transactions of the American Fisheries Society 112: 486–497.

Greenwood, P. H. 1959. The monotypic genera of cichlid fishes in Lake Victoria, Part II. Bulletin of the British Museum of Natural History 5: 165–177.

Griffiths, D. 1994. The size structure of lacustine arctic charr (Pisces: Salmonidae) populations. Biological Journal of the Linnean Society 51: 337–357.

Griffiths, D. 1997. Local and regional species richness in North American lacustrine fishes. Journal of Animal Ecology 66: 49–56.

Grudzien, T. A., and B. J. Turner. 1984a. Direct evidence that the *Ilyodon* morphs are a single biological species. Evolution 38: 402–407.

Hammar, J., J. B. Dempson, and E. Skold. 1989. Natural hybridization between Arctic char (Salvelinus alpinus) and lake char (*S. namaycush*): evidence from northern Labrador. Nordic Journal of Freshwater Research 65: 54–70.

Hammar, J., J. B. Dempson, and E. Verspoor. 1991. Natural hybridization between Arctic charr (Salvelinus alpinus) and brook trout (*S. fontinalis*): evidence from northern Labrador. Canadian Journal of Fisheries and Aquatic Sciences 48: 1437–1445.

Hatfield, T. 1995. Speciation in sympatric stickelbacks: hybridization, reproductive isolation and the maintenance of diversity (PhD thesis). University of British Columbia, Vancouver.

Hatfield, T. 1997. Genetic divergence in adaptive characters between sympatric species of sticklebacks. American Naturalist 149: 1009–1029.

Hindar, K., and B. Jonsson. 1982. Habitat and food segregation of the dwarf and normal Arctic charr (*Salvelinus alpinus*) from Vangsvatnet Lake, western Norway. Canadian Journal of Fisheries and Aquatic Sciences 39: 1030–1045.

Hindar, K., N. Ryman, and G. Stahl. 1986. Genetic differentiation among local populations and morphotypes of Arctic charr, *Salvelinus alpinus*. Biological Journal of the Linnaean Society 27: 269–285.

Hori, M. 1993. Frequency-dependent natural selection in the handedness of scale-eating fish. Science 260: 216–219.

Hubbs, C. L. 1955. Hybridization between fish species in nature. Systematic Zoology 4: 1–20.

Hubbs, C. L. 1961. Isolating mechanisms in the speciation of fishes. Pp. 5–23 *in* W. F. Blair, ed., Vertebrate Speciation. University of Texas Press, Austin.

Klemetsen, A. 1984. The Arctic charr speciation problem as seen from northern Norway. Pp. 65–77 in L. Johnson and B. Burns, eds., Biology of the Arctic Charr. University of Manitoba Press, Winnipeg.

Kurenkov, S. I. 1978. Two reproductively isolated groups of Kokanee Salmon, *Oncorhynchus nerka* Kennerlyi, from Lake Kronotskiy. Journal of Ichthyology 17: 526–534.

Lack, D. 1947. Darwin's Finches. Cambridge University Press, Cambridge.

Lavin, P. A., and J. D. McPhail. 1993. Parapatric lake and stream sticklebacks on northern Vancouver Island: disjunct distribution or parallel evolution? Canadian Journal of Zoology 71: 11–17.

Layzer, J. B., and M. D. Clady. 1987. Phenotypic variation of young-of-year bluegills (*Lepomis macrochirus*) among microhabitats. Copeia 1987: 702–707.

Lee, D. S., C. R. Gilbert, C. H. Hocutt, R. E. Jenkins, D. E. McAllistair, and J. R. Stauffer Jr. 1980. Atlas of North American freshwater fishes. North Carolina State Museum of Natural History, Raleigh.

Lindsey, C. C. 1981. Stocks are chameleons: plasticity in gill rakers of Coregonid fishes. Canadian Journal of Fisheries and Aquatic Sciences 38: 1497–1506.

Lindsey, C. C., and L. F. Kratt. 1982. Jumbo spotted form of least cisco *Coregonus sardinella* in lakes of southern Yukon territory. Canadian Journal of Zoology 60: 2783-2786.

Liou, L. W., and T. D. Price. 1994. Speciation by reinforcement of premating isolation. Evolution 48: 14451–1459.

Magnuson, J. J., J. D. Meisner, and D. K. Hill. 1990. Potential changes in the thermal habitat of the Great Lakes fishes after climate warming. Transactions of the American Fisheries Society 119: 254–264.

Malmquist, H. J., S. S. Snorrason, and S. Skulason. 1985. Bleikjan i Pingvallavatni I. Faeduhaettir. Natturufraedingurinn 55: 195–217.

Mandrake, N. E. 1989. Potential invasion of the Great Lakes by fish species associated with climatic warming. Journal of Great Lakes Research 15: 306–316.

Mann, G. J., and P. J. McCart. 1981. Comparison of sympatric dwarf and normal populations of least cisco (*Coregonus sardinella*) inhabiting Trout Lake, Yukon Territory. Canadian Journal of Fisheries and Aquatic Sciences 38: 240–244.

McCart, P. 1970. Evidence for the existence of sibling species of pygmy whitefish (*Prosopium coulteri*) in three Alaskan lakes. In C. C. Lindsey & C. S. Woods (eds.), Biology of coregonid fishes. Univ. Manitoba Press, Winnipeg. Pp. 81–98.

McAllister, D. E., S. P. Platania, F. W. Schueler, M. E. Baldwin, and D. S. Lee. 1986. Ichthyofaunal patterns on a geographic grid. Pp. 17–52 *in* C. H. Hocutt and E. O. Wiley, eds., The Zoogeography of North American Freshwater Fishes. John Wiley and Sons, New York.

McKinnon, J. S. 1995. Video mate preferences of female threespine sticklebacks from populations with diverent male coloration. Animal Behavior 50: 1645–1655.

McLaughlin, R. L., and J. W. A. Grant. 1994. Morphological and behavioral differences among recently-emerged brook charr, *Salvelinus fontinalis*, foraging in slow- vs. fast-running water. Environmental Biology of Fishes 39: 289–300.

McLaughlin, R. L., J. W. A. Grant, and D. L. Kramer. 1994. Foraging movements in relation to morphology, water-column use, and diet for reently emerged brook trout (*Salvelinus fontinalis*) in still-water pools. Canadian Journal of Fisheries and Aquatic Sciences 51: 268–279.

McPhail, J. D. 1984. Ecology and evolution of sympatric sticklebacks (*Gasterosteus*): morphological and genetic evidence for a species pair in Enos Lake, British Columbia. Canadian Journal of Zoology 62: 1402–1408.

McPhail, J. D. 1992. Ecology and evolution of sympatric sticklebacks (*Gasterosteus*): evidence for a species pair in Paxton Lake, Texada Island, British Columbia. Canadian Journal of Zoology 70: 361–369.

McPhail, J. D. 1993. Ecology and evolution of sympatric sticklebacks (*Gasterosteus*): origin of the species pairs. Canadian Journal of Zoology 71: 515–523.

McPhail, J. D., 1994. Speciation and the evolution of reproductive isolation in the sticklebacks (*Gasterosteus*) of southwestern British Columbia. Pp. 399–437 *in* M. A. Bell and S. A. Foster, eds., Evolutionary Biology of the Threespine Stickleback. Oxford University Press, Oxford.

McPhail, J. D., and C. C. Lindsey 1986. Zoogeography of the freshwater fishes of Cascadia (the Columbia system and rivers north to the Stikine). Pp. 615–638 in C. H. Hocutt and E. O. Wiley, eds., The Zoogeography of North American Freshwater Fishes. John Wiley and Sons, New York.

Meisner, J. D., J. S. Rosenfeld, and H. A. Regier. 1988. The role of ground water in the impact of climate warming on stream salmonids. Fisheries 13: 2–8.

Meyer, A. 1987. Phenotypic plasticity and heterochrony in *Cichlasoma managuense* (Pisces, Cichlidae) and their implications for speciation in cichlid fishes. Evolution 41: 1357–1369.

Meyer, A. 1989. Cost of morphological specialization: Feeding performance of the two morphs in the trophically polymorphic cichlid fish, *Cichlasoma citrinellum*. Oecologia 80: 431–436.

Meyer, A. 1990. Ecological and evolutionary consequences of the trophic polymorphism in *Cichlasoma citrinellum* (Pisces, Cichlidae). Biological Journal of the Linnean Society 39: 279–299.

Meyer, A. 1993. Trophic polymorphisms in cichlid fish: Do they represent intermediate steps during sympatric speciation and explain their rapid adaptive evolution? Pp. 257–266 in J. H. Schroder, J. Bauer, & M. Schartl (eds.), Trends in Ichthyology. Blackwell Scientific.

Miller, R. R., J. D. Williams, and J. D. J. D. Williams. 1989. Extinctions of North American fishes during the last century. Fisheries 14: 22–38.

Noakes, D. L. G., S. Skulason, and S. S. Snorrason. 1989. Alternative life-history styles in salmonine fishes with emphasis on arctic charr, *Salvelinus alpinus*. Pp. 329–346 *in* M. N. Bruton, ed., Alternative life-history styles of animals. Kluwer Academic Publishers, Dordrecht.

Nordeng, H. 1983. Solution to the "char" problem based on arctic char (*Salvelinus alpinus*) in Norway. Canadian Journal of Fisheries and Aquatic Sciences 40: 1372–1387.

Page, L. M., and B. M. Burr. 1991. A field guide to freshwater fishes of North America north of Mexico. Peterson field guide series, 42. Houghton Mifflin Co., Boston, MA.

Pielou, E. C. 1991. After the ice age, Chicago University Press, Chicago.

Raleigh, R. F. 1967. Genetic control in the lakeward migrations of sockeye slamon (*Oncorhynchus nerka*) fry. Fisheries Research Board of Canada 24: 2613–2622.

Reimchen, T. E., E. M. Stinson, and J. S. Nelson. 1985. Multivariate differentiation of parapatric and allopatric populations of threespine stickleback in the Sangan River watershed, Queen Charlotte Islands. Canadian Journal of Zoology 63: 2944–2951.

Rice, W. R., and E. E. Hostert. 1993. Laboratory experiments on speciation: what have we learned in 40 years? Evolution 47: 1637–1653.

Ridgeway, M. S., and J. D. McPhail. 1984. Ecology and evolution of sympatric sticklebacks (*Gasterosteus*): mate choice and reproductive isolation in the Enos Lake species pair. Canadian Journal of Zoology 62: 1813–1818.

Roberts, T. R. 1974. Dental polymorphism and systematics in *Saccodon*, a Neotropical genus of freshwater fishes (Paodontidae, Characoidei). J. Zool. Lond. 173: 303–321.

Robinson, B. W. 1994. Evolutionary ecology of phenotypic polymorphisms in fishes (PhD thesis). Binghamton University, Binghamton, NY.

Robinson, B. W., and D. S. Wilson. 1994. Character release and displacement in fishes: a neglected literature. American Naturalist 144: 596–627.

Robinson, B. W., and D. S. Wilson. 1996. Genetic variation and phenotypic plasticity in a trophically polymorphic population of pumpkinseed sunfish (*Lepomis gibbosus*). Evolutionary Ecology 10: 631–652.

Robinson, B. W., D. S. Wilson, A. S. Margosian, and P. T. Lotito. 1993. Ecological and morphological differentiation by pumpkinseed sunfish in lakes without bluegill sunfish. Evolutionary Ecology 7: 451–464.

Robinson, B. W., D. S. Wilson, and G. O. Shea. 1996. Trade-offs of ecological specialization: an intraspecific comparison of pumpkinseed sunfish phenotypes. Ecology 77: 170–178.

Rundle, H. D., and D. Schluter. 1998. Reinforcement of stickleback mate preferences: sympatry breeds contempt. Evolution 52: 200–208.

Ruzzante, D. E., S. J. Walde, V. E. Cussac, P. J. Macchi, and M. F. Alonso. 1998. Trophic polymorphism, habitat and diet segregation in *Percichthys trucha* in the Andes. Biological Journal of the Linnean Society 65: 191–214.

Sage, R. D., and R. K. Selander. 1975. Trophic radiation through polymorphism in cichlid fishes. Proceedings of the National Academy of Sciences, USA 72: 4669–4673.

Sandlund, O. T., B. Jonsson, H. J. Malmquist, R. Gydemo, T. Lindem, S. Skulason, S. S. Snorrason, and P. M. Jonasson. 1987. Habitat use of arctic charr *Salvelinus alpinus* in Thingvallavatn, Iceland. Environmental Biology of Fishes 20: 263–274.

Schliewen, U. K., D. Tautz, and S. Paabo. 1994. Sympatric speciation suggested by monophyly of crater lake cichlids. Nature 368: 629–632.

Schluter, D. 1993. Adaptive radiation in sticklebacks: size, shape, and habitat use efficiency. Ecology 74: 699–709.

Schluter, D. 1995. Adaptive radiation in sticklebacks: trade-offs in feeding performance and growth. Ecology 76: 82–90.

Schluter, D. 1996. Ecological speciation in postglacial fishes. Philosophical Transactions of the Royal Society of London Series B 351: 807–814.

Schluter, D., and J. D. McPhail. 1992. Ecological character displacement and speciation in sticklebacks. American Naturalist 140: 85–108.

Schluter, D., and L. M. Nagel. 1995. Parallel speciation by natural selection. American Naturalist 146: 292–301.

Schmalhausen, I. I. 1949. Factors of Evolution. University of Chicago Press, Chicago.

Shapiro, J., and J. Chervinski. 1990. The occurrence of two groups of *Saratherodon galilaeus* (Artedi) in Lake Kinneret; a reality. J. Aqua. Trop. 5: 199–208.

Shuter, B. J., and J. R. Post. 1990. Climate, population variability, and the zoogeography of temperate fishes. Transactions of the American Fisheries Society 119: 314–336.

Skaala, O., and G. Naevdal. 1989. Genetic differentiation between freshwater resident and anadromous brown trout, *Salmo trutta*, within watercourses. Journal of Fish Biology 34: 597–605.

Skulason, S., D. L. G. Noakes, and S. S. Snorrason. 1989. Ontogeny of trophic morphology in four sympatric morphs of arctic charr *Salvelinus alpinus* in Thingvallavatn, Iceland. The Biological Journal of the Linnean Society 38: 281–301.

Skulason, S., and T. B. Smith. 1995. Resource polymorphisms in vertebrates. Trends in Ecology and Evolution 10: 366–370.

Skulason, S., S. S. Snorrason, D. Ota, and D. L. G. Noakes. 1993. Genetically based differences in foraging behaviour among sympatric morphs of arctic charr (Pisces: Salmonidae). Animal Behaviour 45: 1175–1192.
Smith, G. R., and T. N. Todd 1984. Evolution of species flocks of fishes in north temperate lakes. Pp. 45–68 *in* A. A. Echelle and I. Kornfield, eds., Evolution of Fish Species Flocks. University of Maine at Orono Press, Orono.
Smith, T. B. 1987. Bill size polymorphism and intraspecific niche utlization in an African finch. Nature 329: 717–719.
Smith, T. B. 1990a. Natural selection on bill characters in the two bill morphs of the African finch *Pyrenestes ostrinus*. Evolution 44: 832–842.
Smith, T. B. 1990b. Resource use by bill morphs of an African finch: evidence for intraspecific competition. Ecology 71: 1246–1257.
Smith, T. B., and S. Skulason. 1996. Evolutionary significance of resource polymorphisms in fish, amphibians and birds. Annual Review of Ecology and Systematics 27: 111–133.
Salmonidae): morphological divergence and ontogenetic niche shifts. Biological Journal of the Linnean Society 52: 1–18.
Snorrason, S. S., S. Skulason, O. T. Sandlund, H. J. Malmquist, B. Jonsson, and P. M. Jonasson. 1989. Shape polymorphism in arctic charr, *Salvelinus alpinus*, in Thingvallavatn, Iceland. Pp. 393–404 in H. Kawanabe, F. Yamazaki, and D. L. G. Noakes, eds., Biology of Chars and Masu Salmon. Kyoto University Press, Kyoto, Japan.
Stahl, G. 1987. Genetic population structure of Atlantic salmon. Pp. 121-143 *in* N. Ryman and F. Utter, eds., Population Genetics and Fishery Management. University of Washington Press, Seattle.
Svardson, G. 1961. Young sibling fish species in northwestern Europe. Pp. 498–513 in W. F. Blair, ed., Vertebrate Speciation. University of Texas Press, Austin.
Svardson, G. 1970. Significance of introgression in coregonid evolution. Pp. 33–60 in C. C. Lindsey and C. S. Woods, eds., Biology of Coregonid fishes. University of Manitoba Press, Winnipeg.
Svedang, H. 1990. Genetic basis of life-history variation of dwarf and normal Arctic charr, *Salvelinus alpinus* (L.), in Stora Rosjon, central Sweden. Journal of Fish Biology 36: 917–932.
Swain, D. P., and L. B. Holtby. 1989. Differences in morphology and behavior between juvenile coho salmon (*Oncorhynchus kisutch*) rearing in a lake or in its tributary stream. Canadian Journal of Fisheries and Aquatic Sciences 46: 1406–1414.
Taylor, E. B., and P. Bentzen. 1993a. Evidence for multiple origins and sympatric divergence of trophic types of smelt (Osmerus) in Northeastern North America. Evolution 47: 813–832.
Taylor, E. B., and P. Bentzen. 1993b. Molecular genetic evidence for reproductive isolation between sympatric populations of smelt *Osmerus* in Lake Utopia, south-western New Brunswick, Canada. Molecular Ecology 2: 345–357.
Taylor, E. B., C. J. Foote, and C. C. Wood. 1996. Molecular genetic evidence for parallel life-history evolution within a Pacific salmon (sockeye salmon and kokanee, *Oncorhynchus nerka*). Evolution 50: 401–416.
Taylor, E. B., S. Harvey, S. Pollard, and J. Volpe. 1997. Postglacial genetic differentiation of reproductive ecotypes of kokanee *Oncorhynchus nerka* in Okanagan Lake, British Columbia. Molecular Ecology 6: 213–227.
Thompson, C. E., E. B. Taylor, and J. D. McPhail. 1998. Parallel evolution of stream-lake pairs of threespine ticklebacks (Gasterosteus) inferred from mitochondrial DNA variation. Evolution 51: 1955–1965.

Todd, T. N., G. R. Smith, and L. Cable. 1981. Environmental and genetic contributions to morphological differentiation in ciscoes (Coregoninae) of the Great Lakes. Canadian Journal of Fisheries and Aquatic Sciences 38: 59–67.

Tonn, W. M. 1990. Climate change and fish communities: a conceptual framework. Transactions of the American Fisheries Society 119: 337–352.

VanValen, L. 1965. Morphological variation and the width of the ecological niche. The American Naturalist 99: 377–390.

Via, S., R. Gomulkiewicz, G. DeJong, S. M. Scheiner, C. D. Schlichting, and P. H. Van Tienderen. 1995. Adaptive phenotypic plasticity: consensus and controversy. Trends in Ecology and Evolution 10: 212–217.

Vrijenhoek, R. C. 1978. Coexistence of clones in a heterogenous environment. Science 199: 549–552.

Vrijenhoek, R. C. 1994. Unisexual fish: Model systems for studying ecology and evolution. Annual Review of Ecology and Systematics 25: 71–96.

Vuorinen, J., R. A. Bodaly, J. D. Reist, L. Bernatchez, and J. J. Dodson. 1993. Genetic and morphological differentiation between dwarf and normal size forms of lake whitefish (*Coregonus clupeaformis*) in Como Lake, Ontario. Canadian Journal of Fisheries and Aquatic Sciences 50: 210–216.

Wilmot, R. L., and C. V. Burger. 1985. Genetic differences among populations of Alaskan sockeye salmon. Transactions of the American Fisheries Society 114: 236–243.

Wilson, D. S. 1989. The diversification of single gene pools by density- and frequency-dependent selection. Pp. 366–385 *in* D. Otte and J. A. Endler, eds., Speciation and Its Consequences. Sinauer Associates, Sunderland, MA.

Wimberger, P. 1991. Plasticity of jaw and skull morphology in the neotropical cichlids *Geophagus brasiliensis* and *G. steindachneri*. Evolution 45: 1545–1563.

Wimberger, P. 1992. Plasticity of fish body shape: the effects of diet, development, family and age in two species of Geophagus (Pisces: Cichlidae). Biological Journal of the Linnean Society 45: 197–218.

Wimberger, P. H. 1994. Trophic polymorphisms, plasticity, and speciation in vertebrates. Pp. 19–43 in D. J. Stouder, K. Fresh, and R. J. Feller (eds.), Theory and application in fish feeding ecology. University of South Carolina Press, Columbia, S. Carolina.

Wood, C. C., and C. J. Foote. 1990. Genetic differences in the early development and growth of sympatric Sockeye salmon and kokanee (*Oncorhynchus nerka*), and their hybrids. Canadian Journal of Fisheries and Aquatic Sciences 47: 2250–2260.

Wood, C. C., and C. J. Foote. 1996. Evidence for sympatric genetic divergence of anadromous and nonanadromous morphs of sockeye salmon (*Oncorhynchus nerka*). Evolution 50: 1265–1279.

4

Understanding Natural Selection on Traits That Are Influenced by Environmental Conditions

RUEDI G. NAGER, LUKAS F. KELLER, AND
ARIE J. VAN NOORDWIJK

To predict microevolutionary responses to changing environments, genetic effects on a trait need to be separated from those of environmental factors. Resemblance between family members in ecologically important traits can be estimated by long-term studies of populations with marked individuals of known decent. The observed resemblance, however, can be caused by genetic and environmental factors, and determining their relative importance requires a detailed understanding of the way genotype and environment influence the phenotype. Phenotypic plasticity provides a useful theoretical framework in which to study the influence of genes and environment on the phenotype. Field experiments have limited ability to measure genotype-by-environment interactions because experimentally induced environmental variation is usually small relative to natural variation. To study the proximate control of the phenotype, we propose ecophysiological experiments combined with measurements of environmental conditions at the individual scale. To understand the evolution of a trait, we must consider the pattern of environmental heterogeneity, the form of natural selection within each environment, and possible effects of past environments. We present a framework for such studies and illustrate it with the timing of reproduction in great tits (*Parus major*). The expression of laying date depends strongly on local environmental conditions and the correct timing of the bird's nestling period in relation to food availability has important fitness consequences. The response of laying date to environmental conditions early in the spring may allow the bird to anticipate whether caterpillars are likely to appear early or late in the season. However, there is no correlation between early and late weather conditions, and the caterpillars' peak date cannot be predicted from conditions in early spring. As a consequence, direction of selection on laying date varies unpredictably between years. In other words, two different, uncorrelated environmental conditions affect the expression of the laying date and the natural selection on that trait. The lack of predictability in environmental conditions will favor the presence of a large variability in that trait, so that at least a few phenotypes can breed successfully rather than an adaptive phenotypic plasticity.

4.1 Phenotypic Variation and Evolution

A distinguishing feature of organisms is that they can evolve if the environment they live in changes. There are good case studies that show measurable changes in the properties of organisms over a few generations in response to changes in the environment (e.g., Grant and Grant 1995, Losos et al. 1997, Reznick et al. 1997). Any property of an individual that can be measured or attributed to a discrete category can be regarded as a trait, and this, in turn, can be characterized by a frequency distribution. If this frequency distribution shifts between generations, we have evolution. Traits can evolve only if two conditions are fulfilled: there must be heritable variation in the trait, and individuals with different traits must vary in fitness (Endler 1986, Stearns 1992). Organisms typically live in heterogeneous environments where they face the problem of how to maximize fitness under variable conditions. To overcome such problems, a given organism may express different phenotypes depending on the environmental conditions. Observed values of a trait (the phenotype) are thus produced through genetic factors (the genotype), environmental influences, or a combination of the two. It is on the product of genetic and environmental influences that natural selection acts (Lewontin 1974, Stearns 1989, West-Eberhard 1989). Environmental heterogeneity therefore has important implications for our understanding of the evolution of traits.

Phenotypic variation is the focus of evolutionary biology because it fuels evolutionary change. But proximate processes that generate variation have received little attention in evolutionary theory until recently. To understand evolution we need to know the processes generating phenotypic variation in a trait, how it is connected to genetic variation, and how it interacts with other traits and the environment to determine fitness (Stearns 1989). The various aspects of the phenotype are usually studied by different fields of biology. Quantitative genetics studies the connection between the genotype and the phenotype, and life-history theory studies how phenotypic traits interact to determine fitness, but neither examines how phenotypes actually arise. We must therefore combine proximate questions concerning the causal mechanism and ultimate questions about the functional significance of phenotypic variation. The concept of phenotypic plasticity provides a useful framework that enables us to bring together such disparate fields as genetics, development, ecophysiology, and ecology (Stearns 1989, van Noordwijk 1989, Pigliucci 1996).

The aims of this chapter are to (1) consider how we can disentangle the relevant genetic effects from environmental effects on the observed phenotypic variation, (2) discuss the relevance of phenotypic plasticity to evolutionary studies, and (3) emphasize the importance of understanding the interactions between phenotypic variation and environmental conditions. We then illustrate the usefulness of this approach with a case study of variation in timing of breeding in a natural population of great tits (*Parus major*).

4.2 Genetic and Environmental Causes of Phenotypic Variation

The extent to which differences in phenotypes between two individuals are genetically determined depends on the relatedness of the individuals and the relative contribution of genes to phenotypic variation. With information on relatedness and quantitative genetics, one can estimate heritabilities—that is, the proportion of phenotypic variation attributable to the additive effects of genes (Falconer 1989, Falconer and Mackay 1996, Roff 1997). Many wild populations have been shown to have significant genetic variation, although most studies are from vertebrates (Mousseau and Roff 1987, Weigensberg and Roff 1996). This is because detailed studies of marked individuals over several generations are typically necessary to know the relatedness between individuals. These have mainly been carried out in vertebrates due to the ease in following marked individuals over several generations. Relatedness can also be estimated from molecular data (e.g., Queller and Goodnight 1989), and these estimates will become cheaper and easier to obtain with advances in the field of molecular biology. Estimates of relatedness based on molecular markers from a wide variety of species are emerging rapidly (e.g., Gagneux et al. 1997, Peacock and Smith 1997, Ritland, this volume). Molecular methods could soon be used to estimate heritabilities in invertebrates and in vertebrate species with significant amounts of extrapair fertilization.

Classic quantitative genetics theory assumes that the genotype relates to the environment in a simple way. There are, however, complex relationships between genotype and environment in forming the observed phenotypic distribution. Two main relationships between genotype and environment need to be considered in evolutionary studies: genotype–environment correlations and genotype × environment interactions. The former concerns the nonrandom distribution of genotypes over environments, and the latter refers to different reactions by different genotypes to the same environmental conditions. We will deal in the following section with genotype–environment correlations and in the next section with genotype × environment interactions.

4.2.1 Genotype–environment correlations

Relatives show, on average, more similar phenotypes than nonrelatives not only because they share genes that descend from their common ancestors, but because they may also experience similar environmental conditions. Body weight in great tits provides a good example of environmentally caused resemblance (Gebhardt-Henrich and van Noordwijk 1991). The body weight of cross-fostered, full-grown fledglings correlated positively with the body weight of their foster parents during the breeding period, particularly under poor environmental conditions. There was no correlation, however, with the foster parent's body weight in winter. The correlation between breeding weight of the foster parent and the fostered young was not due to shared genes, but lighter parents reared lighter young because they shared the same poor environmental conditions.

Because relatives in natural populations are likely to experience similar environmental conditions we need to separate environmental and genetic causes of resemblance. This can be achieved by experimental manipulations. Cross-fostering, exchanging young between nests, allows us to compare the resemblance between relatives experiencing the same environment and relatives experiencing a different environment and to test for the effects of a common environment (Boag and van Noordwijk 1987). Alternatively, experimental manipulations of the environmental conditions can separate genetic and environmental effects (Gebhardt-Henrich and van Noordwijk 1991, Merilä 1996). If experimental manipulations are impractical, one can separate genetic and environmental causes of resemblance between relatives by comparing repeatabilities of a given trait (Falconer and Mackay 1996) between suitable groups of individuals. Various comparisons have been used to shed light on the relative effects of genes and the environment (1) between different pairs occupying the same environment (i.e., territory), (2) between unrelated individuals such as neighbors sharing the same local environment (van Noordwijk et al. 1980, van Noordwijk 1984), and (3) between two groups of relatives that are unlikely to share a common environment, such as maternal versus paternal grandparents (van Noordwijk et al. 1981a).

The presence of genotype–environment correlations can have important implications for our understanding of the evolution of a trait. As a result of genotype–environment correlations, the genetic similarity between individuals may be confounded with a similarity due to common environment. For some traits natural selection may act on the nonheritable environmental component rather than on the genetic component (Price et al. 1988). We would then observe no phenotypic change. Thus, if the positive correlation between the trait and fitness is exclusively due to a nonheritable environmental component, the trait may appear to be under directional selection and yet not evolve. Such cases have been suggested for life-history traits (Price et al. 1988, Price and Liou 1989) and morphological characters (van Noordwijk et al. 1988, Alatalo et al. 1990). This has important consequences for our ability to predict phenotypic changes from measured natural selection.

4.2.2 Phenotypic plasticity and genotype × environment interactions

A single genotype can express different phenotypes across different environmental conditions. This is referred to as "phenotypic plasticity" (Stearns et al. 1991). Genotypes can be treated as developmental and physiological programs that incorporate information about the environment into the expression of the phenotype (Falconer 1989). The direction and the extent of the plasticity of a phenotype can be summarized as a continuous function of the phenotypic expressions in specific environmental conditions (fig. 4.1a). This function has been called a "reaction norm" (Stearns 1989) and is a property of the genotype. The quantitative genetics approach treats phenotypic plasticity either as one trait in two discrete environments as if they were two traits (e.g., Via and Lande 1985) or as a continuous function (e.g., de Jong 1990, Gomulkiewicz and Kirkpatrick 1992). Reaction norms may differ between genotypes (fig. 4.1b).

If a single genotype can be represented by a reaction norm, then we can represent a population as the bundle of reaction norms of all genotypes forming that population (van Noordwijk 1989). In the absence of genetic variation in plasticity, all phenotypic variation would be due to differences in environment, and all individuals would have parallel reaction norms (fig. 4.1c) or, in the absence of genetic variation in the trait itself, the same reaction norm. Heritable variation in the trait means that different genotypes produce different phenotypes under the same environmental conditions. The width of the bundle of reaction norms can therefore be interpreted as the genetic variation present in the average environment. If there is genetic variation in phenotypic plasticity between genotypes, the population is represented by a bundle of crossing reaction norms (fig. 4.1d, e). The crossing between reaction norms is called a genotype × environment interaction. This graphical model of genotype × environment interaction corresponds to the more formal treatment of the interaction as a statistical interaction term when phenotypes are measured under two discrete environmental conditions (Fry 1992, Via 1993, 1994). As a consequence of genotype × environment interaction, the width of the bundle, and therefore the genetic variance present in the population, varies between environmental conditions (Turelli 1988, van Noordwijk 1989). The genetic variance

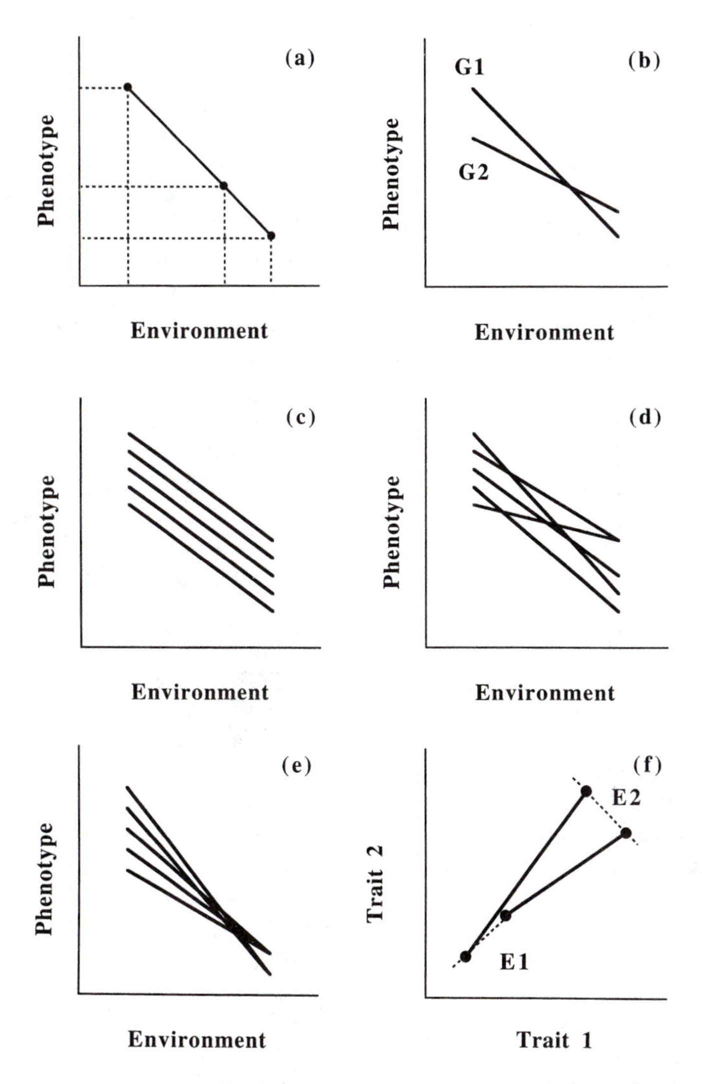

Figure 4.1 Reaction norms of individual genotypes and of populations of genotypes illustrate changes in genetic parameters with environmental conditions. (a) The reaction norm translates environmental variation into phenotypic variation for a given genotype. (b) Two genotypes (G1 and G2) may differ in their sensitivity to environmental changes, discussed as genotype × environment interactions (c–e). A population consists of several genotypes and therefore can be represented by a bundle of reaction norms. The phenotypic variation in that population is affected by environmental factors (slope of the reaction norm) and genetic factors (width of the bundle of reaction norms). (c) The genotypes within a population do not differ in plasticity. (d) The genotypes differ in plasticity, but there is the same amount of genetic variation over all environments encountered. (e) Due to genotype × environment interactions, genetic variation changes over environments. (f) Bivariate reaction norms. The values of two traits are plotted on each other for two environments (E1 and E2). The dotted lines indicate the change in direction of the genetic correlation between the two traits in different environments.

in the trait then needs to be estimated from analyses in which observations from each environment are analyzed separately.

If phenotypic plasticity is considered to evolve, then it must have a genetic basis. Although the mechanism of genetic control has important consequences on how phenotypic plasticity evolves, the question of which mechanism actually applies is still controversial (Scheiner 1993, Schlichting and Pigliucci 1993, Via 1993, Roff 1997). Some workers proposed that plasticity is a trait with a genetic basis that natural selection acts directly upon (Bradshaw 1965, Schlichting 1986, Scheiner and Lyman 1989, 1991). In the quantitative genetic view, plasticity is thought to evolve as a byproduct of stabilizing selection on the mean values of the character states expressed in each environment (Via and Lande 1985, Via 1993). In the latter view, genetic variation in plasticity is described as the genetic independence of the expression of a trait in different environments. Unfortunately, the genetic independence does not increase monotonically to the genotype × environment interaction variance (Via 1984, 1994). Genetic independence across environments is better estimated using the genetic correlation between character states expressed at different environmental conditions (see Via 1984, Fry 1992, Gomulkiewicz and Kirkpatrick 1992). Genetic correlations can be represented in our graphical model by plotting the environmental reaction of two traits, or expressions of a trait in two environments, against one another (fig. 4.1f). Here the environmental conditions then vary along the norm rather than along one of the axes.

If variation in plasticity is heritable and individuals with different plasticities vary in fitness, plasticity can evolve. Environmental conditions interact with genes to cause phenotypic variation as well as playing an important role in selection upon phenotypic variation. Many organisms live in heterogeneous environments that are highly variable, and environmental conditions during the formation of a phenotype and during selection upon it may therefore differ. One should therefore measure the environments during phenotype formation and phenotypic selection separately. Van Noordwijk and Müller (1994) have proposed a general framework for the study of phenotypic plasticity in heterogeneous environments (fig. 4.2). Consider an organism that modifies its phenotype in relation to environmental conditions, which is then favored by natural selection (adaptive plasticity; van Noordwijk and Müller 1994, Gotthard and Nylin 1995). This scenario assumes that the environment during the formation of the phenotype (time T1) is the same as the environment affecting selection on that trait (time T2). In many situations, however, phenotypic selection occurs some time after the phenotype has been formed or the organism has moved to another environment. Conditions at T2 can then be unrelated to conditions at T1, and this has important implications for the evolution of plasticity in that trait. In this case, phenotypes caused by environmental conditions at T1 are randomly distributed among environments at T2, and therefore there is no consistent selection pressure favoring particular phenotypic changes in relation to environmental conditions at T1. The separation of environmental conditions that cause phenotypic variation from those that affect selection on that trait and the relationship between the two plays an important role in our understanding of evolution in heterogeneous environments. This basic model can be further expanded (van Noordwijk and Müller 1994). For example, one can incorporate other potential influences on phenotype formation or selection, such as delayed effects of previously experienced environments, maternal effects, or the effects of other genotypes (e.g., the effects of the mate's genotype).

In this section we wanted to stress that natural selection on traits in environments is a complex process. Environmental influences on phenotype formation need to be separated

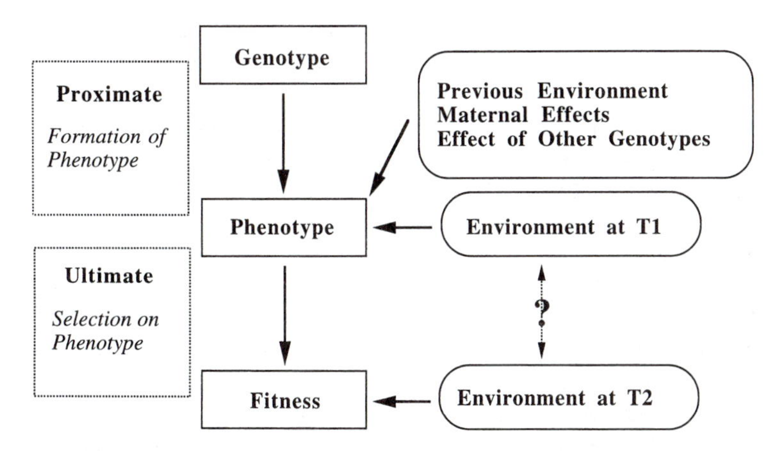

Figure 4.2 A conceptual model of phenotypic plasticity in a heterogeneous environment. The formation of a phenotype can be influenced by the genotype, the environment during the formation of that phenotype at T1, and other effects (e.g., previously experienced environment, maternal effect, and effect of other genotypes). The selection on the phenotype, however, takes place in the environment at T2, which is separated in time and/or space from the environment at T1. Thus, the degree to which environment at T2 can be predicted from the environment at T1 has important implications for the evolution of adaptive phenotypic plasticity (after van Noordwijk and Müller 1994).

from genetic factors. Models of phenotypic selection also often assume that genetic parameters are constant over all environments. We have seen that deviations from this assumption have important consequences for our ability to predict microevolutionary patterns. Furthermore, different environmental factors may affect phenotype formation than establish fitness consequences on that phenotype, and this has evolutionary consequences. Via et al. (1995) list the following requirements to describe natural selection on phenotypically plastic traits in heterogeneous environments: (1) determine how a trait responds to a particular environmental factor (proximate control), (2) describe the form of selection within each environment (ultimate function), (3) assess the pattern of environmental heterogeneity (spatial and temporal variation), and (4) determine possible biological carryover effects (e.g., earlier experience, maternal effects). We now illustrate these points by a field study on the timing of breeding in the great tit.

4.3 The Timing of Breeding in Great Tits: Adaptive Phenotypic Plasticity?

The great tit (*Parus major*) is a common monogamously breeding bird of European woodlands. The birds readily accept artificial nest-boxes for breeding, and if the number of natural cavities is limited, they will use nest-boxes exclusively, allowing an entire population to be studied easily. It is possible to ring most parents and all young in the nest. This allows us to study populations of marked individuals. There is a long tradition of population studies on this species, going back to the 1930s and still continuing (e.g., Kluyver 1951, Perrins 1991).

Great tits have an annual survival rate between 0.40 and 0.44 (Clobert et al. 1988) and show relatively high site-fidelity (Perrins 1979). Thus a fair proportion of the breeding birds can be followed over two or more breeding seasons. During breeding both parents feed the young. Caterpillars and leaf-feeding larvae of Lepidoptera and sawflies, compose 80–90% of the nestlings' diet (van Balen 1973). Because caterpillar abundance can be readily measured (Zandt 1994), we can link the bird's biology to its food supply.

The timing of great tits' breeding varies considerably between years (e.g., fig. 4.3) and habitats. Laying date of individual birds also differs between years. Differences in laying date have been related to ambient temperature (Perrins 1979, Nager 1990, Nager and van Noordwijk 1995), food availability (Källander 1974, Clamens and Isenmann 1989, Nager et al. 1997), and genetic factors (van Noordwijk et al. 1981b). To raise young successfully, birds are thought to breed when food is most plentiful (Lack 1968). Laying date largely determines when the birds have their young in the nest. For great tits this is about 3–4 weeks after the decision to start breeding has been made, and nestling rearing lasts about 3 weeks in this species. Caterpillars, the main prey of breeding great tits, are abundant for only a short period of about 2 weeks, however, and the time of caterpillar's peak abundance varies between years (Perrins 1979). This means that great tits cannot delay breeding until food is abundant but must start some time before food becomes plentiful. The question then arises, how do great tits time their breeding so that they have young in the nest when food is plentiful?

There are two main explanations for the timing of breeding. (1) The breeding season is believed to be adapted to the timing of caterpillar peak abundance when food for the young is most plentiful. Lack (1954) suggested that an environmental factor prior to egg laying provides information on conditions for rearing young and is the proximate factor of timing of breeding (the anticipation hypothesis). (2) Egg production is a demanding process in terms of energy and nutrients (Carey 1996). Eggs are typically laid before caterpillars appear, when food abundance is still low, and food supply at that time may constrain egg production (the constraint hypothesis; Lack 1966, Perrins 1970). Females that happen to occupy the territories with highest food availability will be the first to reach a condition allowing egg formation. Most females have to wait until food abundance increases before they can start laying, and this delays their breeding until after the best time for chick rearing (Perrins 1965, 1970, 1991, Verhulst and Tinbergen 1991, Verhulst et al. 1995, but see van Noordwijk et al. 1981b, Nager and van Noordwijk 1995). The proximate factor of laying date may therefore be either the relative changes in some cues about the timing of food supply (anticipation) or passing a threshold of food supply needed for egg formation (constraint).

These different proximal causes of timing of breeding have different evolutionary consequences. Under the constraint hypothesis selection would act on the environment-dependent condition rather than on genetic variation for laying date (Price et al. 1988). As a consequence, there would be no evolutionary response in laying date, and genetic variation in that trait would be maintained. If, on the other hand, the birds can anticipate the most favorable time to raise young, stabilizing selection for the optimal phenotypic plasticity in laying date is expected.

In summary, the timing of breeding of great tits shows all the properties of a trait that should be evolving. It is reported to be heritable, and natural selection seems to favor earlier laying, but yet earlier breeding has not evolved (Lack 1968). Timing of breeding shows high phenotypic variability despite a considerable part of the variance being determined by genetic factors. These contradictions show that there is much left to be understood about the evolution of this trait. In the following sections we explore these issues in the great tit by looking at (1) how tits respond to variability in environmental conditions during the laying period, (2) how the environment influences individual fitness, and (3) if the responses in laying date to the environmental conditions are adaptive.

4.3.1 Ecophysiological considerations of laying date

The constraint hypothesis predicts that food availability increases in spring and that birds will start to breed only when sufficient food is available to sustain egg production. Great tits usually lay earlier in years or territories experiencing higher spring temperatures (e.g., Perrins 1965, Nager 1990), presumably because of the reduced thermoregulation cost that increases energy available for egg production. Tits also lay earlier when food availability is more favorable (Källander 1974, Perrins 1979, Clamens and Isenmann 1989, Nager et al. 1997). These findings are in agreement with the constraint hypothesis. We tested this hypothesis further by supplementarily feeding great tits from 3 to 4 weeks before laying until clutch completion in a rich oak forest near Basel, in northwestern Switzerland (Nager et al. 1997).

We provided two groups of birds with one of two food types to test the effects of energy and nutrient availability on timing of breeding. Each food type had equal energy content but a different protein content. The birds were given more energy than they needed for successful egg formation. If the limit to date of egg laying was purely energy, we would expect both experimental groups to be advanced and to lay earlier than the earliest control birds. The earliest observed laying date in our study area was mid-March (Nager 1993), and thus all birds must have been hormonally able to lay shortly after the start of feeding. Although we found a significant advancement in average laying date in the two groups of supplemented birds compared to control birds ($P < .001$; Nager et al. 1997), supplemented individuals did not start to lay before the earliest control birds started egg laying (fig. 4.3; see also Högstedt 1981). This is not what the constraint hypothesis predicts.

It might be that egg production in birds is constrained by the supply of some critical nutrients rather than by energy, and that the supplementary food provided an inappropriate supply of nutrients. Protein availability is usually believed to be a critical resource for egg production (Jones and Ward 1976, Ankney and McInnes 1978, Houston et al. 1983). Egg production might thus be constrained by protein availability rather than by energy supply. We could find no difference, however, in laying date between tits fed a protein-poor and a protein-rich supplement ($P = .50$; fig. 4.3). In the same experiment supplemented tits laid significantly more but not larger eggs (Nager et al. 1997), and inadequate quality of the food supplements therefore seem to be an unlikely explanation for the relatively small advancement of laying in supplementarily fed birds. Thus, food availability may contain more information for breeding great tits than only the level of currently available resources.

Alternatively, instead of directly increasing a bird's energy supply for egg production by supplementary feeding, the bird's maintenance costs can also be manipulated and thus indirectly affect the energy available for egg production. In the prelaying and laying period the females spend their nights in nest-boxes. By heating and cooling the nest-boxes during the night, we can manipulate the overnight energy expenditure for thermoregulation and thus the energy available for egg production (Nager and van Noordwijk 1992). This approach has

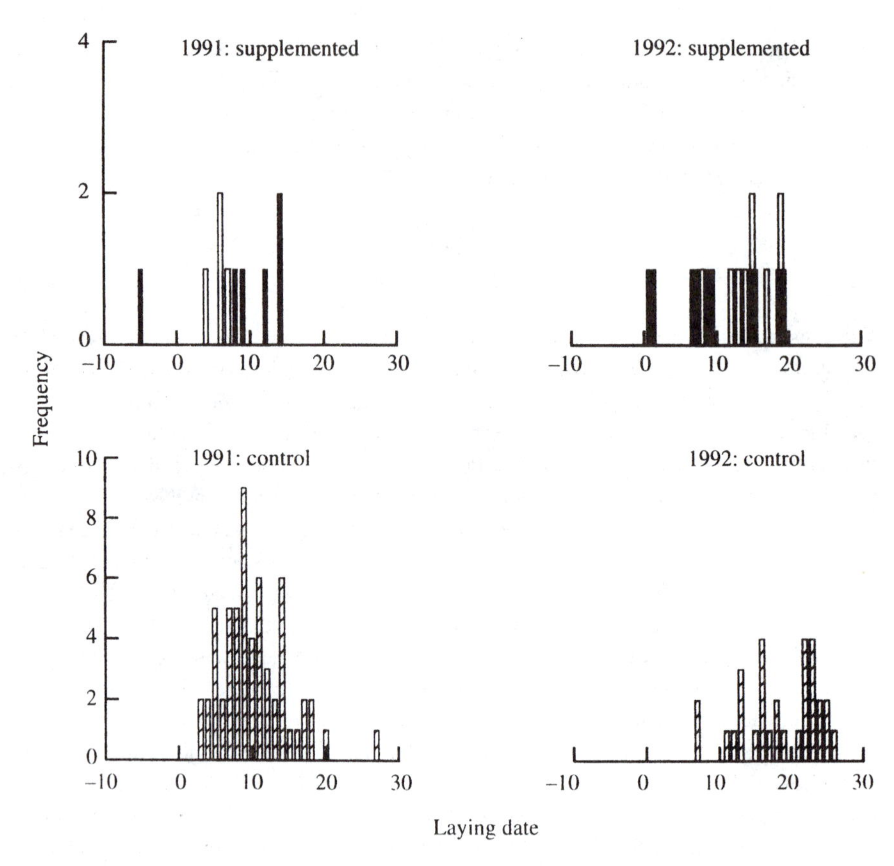

Figure 4.3 Supplementary feeding experiment and timing of breeding in great tits. During the prelaying and laying period of 1991 and 1992 at the Hard study site, great tits were provided with a protein-rich food supplement consisting of mealworms (top, filled bars), protein-poor food consisting of sunflower seeds (top, open bars), or were provided with no food and served as controls (bottom, hatched bars). The two food types have a similar energy content, but compared to mealworms sunflower seeds contain 50% less protein and have much reduced levels of sulfur amino acids, which are important components of eggs (Nager et al. 1997). The feeding treatment between the two supplemented groups was reversed between 1991 (left) and 1992 (right). Shown are the frequency distributions of laying dates for the different treatments and years (day 1 = 1 April). Laying started 11 days earlier in 1991 than in 1992 ($P < .001$; Nager et al. 1997). The type of food had no effect on the degree of advancement. Although, overall, supplemented birds started egg laying 5.6 days earlier than control birds, there is a large overlap in laying dates between supplemented and control birds. Reproduced with permission from Nager et al. (1997).

the advantage that it does not affect food availability. If energy available for egg formation is the sole proximate factor, then cooling should increase the costs for body maintenance at the expense of egg formation, and egg laying should be delayed. In contrast, heating increases the energy available for egg formation, and the birds are expected to advance egg laying. During 3 years we heated and cooled nest-boxes during the nights, starting at least 2 weeks before laying until clutch completion. Cooling reduced nighttime temperature by 1.5°C, whereas heating increased nest-box temperature by 1.5°C (Nager and van Noordwijk 1992). The temperature manipulation did not affect laying date (fig. 4.4a). The experiment still had an impact on the birds because females nesting in heated nest-boxes laid larger eggs (mean volume of 1.650 cm^3) than birds from cooled nest-boxes (mean volume of 1.422 cm^3; $P = .001$; Nager and van Noordwijk 1992), but there was no significant difference in the number of eggs laid (unpublished data). The expected mean difference in egg size between temperature treatments was 0.32 cm^3 (Nager and van Noordwijk 1992), close to the observed difference, indicating that the entire effect of the temperature manipulation was translated into egg-size differences.

The temperature manipulation was also expected to result in a difference in laying dates between treatment groups of 6 days. The expected effect is relatively small compared to the naturally occurring variation in laying date (see fig. 4.3). If the magnitude of natural environmental variation between nests was much larger than the experimentally created differences, then the experimental approach would be less powerful (Gebhardt-Henrich and van Noordwijk 1991). Under these conditions an alternative, more powerful approach is to measure phenotypes produced by the same individual over a range of environments. This approach works for traits that are formed repeatedly during a lifetime such as laying date, and the relevant environment can be experimentally manipulated. Taking only individuals breeding twice, once as control and once as treatment, each individual can generate its own null expectation. Individual great tit females show high repeatabilities of their laying dates, corrected for differences between years (van Noordwijk et al. 1981b, Nager and van Noordwijk 1995). Temperature treatments had no effect on the within-individual variation in timing of breeding (fig. 4.4b). Thus, variation in nighttime temperature, independent of food availability, does not cause differences in timing of breeding.

In summary, great tits lay earlier in warmer springs and when more food is available, but often these two factors correlate with each other. Our experiments attempted to disentangle the effects of energy, nutrients, and temperature on laying date. All experiments were carried out in the same breeding population and over the same time period. We found no effects of food quality on laying date and egg production. If energy availability for egg formation is the sole proximate factor determining laying date, then both experiments, supplementary feeding and temperature manipulation, should have affected timing of breeding. Direct manipulation of energy availability for egg formation by supplementary feeding significantly affected timing of breeding, whereas indirect manipulation of energy availability for egg formation did not affect timing of breeding. The absence of a direct temperature effect on laying date makes it unlikely that temperature is the causal factor of laying date. The observed correlation between laying date and temperature may instead arise through an indirect effect of temperature on food availability (Avery and Krebs 1984). Variation in timing of breeding thus seems to be best explained by food availability.

Food availability can affect laying date either directly through the energy necessary for egg formation (constraint hypothesis) or indirectly by giving information on the forthcoming food condition (anticipation hypothesis). It is difficult to distinguish between these two possibilities, but they have different implications for the evolution of this trait, as discussed later. We have, however, two indications that food may act as a clue rather than as a constraint. First, supplementarily fed great tits did not lay earlier than the earliest control birds even though they had plenty of food available to start laying much earlier. Other factors must

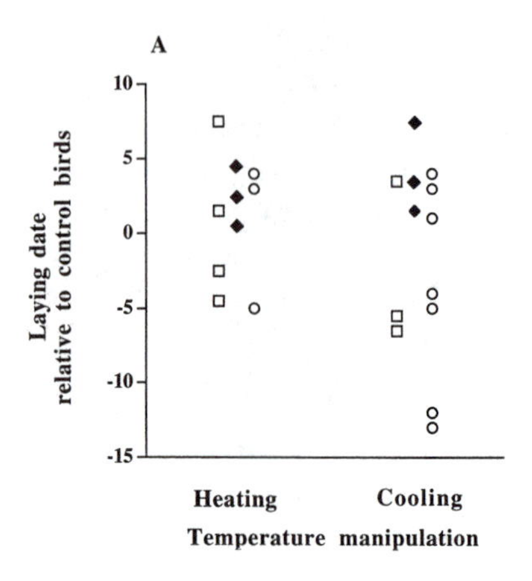

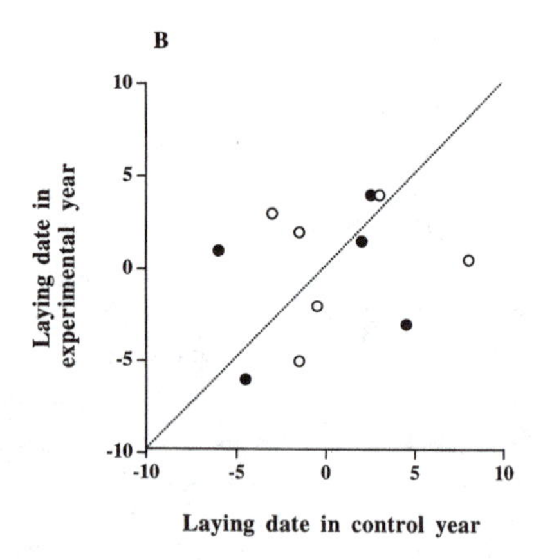

Figure 4.4 Temperature manipulation and variation in timing of breeding in great tits. During the prelaying and laying period of 1990 to 1992 at the Hard study site, nighttime temperatures of nest-boxes were increased by 1.5°C (heated) or decreased by 1.5°C (cooled). These experiments were carried out in the same population as the experiments described in figure 4.3. Every year different groups of nest-boxes were heated and cooled. (A) Between-individual variation in laying date expressed relative to the median laying date of all unmanipulated birds in that year (= day 0), separate for the 3 years of study (open squares = 1990, filled diamonds = 1991, open circles = 1992). There was no effect of temperature manipulation on laying date (ANOVA; effect of treatment, $F_{1,24} = 0.34$, $P = .56$; effect of year, $F_{2,25} = 2.26$, $P = .13$, no interaction). (B) Within-individual changes in timing in relation to temperature manipulation during the egg formation period. Laying dates are given as dates relative to the median laying date of unmanipulated birds in the corresponding year (e.g., -5 means that female laid 5 days before the median date) and shows his repeatability between years (see text). The dotted line shows equal relative laying dates in the 2 years. If cooling a nest-box (filled circles) results in delayed laying, the laying date in the experimental year is expected to be above the dotted line. If the nest-box was heated (open circles), the laying date in the experimental year is expeced to be below the line. Temperature treatment, however, did not affect timing of breeding (ANCOVA, effect of treatment corrected for laying date in control year, $F_{1,8} = 0.23$, $P = .64$, no interaction). Data from Nager and van Noordwijk (1992) and R.G. Nager (unpublished).

therefore also play an important role. Second, in a nearby population we also followed the laying date of individual great tits over several years and correlated variation in laying date with measured differences in local environmental conditions (Nager and van Noordwijk 1995). In a multiple regression both local temperature before laying and local food availability at laying significantly influenced laying date. At the time of laying, caterpillars are responsible for only 7% of the total food biomass available (Nager and van Noordwijk 1995). Only caterpillar abundance, but not overall food availability, significantly affected laying date. In particular, in a total of 33 cases we never observed great tits starting to lay before the first small caterpillar appeared in their territory (Nager and van Noordwijk 1995). This might suggest that food does not act as a constraint on laying date, but that the appearance of the first caterpillars acts as a reliable signal for the start of breeding.

4.3.2 Environmental influences on fitness

We now turn to ultimate considerations of timing of breeding. According to Lack's (1968) idea, birds that rear their young when food is most plentiful are expected to be the most successful breeders. To test this idea we measured weekly caterpillar abundance throughout the breeding season over a 5-year period. We made these measurements at the Blauen, about 15 km away from our experimental plot. Blauen is a mixed deciduous forest, where we selected eight sites differing in altitude, exposure, and tree composition. We observed natal dispersal between sites, and birds of different sites are therefore not genetically isolated from each other. Caterpillar abundance showed a pronounced seasonal pattern with a short period of peak abundance (fig. 4.5). The date of peak caterpillar abundance varied both between years and within years between sites. Between years the average caterpillar peak date ranged from 4 May (in 1989) to 21 May (in 1991). Within-years there was a significant difference in caterpillar peak date between sites; at some sites caterpillar density always peaked earlier than at other sites. At the earliest site caterpillars peaked on average 10 days earlier than at the latest site; thus the within-year variation is of similar magnitude as the between-year variation. Thus within a population tits face considerable variation in environmental conditions.

The chick-rearing capacity of great tit parents (chick growth and fledging success) is strongly affected by food availability (Keller and van Noordwijk 1994, Nager and van Noordwijk 1995). The workload of breeding great tits reaches its maximal level when the young are about 1 week old (van Balen 1973). Pairs with 1-week-old young in the nest when caterpillars were most abundant fledged more young and young of better quality (fig. 4.6). A similar pattern has also been found for blue tits (*Parus caeruleus*) (Dias and Blondel 1996). Thus, in agreement with Lack's prediction, the birds that reared their young when food was most abundant produced more and better offspring. According to the constraint hypothesis we would predict that the earliest birds are always well synchronized with the caterpillar peak. We found, however, that only in some years did the earliest birds raise their young when the caterpillars were most abundant; there were other years when intermediately laying birds or even late-laying birds raised young when caterpillars were most abundant (fig. 4.6). Thus, instead of finding consistent directional selection for early laying, we find between-year variation in the direction of selection on laying date. Such oscillating selection on laying date in tits was also observed in other studies (van Noordwijk et al. 1995, Svensson 1997).

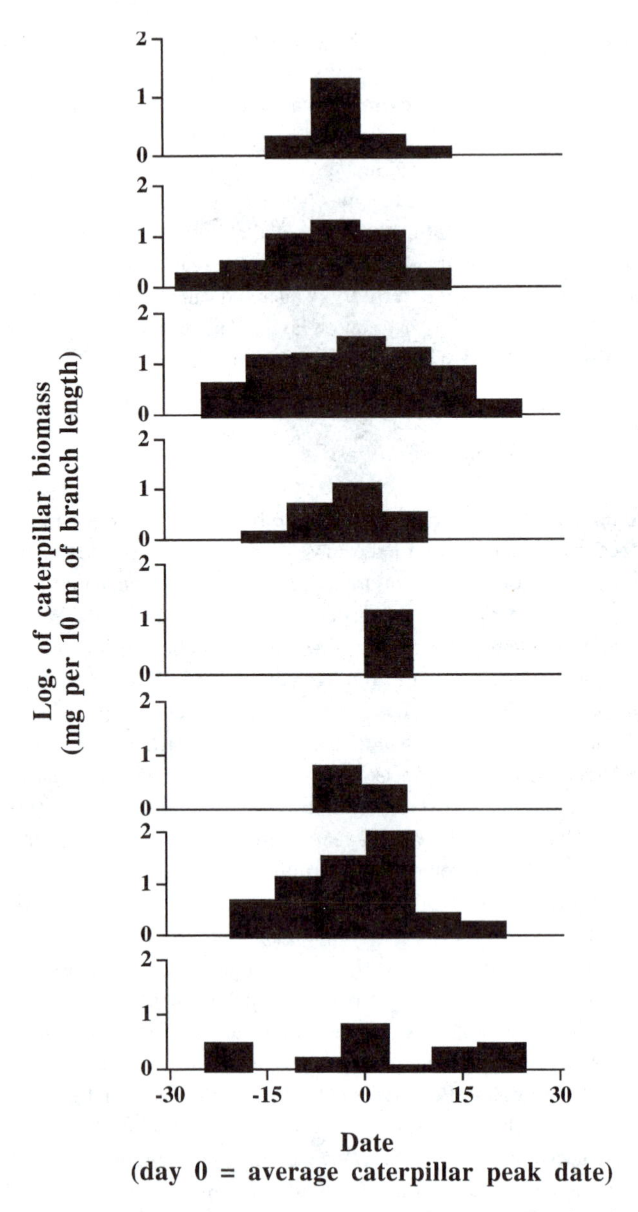

Figure 4.5 Variation in caterpillar abundance at the Blauen study site. Caterpillars were dominant in biomass at the time of maximum food abundance, but accounted for only about 7% of the food biomass during egg laying. Shown are the average seasonal patterns of caterpillar abundance for eight sites over a study period of 5 years. Day 0 is the overall mean caterpillar peak date (date of maximal caterpillar abundance), and local peak dates are expressed as the average deviations from the overall mean. The caterpillar peak date differs between the eight sites ($P = .004$; Nager and van Noordwijk 1995). Modified from Nager and van Noordwijk (1995).

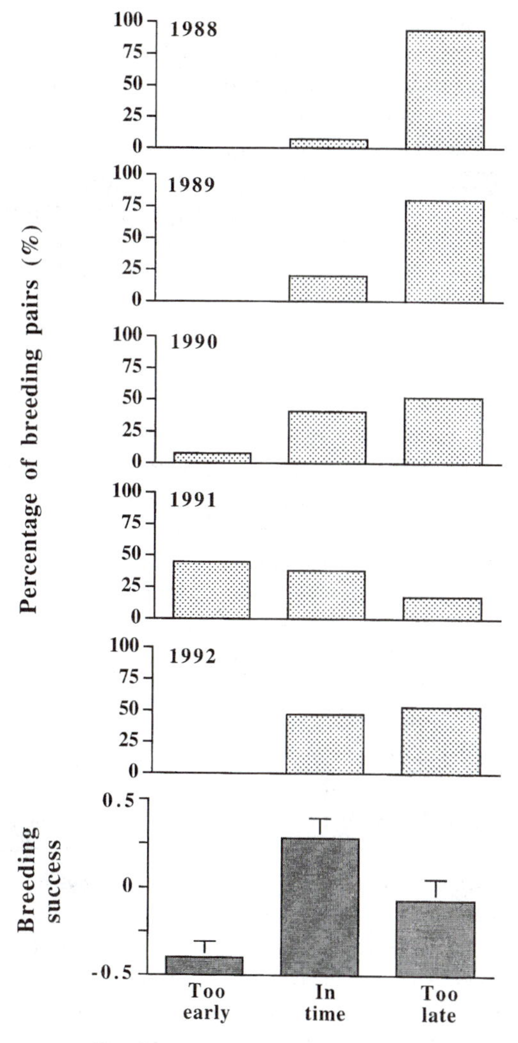

Figure 4.6 Variation in selection pressures on laying date over a 5-year period at the Blauen study area (northwest Switzerland). The top five figures show the frequency distribution of tits laying "in time" (raising 1-week-old young ± 7 days of caterpillar peak abundance), "too early" (raising young before caterpillars peaked) and "too late" (raising young after caterpillars peaked). The bottom figure shows breeding success (number of young expected to breed the following year) in relation to the timing of peak caterpillar abundance. Given are standardized values (corrected for differences among years) ± SE. Birds that bred "in time" had a higher breeding success compared with birds that laid earlier or later ($P = .019$; Nager and van Noordwijk 1995). In the years 1988–1990 birds laid, on average, too late and the earliest laying birds were favored. However, in 1991 and 1992 more birds laid too early, and birds that laid later in those years had a selective advantage. Selection pressure on laying date differed significantly between years ($P = .0001$; Nager and van Noordwijk 1995). Modified from Nager and van Noordwijk (1995).

It is the timing of tits' breeding in relation to the time of caterpillar peak abundance that proved to be crucial to breeding success. Both appearance and development of caterpillars are strongly influenced by temperature (Perrins 1979). In contrast to the insects, the timing of peak food demand of tits is largely fixed by laying date, and temperatures after the birds started laying only influence the development of caterpillars but not of birds. As a consequence, the correlation coefficient between early spring temperature and the date of peak caterpillar abundance is low and not significant ($r = .025$; Nager and van Noordwijk 1995). Thus, at the time laying date is expressed, the environment under which selection on laying date will occur cannot be predicted.

4.3.3 The evolution of timing of breeding

The data clearly show that the time when caterpillars are most abundant is the key factor in the timing of great tits' breeding; variation in this environmental factor has important consequences on the bird's fitness. The timing of peak caterpillar abundance is also variable, both between years and between territories, but only the spatial variation pattern is predictable. Laying date in great tits is a plastic trait and responds to changes in environmental conditions in early spring. Given these changing environmental conditions, great tits can either respond to these changing conditions to maximize fitness (adaptive phenotypic plasticity) or adopt a strategy to express many different phenotypes so that at least some phenotypes will get the timing right (bet-hedging or spreading the risk). Although we do not yet fully understand the proximate mechanism responsible for the timing of breeding, we are still able to consider whether the observed variation in laying date represents an adaptive phenotypic plasticity.

To answer the question about adaptive phenotypic plasticity, we would have to compare the effect on fitness of a particular plasticity relative to other plasticities. This involves measurements of fitness of phenotypes in each environment and the knowledge of the frequency distribution of environments. From that an overall fitness associated with each plasticity can be estimated (Newman 1992). For the Blauen study area we found both temporal and spatial variation in the timing of peak caterpillar abundance. Birds using an adaptive strategy would respond to environmental variation in such a manner that they would breed more successfully than when using any other strategies. We found that great tits are able to respond to the spatial variation in insect densities (Nager and van Noordwijk 1995). Because the same territory has, for example, repeatable caterpillar peaks, and great tits usually remain faithful to their breeding territory, they are expected to incorporate previous experience into current decisions. Each year individual great tits lay at similar environmental conditions ($R^2 = .29$, $P = .0003$; Nager and van Noordwijk 1995), and these environmental conditions can be measured on a local scale. If, for example, a bird has bred too late the previous year and raised its young after the caterpillar peak, it is expected to lay at earlier environmental conditions in order to raise young at the right time next year. We found a strong correlation of between-year changes in laying date that could not be accounted for by environmental conditions with the difference between the bird's and the caterpillars' timing in the previous year ($r = -.559$, $P = .0006$; Nager and van Noordwijk 1995). This means that great tits that raised their young after the caterpillar peak date in one year started to lay earlier the next year. Individuals that raised their young before the caterpillar peak in a given year likewise delayed laying to later in the season. Individuals that incorporated previous experience in their decisions raised their young when food was more abundant than birds adopting a hypothetical strategy to lay repeatedly under the same environmental conditions.

These results show that great tits have the sensory capacities necessary to respond adaptively to changing environmental conditions if the environment is predictable, such as the spatial variation in timing of caterpillar abundance. We found, however, that at the onset of laying, between-year differences in date of peak caterpillar abundance are unpredictable. Conditions during laying are not correlated with conditions during chick rearing, and thus the bird cannot predict when food will be most plentiful. Lack of predictability in future conditions will favor the evolution of bet-hedging strategies rather than adaptive phenotypic plasticity (Roff 1992). We also found no between-individual differences in the slope of plasticities, which implies that this plasticity cannot evolve but has been fixed.

Instead, under unpredictable conditions selection is expected to favor maximization of the geometric average of fitness (Roff 1992). This, for example, has been applied to clutch size in great tits where the penalty for laying too many eggs is bigger than laying too few. Therefore, tits were predicted to lay a clutch smaller than the most productive clutch (Boyce and Perrins 1987). In our great tit population the disadvantage of laying too early tended to be larger than the disadvantage of laying too late in relation to the caterpillar peak (fig. 4.6). Similarly, greater costs of being too early were suggested for blue tits breeding in the Mediterranean region (Dias and Blondel 1996) and in Sweden (Nilsson 1994). The reason for this is probably that there will always be some late-growing, large caterpillar species available. But early in the season before the caterpillar biomass peaks, only small larvae are available, which cannot be exploited in an efficient way. This leads to an asymmetric fitness curve skewed toward later laying. We therefore expect a plasticity that results, on average, in a later laying date than the most productive one.

Between-individual variation in the response to environmental cues can itself be adaptive. For example, in desert plants variation in diapause among the offspring of a single parent has been suggested to be adaptive (Cohen 1966, 1967). Thus variation in response to environmental conditions concerning when to breed within a population of great tits may simply ensure that under any given environmental condition, at least some individuals will breed successfully. Each individual has a different environmental threshold level at which it starts to lay to ensure that at least some raise young when food is most abundant. Unlike the constraint hypothesis, individuals that get the timing right are not always the earliest ones; our study shows that the most productive laying date differs between years. This results in oscillating selection pressures on laying date, mainly because of fluctuating weather conditions after the birds started to lay (resulting in significant changes of the caterpillars' timing of peak abundance). A similar pattern of variation in selection pressures between years has also been found in other studies of timing in tits (van Noordwijk et al. 1995, Svensson 1997), as well as in Darwin's finches on Galapagos (Gibbs and Grant 1987). Such a selection pattern does not select a particular value of laying date but favors the large variation observed in laying date. Timing of breeding in great tits thus does not represent a classic adaptive plasticity. It is also not a nonadaptive explanation because the variation in laying date itself is adaptive. Due to the unpredictable nature of the bird's food supply, no single laying date or single strategy of phenotypic plasticity will result in repeatedly correct timing in relation to caterpillar densities. Instead, natural populations of birds rely on a large variability to ensure that despite an unpredictable environment at least some great tits breed successfully each year.

Acknowledgments We thank David Houston, Kathy Jones, Graham Ruxton, and two anonymous referees for useful comments on an earlier draft. Ideas and data presented here originated when all three authors worked at the Zoology Institute of the University of Basel, Switzerland, and the research has been supported by the Swiss Science Foundation.

References

Alatalo, R. V., L. Gustafsson, and A. Lundberg. 1990. Phenotypic selection on heritable size traits: environmental variance and genetic response. Am. Nat. 135:464–471.

Ankney, C. D., and C. D. McInnes. 1978. Nutrient reserves and reproductive performance of female lesser snow geese. Auk 95:459–471.

Avery, M. I., and J. R. Krebs. 1984. Temperature and foraging success of great tits (*Parus major*) hunting for spiders. Ibis 126:33–38.

Boag, P. T., and A. J. van Noordwijk. 1987. Quantitative genetics. Pp. 45–78 *in* F. Cooke and P. A. Buckley, eds., Avian Genetics: A Population and Ecological Approach. Academic Press, London.

Boyce, M. S., and C. M. Perrins. 1987. Optimizing great tit clutch size in a fluctuating environment. Ecology 68:142–153.

Bradshaw, A. D. 1965. Evolutionary significance of phenotypic plasticity in plants. Adv. Genet. 13:115–155.

Carey, C. 1996. Female reproductive energetics. Pp. 324–374 *in* C. Carey, ed., Avian Energetics and Nutritional Ecology. Chapman & Hall, New York.

Clamens, A., and P. Isenmann. 1989. Effects of supplemental food on the breeding of blue and great tits in the Mediterranean habitats. Ornis Scand. 20:36–42.

Clobert, J., C. M. Perrins, R. H. McCleery, and A. G. Gosler. 1988. Survival rate in the great tit *Parus major* in relation to sex, age, and immigration status. J. Anim. Ecol. 57:287–306.

Cohen, D. 1966. Optimizing reproduction in a randomly varying environment. J. Theor. Biol. 12:119–129.

Cohen, D. 1967. Optimizing reproductionin a randomly varying environment when a correlation may exist between the conditions at the time a choice has to be made and the subsequent outcome. J. Theor. Biol. 16:1–14.

de Jong, G. 1990. Quantitative genetics of reaction norms. J. Evol. Biol. 3:447–468.

Dias, P. C., and J. Blondel. 1996. Breeding time, food supply and fitness components of blue tits *Parus caeruleus* in Mediterranean habitats. Ibis 138:644–649.

Endler, J. A. 1986. Natural Selection in the Wild. Princeton University Press, Princeton, NJ.

Falconer, D. S. 1989. Introduction to Quantitative Genetics, 3rd ed. Longman, Essex, UK.

Falconer, D. S., and T. F. C. Mackay. 1996. Introduction to quantitative genetics, 4th ed. Longman, Essex, UK.

Fry, J. D. 1992. The mixed model analysis of variance applied to quantitative genetics: biological meaning of the parameters. Evolution 46:540–550.

Gagneux, P., D. S. Woodruff, and C. Boesch. 1997. Furtive mating in female chimpanzees. Nature 387:358–359.

Gebhardt-Henrich, S., and A. J. van Noordwijk. 1991. Nestling growth in the great tit I. Heritability estimates under different environmental conditions. J. Evol. Biol. 4:341–362.

Gibbs, H. L., and P. R. Grant. 1987. Oscillating selection on Darwin's finches. Nature 327:511–513.

Gomulkiewicz, R., and M. Kirkpatrick. 1992. Quantitative genetics and the evolution of reaction norms. Evolution 46:390–411.

Gotthard, K., and S. Nylin. 1995. Adaptive plasticity and plasticity as an adaptation: a selective review of plasticity in animal morphology and life-history. Oikos 74:3–17.

Grant, P. R., and B. R. Grant. 1995. Predicting microevolutionary responses to directional selection on heritable variation. Evolution 49:241–251.

Högstedt, G. 1981. Effect of additional food on reproductive success in the magpie (*Pica pica*). J. Anim. Ecol. 50:219–229.

Houston, D. C., P. J. Jones, and R. M. Sibley. 1983. The effect of female body condition on egg laying in lesser black-backed gulls *Larus fuscus*. J. Zool. 200:509–520.

Jones, P. J., and P. Ward. 1976. The level of reserve protein as the proximate factor controlling the timing of breeding and clutch size in the red-billed quelea *Quelea quelea*. Ibis 118:547–574.

Källander, H. 1974. Advancement of laying of the great tit by the provision of food. Ibis 116:365–367.

Keller, L. F., and A. J. van Noordwijk. 1994. Effects of local environmental conditions on nestling growth in the great tit *Parus major*. Ardea 82:349–362.

Kluyver, H. N. 1951. The population ecology of the great tit *Parus major* L. Ardea 19:1–135.

Lack, D. 1954. The Natural Regulation of Animal Numbers. Clarendon, Oxford.

Lack, D. 1966. Population Studies of Birds. Oxford University Press, Oxford.

Lack, D. 1968. Ecological Adaptations for Breeding in Birds. Methuen, London.

Lewontin, R. C. 1974. The Genetic Basis of Evolutionary Change. Columbia University Press, New York.

Losos, J. B., Warheit, K. I., and Schoener, T. W. 1997. Adaptive differentiation following experimental island colonization in *Anolis* lizards. Nature 387:70–73.

Merilä, J. 1996. Genetic variation in offspring condition: an experiment. Funct. Ecol. 10: 465–474.

Mousseau, T. A., and D. A. Roff. 1987. Natural selection and the heritability of fitness components. Heredity 59:181–197.

Nager, R. G. 1990. On the effects of small scale variation in temperature and food availability on laying date and egg size in great tits (*Parus major*). Pp. 187–197 *in* J. Blondel, A. Gosler, J.-D. Lebreton, and R. McCleery, eds., Population Biology of Passerine Birds. Springer, Berlin.

Nager, R. G. 1993. On the timing of breeding of great tits (*Parus major*) in relation to local environmental conditions. Ph.D. Thesis, University of Basel, Basel.

Nager, R. G., C. Rüegger, and A. J. van Noordwijk. 1997. Nutrient or energy limitation on egg formation: a feeding experiment in great tits. J. Anim. Ecol. 66:495–507.

Nager, R. G., and A. J. van Noordwijk. 1992. Energetic limitation in the egg laying period of great tits. Proc. R. Soc. Lond. B 249:259–263.

Nager, R. G., and A. J. van Noordwijk. 1995. Proximate and ultimate aspects of phenotypic plasticity in timing of great tit breeding in a heterogeneous environment. Am. Nat. 146:454–474.

Newman, R. A. 1992. Adaptive plasticity in amphibian metamorphosis. BioScience 42: 671–678.

Nilsson, J.-Å. 1994. Energetic bottle-necks and the reproductive cost of being early. J. Anim. Ecol. 63:200–208.

Peacock, M. M., and A. T. Smith. 1997. Nonrandom mating in pikas *Ochotona princeps*: evidences for inbreeding between individuals of intermediate relatedness. Mole. Ecol. 6:801–811.

Perrins, C. M. 1965. Population fluctuations and clutch size in the great tit (*Parus major*). J. Anim. Ecol. 34:601–647.

Perrins, C. M. 1970. The timing of birds' breeding season. Ibis 112:242–255.
Perrins, C.M. 1979. British Tits. Collins, London.
Perrins, C. M. 1991. Tits and their caterpillar food supply. Ibis 133 (suppl. 1):49–54.
Pigliucci, M. 1996. How organisms respond to environmental changes: from phenotypes to molecules (and vice versa). Trends Ecol. Evol. 11:168–173.
Price, T., M. Kirkpatrick, and S. J. Arnold. 1988. Directional selection and the evolution of breeding date in birds. Science 240:798–799.
Price, T., and L. Liou. 1989. Selection on clutch size in birds. Am. Nat. 134:950–959.
Queller, D. C., and K. F. Goodnight. 1989. Estimating relatedness using genetic markers. Evolution 43:258–275.
Reznick, D. N., F. H. Shaw, F. H. Rodd, and R. G. Shaw. 1997. Evaluation of the rate of evolution in natural populations of guppies (*Poecilia reticulata*). Science 275:1934–1937.
Roff, D. A. 1992. The Evolution of Life Histories. Chapman and Hall, New York.
Roff, D. A. 1997. Evolutionary Quantitative Genetics. Chapman and Hall, New York.
Scheiner, S. M. 1993. Plasticity as a selectable trait: reply to Via. Am. Nat. 142:371–373.
Scheiner, S. M., and R. F. Lyman. 1989. The genetics of phenotypic plasticity I. Heritability. J. Evol. Biol. 2:95–107.
Scheiner, S. M., and R. F. Lyman. 1991. The genetics of phenotypic plasticity II. Response to selection. J. Evol. Biol. 3:23–50.
Schlichting, C. D. 1986. The evolution of phenotypic plasticity in plants. Annu. Rev. Ecol. Syst. 17:667–693.
Schlichting, C. D., and M. Pigliucci. 1993. Control of phenotypic plasticity via regulatory genes. Am. Nat. 142:366–370.
Stearns, S. C. 1989. The evolutionary significance of phenotypic plasticity. BioScience 39:436–445.
Stearns, S. C. 1992. The Evolution of Life Histories. Oxford University Press, Oxford.
Stearns, S. C., G. de Jong, and R. A. Newman. 1991. The effects of phenotypic plasticity on genetic correlations. Trends Ecol. Evol. 6:122–126.
Svensson, E. 1997. Natural selection on avian breeding time: causality, fecundity-dependent, and fecundity-independent selection. Evolution 51:1276–12863.
Turelli, M. 1988. Phenotypic evolution, constant covariances, and the maintenance of additive variance. Evolution 42:1342–1347.
van Balen, J. H. 1973. A comparative study of the breeding ecology of the great tit *Parus major* in different habitats. Ardea 61:1–93.
van Noordwijk, A. J. 1984. Problems in the analysis of dispersal and a critique on its 'heritability' in the great tit. J. Anim. Ecol. 53:533–544.
van Noordwijk, A. J. 1989. Reaction norms in genetical ecology. BioScience39:453–458.
van Noordwijk, A. J., R. H. McCleery, and C. M. Perrins. 1995. Selection for the timing of great tit breeding in relation to caterpillar growth and temperature. J. Anim. Ecol. 64:451–458.
van Noordwijk, A. J., and C. B. Müller. 1994. On adaptive plasticity in reproductive traits, illustrated with laying date in the great tit and colony inception in a bumble bee. Physiol. Ecol. Jpn. 29:180–194.
van Noordwijk, A. J., J. H. van Balen, and W. Scharloo. 1980. Heritability of ecologically important traits in the great tit. Ardea 68:193–203.
van Noordwijk, A. J., J. H. van Balen, and W. Scharloo. 1981a. Genetic and environmental variation in clutch size of the great tit. Neth. J. Zool. 31:342–372.
van Noordwijk, A. J., J. H. van Balen, and W. Scharloo. 1981b. Genetic variation in the timing of reproduction in the great tit. Oecologia 49:158–166.

van Noordwijk, A. J., J. H. van Balen, and W. Scharloo. 1988. Heritability of body size in a natural population of the great tit and its relation to age and environmental conditions during growth. Genet. Res. 51:149–162.

Verhulst, S., and J. M. Tinbergen. 1991. Experimental evidence for a causal relationship between timing and success of reproduction in the great tit (*Parus major*). J. Anim. Ecol. 60:269–282.

Verhulst, S., J. H. van Balen, and J. M. Tinbergen. 1995. Seasonal decline in reproductive success of the great tit: variation in time or quality? Ecology 76:2392–2403.

Via, S. 1984. The quantitative genetics of polyphagy in an insect herbivore II. Genetic correlations in larval performance within and across host plants. Evolution 38:896–905.

Via, S. 1993. Adaptive phenotypic plasticity: a target or by-product of selection in a variable environment? Am. Nat. 142:352–365.

Via, S. 1994. The evolution of phenotypic plasticity: what do we really know? Pp. 35-57 *in* L. A. Real, ed., Ecological Genetics. Princeton University Press, Princeton, NJ.

Via, S., R. Gomulkiewicz, G. de Jong, S. M. Scheiner, C. D. Schlichting, and P. H. van Tienderen. 1995. Adaptive phenotypic plasticity: consensus and controversy. Trends Ecol. Evol. 10:212–217.

Via, S., and R. Lande. 1985. Genotype-environment interaction and the evolution of phenotypic plasticity. Evolution 39:505–523.

Weigensberg, I., and D. A. Roff. 1996. Natural heritabilities: can they be reliably estimated in the laboratory? Evolution 50:2149–2157.

West-Eberhard, M. J. 1989. Phenotypic plasticity and the origins of diversity. Annu. Rev. Ecol. Syst. 20:249–278.

Zandt, H. S. 1994. A comparison of three sampling techniques to estimate the population size of caterpillars in trees. Oecologia 97:399–406.

5

Adaptive Evolution and Neutral Variation in a Wild Leafminer Metapopulation

SUSAN MOPPER, KELI LANDAU, AND PETER VAN ZANDT

Allozyme and DNA markers reveal substantial neutral genetic structure within natural insect populations, but there is little information about adaptive structure in these systems. We combine genetic analyses and field experiments to measure gene flow and neutral and adaptive structure in a wild leafminer population. This integrative approach reveals temporal and spatial variation in population structure and provides a more accurate and comprehensive understanding of the evolutionary ecology of wild insect populations.

5.1 Spatiotemporal Structure of Wild Insect Populations

Insect populations are ideal for investigating evolutionary processes because they reside in patchy habitats that promote geographic structure (reviewed by Roderick 1996, McCauley and Goff 1998, Peterson and Denno 1998). In the last decade, insect models have played a central role in the development of metapopulation theory, helping to reveal how colonization, extinction, and migration influence the genetic structure and evolutionary ecology of populations (see reviews in Hassell et al. 1991, Hastings and Harrison 1994, Hanski and Gilpin 1997). Phytophagous insects are particularly intriguing models for investigating evolutionary processes because of their "parasitic" relationship with long-lived plants, on which they may feed and reside for many generations (Price 1980). The genetic heterogeneity of host plant populations (Edmunds and Alstad 1978, Schmitt and Gamble 1990, Zangerl and Berenbaum 1990, Mopper et al. 1991, Michalakis et al. 1993, Sork et al. 1993, Berg and Hamrick 1995) can promote local differentiation and adaptive evolution in phytophagous insects at very fine spatial scales (reviewed in Mopper and Strauss 1998).

5.1.1 Sympatric race formation

One of the most important examples of local adaptive structure is the formation of insect host races. Differential selection imposed by distinct plant species results in genetic differentiation, reproductive isolation, and local evolution in a diverse array of insects (Futuyma and Peterson 1985, Barton et al. 1988, Via 1990, Bush 1994). Some of the best known examples are the hymenopteran shoot-galling sawfly, *Euura atra* (Roininen et al. 1993), lepidopteran fall armyworm *Spodoptera frugiperda* (Pashley 1988) and fall cankerworm, *Alsophila pometeria* (Mitter et al. 1979), homopteran pea aphid, *Acyrthosiphon pisum* (Via 1991), dipteran gall fly, *Eurosta solidaginis* (Craig et al. 1997), leafminer, *Liriomyza brassicae* (Tavormina 1982), and coleopteran pine beetle, *Dendroctonus ponderosae* (Sturgeon and Mitton 1986).

Several host races have evolved in response to recently introduced exotic plant species. The classic example is *Rhagoletis pomonella*, which has diversified in the past 130 years since the introduction of European apple, *Malus pumila*, to North America. Now two races of fly exist sympatrically, one that prefers the native hawthorn host (*Crataegus* spp.), and one that specializes on the exotic apple (McPheron et al. 1988, Feder et al. 1990, 1998). The native soapberry bug, *Jadera haematoloma*, has rapidly differentiated following the introduction of *Koelreuteria elegans* into North America from southeast Asia (Carroll and Boyd 1992, Carroll et al. 1997). In less than 50 years, plant-related selection on beak length has resulted in bug populations locally adapted to new host species. Rapid evolution is also evident in the checkerspot butterfly, *Euphydryas editha*, which is currently expanding from its native host plant, *Collinsia parviflora*, to the European *Plantago lanceolata*, introduced into California in the last 15–20 years (Singer et al. 1993, Thomas and Singer 1998). Evolutionarily poised between demes and incipient species (Bush 1969), host races exhibit phenotypic variation and partial or complete reproductive isolation and are excellent models for investigating the spatial patterns and temporal dynamics of genetic isolation and adaptive differentiation at local levels. The many examples of host-race formation have helped change how biologists view evolution, from a strictly allopatric process to one that can occur in close physical proximity (Bush 1994).

5.1.2 Population subdivision

Distinct host species represent potent sources of differential selection, even for sympatrically distributed insect herbivores. Evidence is also mounting for fine-scale genetic structure within populations of insects that feed on only one host. Peterson and Denno (1998) reviewed more than 200 studies to explore how insect life histories influence population structure at local levels. Not surprisingly, in many (but not all) populations, there was a negative relationship between genetic structure and insect dispersal ability and putative gene flow. But regardless of mobility, insects were much more likely to exhibit population structure on long-lived woody plants than on herbaceous hosts. Host attributes such as longevity can affect the evolutionary potential of the entire population because they promote genetic heterogeneity and population genetic structure (Levin 1995).

Migration between, colonization of, and extinction on host plants are important forces in phytophagous insect populations and are central theoretical issues in the study of metapopulation dynamics (Slatkin 1987, Wade and McCauley 1988, Olivieri et al. 1995). Founder effects and genetic drift can create age-structured populations in which newly colonized habitat patches exhibit greater genetic variance than long-inhabited patches. For example, Nürnberger and Harrison (1995) observed high genetic variance in mtDNA markers owing to genetic drift in ponds recently colonized by whirligig beetles (*Dineutus assimilis*). Using allozymes, Whitlock (1992) observed a similar pattern in fungus beetles: incipient populations on newly colonized decaying logs exhibited significantly greater genetic variance than older populations. Similarly, a long-term colonization experiment provided indirect support for age-structured neutral genetic variance in local populations of milkweed beetle, *Tetraopes tetraophthalmus* (McCauley and Eanes 1987, McCauley 1989). In fragmented populations, random events can interact with gene flow to sustain high levels of genetic variation upon which natural selection acts, with profound effects on the evolutionary potential of the population (Wright 1977).

5.1.3 Local adaptation

Edmunds and Alstad (1978) proposed the adaptive deme formation hypothesis (ADF), inspired by their long-term studies of black pineleaf scale infestation of ponderosa pine (*Nuculaspis californica* and *Pinus ponderoseae*, respectively). The proposition that insect demes evolved in response to individual pine trees initiated a debate among evolutionary ecologists that continues today (Alstad 1998, Boecklen and Mopper 1998, Cobb and Whitham 1998). Deme formation is the development of genetically isolated groups of conspecific organisms at local spatial scales (Alstad and Corbin 1990, Costa and Ross 1993, Mopper 1996, Hanks and Denno 1998, McCauley and Goff 1998). Demes can arise from random events such as genetic drift or from adaptation in response to selection pressures. The ADF hypothesis predicts that insect populations will subdivide into locally adapted demes if (1) host plants are long-lived relative to insects, (2) individual host plants exert distinct selection pressures, and (3) insect migration between hosts is limited.

Obligate or facultative host fidelity is necessary for the evolution of locally adapted demes in phytophagous insect populations because genetic "viscosity" is a key isolating mechanism (Edmunds and Alstad 1978, Peterson and Denno 1998). In other words, sessile insects should have greater potential for adaptive deme formation than dispersive organisms with higher rates of interplant dispersal (Alstad 1998). The extremely sessile black pineleaf scale was the first insect model for adaptive deme formation (Edmunds and Alstad 1978), and most subsequent studies of adaptive structure have focused on insects with limited mobility. However, the relationship between adaptive structure and gene flow has proven unpredictable at local levels (Mopper 1996, Van Zandt and Mopper 1998). Although no longer viewed as a barrier to adaptive evolution (Ehrlich and Raven 1969, Endler 1977, Slatkin 1987, Holt and Gomulkiewicz 1997), the role of gene flow in local adaptation is relatively untested empirically (but see Roininen et al. 1993, Bossart and Scriber 1995). Although consistent at regional scales, the relationship between mobility and population subdivision appears unpredictable at local spatial scales (Peterson and Denno 1998). Our empirical studies in oak leafminer populations are currently focused on quantifying gene flow and determining its influence on neutral and adaptive population structure.

5.1.4 A Meta-analysis of adaptive deme formation

One of our primary interests is to discover the forces associated with local adaptive structure. One approach is meta-analysis, the quantitative analog of a narrative literature review (Gurevitch and Hedges 1993). By synthesizing the results of independent studies and weighting them based on experimental power, meta-analysis is a robust and objective tool for drawing general inferences about biological processes (Hedges and Olkin 1985, Rosenberg et al. 1997). Meta-analysis is particularly helpful when studies testing a specific hypothesis are divided, as in the insect deme formation literature, but it can also be used to compare different experiments within the same study (Hechtel and Juliano 1997), which we demonstrate later in this chapter.

Meta-analysis of published experiments indicates that diverse insect taxa exhibit local adaptation in association with long-lived host plants (Van Zandt and Mopper 1998). However, there is no apparent relationship between insect dispersal ability and local adaptation, as was predicted by the original hypothesis (Edmunds and Alstad 1978). Some dispersive insect species such as leafminers (Mopper et al. 1995), aphids (Akimoto 1990, Komatsu and Akimoto 1995), and gall-forming midges (Stiling and Rossi 1998) exhibit adaptive population structure, whereas some highly sessile scale insects do not (Unruh and Luck 1987, Cobb and Whitham 1993). Ironically, the subdivision observed within black pineleaf scale populations is now attributed to stochastic forces rather than to natural selection (Alstad 1998). It seems *N. californica* is too sessile to accumulate enough genetic variation for selection to create individual demes; its population structure is more influenced by founding events and genetic drift than local adaptation as originally believed.

5.2 Empirical Background

5.2.1 Natural history of an oak leafminer

Stilbosis quadricustatella (Hodges) is a microlepidopteran leafminer (Cosmopterigidae) whose primary host is *Quercus geminata* (sand live-oak), a shrubby, semievergreen species inhabiting the coastal areas and sandy inland soils of central and north Florida, USA. Since 1985 we have studied a monospecific population of mixed-age *Q. geminata* distributed as a narrow band of trees surrounding a small freshwater lake 13 km south of Tallahassee, Florida (fig. 5.1a,b). Many Lost Lake *Q. geminata* are heavily infested by *S. quadricustatella* leafminers, although insect densities vary widely and consistently among individual trees (Mopper and Simberloff 1995). *S. quadricustatella* leafminers produce one generation per year. In late spring, the adults, which do not feed, emerge from leaf litter and fly to the canopy to mate and oviposit on newly emerged foliage. Females deposit single eggs among the dense trichomes of the lower leaf surface. When eggs hatch, the 1-mm larvae wander the leaf surface to locate a site along the mid-vein to excavate an interior mine.

There are three important sources of larval mortality: host plants, parasitoids, and predators (Mopper et al. 1984, 1995). Mortality sources are easily identified by characteristic features such as intact dead larvae in the mines of attached or abscised leaves (plant-related mortality); tiny circular exit holes and pupal remains (hymenopteran parasitism); and opened mines with missing larvae (hymenopteran and avian predation). First and second instars

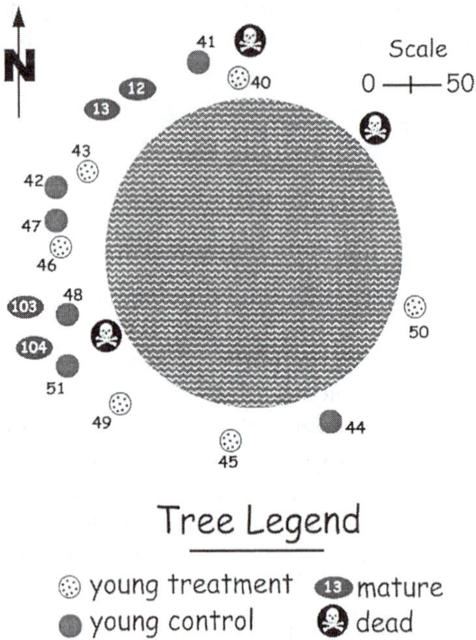

Figure 5.1 Photograph (A) and map (B) of the inland *Quercus geminata* population. Lost Lake is 203 m in diameter and surrounded by a narrow band of *Q. geminata* trees of different age classes. After Mopper et al. (2000).

incur the highest mortality, primarily by failing to successfully excavate or sustain viable mines. Intermediate larval instars suffer parasitism by wasps, and late instars are killed primarily by wasp, ant, and bird predators. Larvae do not have physical contact nor do mines coalesce, and there are no apparent antagonistic interactions (fig. 5.2). Leafminer larvae feed from June through September, then create a characteristic sickle-shaped emergence hole in the lower leaf surface, drop to the ground, and pupate beneath the tree until spring. In late summer and fall, premature leaf abscission can cause substantial mortality of late-instar larvae (Simberloff and Stiling 1987), which occurs at high rates on heavily infested trees (Mopper and Simberloff 1995). Unique leaf scars indicate the source of leafminer mortality and enable us to accurately determine the fates of individual larvae.

5.2.2 *Extinction experiment*

We conducted an extinction experiment to assess leafminer gene flow and population structure on 10-year-old trees. In August 1990 we selected 12 *Q. geminata* trees and hand-removed all late-instar leafminers from 6 of them chosen at random. *S. quadricustatella* is univoltine; therefore, removing all larvae from the treatment trees created vacant hosts for the 1991 Lost Lake population. Leafminers were allowed to colonize trees naturally in 1991 and subsequent years. In 1994 we determined the fate of individual leafminers on the control and treatment trees. The 1994 cohort had inhabited treatment trees for a maximum of 4 generations, whereas leafminers had a maximum of 10 generations on control trees.

5.2.3 *Reciprocal transfer experiment*

A reciprocal transfer experiment was conducted to assess adaptive structure on mature 40-year-old trees. In 1992, Mopper et al. (1995) reciprocally transferred leafminer eggs among four mature oak trees. One hundred and sixty leafminer eggs were collected from each tree and randomly distributed back either to the natal host or to one of the three novel trees. All transfers were monitored until mines were completed and leafminer fates determined. Leafminers transferred back to their natal host had a maximum residence of 40 generations, and leafminers transferred to novel mature hosts were in their first year of colonization.

For simplicity, we use the term "migrants" to indicate fourth-generation leafminers colonizing young extinction-treatment trees and first-generation leafminers transferred to novel mature trees. We use "residents" to indicate the 10th-generation leafminers on young control trees and the 40th-generation leafminers transferred back to their natal mature trees.

5.3 Hypotheses

5.3.1 *Gene flow*

Gene flow can be difficult to measure in natural systems, and the benefits of demographic versus genetic estimates are under debate (Slatkin and Barton 1989, Mitton 1994, Neigel 1997, Peterson and Denno 1998). The direct demographic method provides an ecological perspective, whereas the indirect genetic method has a historical bias, and both are prone to mul-

Figure 5.2 Two distinct late-instar *S. quadricustatella* larval leafminers feeding within a *Quercus nigra* leaf (after Mopper 1996).

tiple interpretations (Bossart and Prowell 1998). We used both the demographic and the genetic methods to estimates rates of leafminer gene flow between *Q. geminata* host trees.

5.3.1.1 Demographic estimate of gene flow *S. quadricustatella* leafminers are winged and fly readily, suggesting that gene flow could be high between trees. Nonetheless, they exhibit adaptive structure at local and regional spatial scales (Mopper et al. 1995). We used colonization and survival data from the extinction experiment to estimate the rate of gene flow between trees with the equation

$$M = \xi K$$

where M = the number of leafminers migrating between host trees, K = number of leafminers colonizing new host trees, and ξ = survival rate of leafminers through the larval stage (notation after Slatkin 1993, Ingvarsson et al. 1997).

Colonization of the vacant hosts created by the extinction treatment indicates substantial inter-tree dispersal (fig. 5.3), and combined with the survival data ($M = \xi K$), suggests that gene flow ranges from 0 to 249 migrants between young trees each generation (fig. 5.4). This is a sizable portion of the local deme, assuming that the young trees contain approximately 3000 leaves (E. Connor, D. Simberloff, and P. Stiling, personal observations) and leafminer densities average 10 ± 3% of total leaf number (based on 1994 census data). The extinction experiment also demonstrates that leafminer colonization rates vary widely. For example, one extinction treatment tree was not colonized at all in 1991, whereas another treatment tree was colonized by 249 leafminers in 1991 and had doubled in density of leafminers by 1994.

5.3.1.2 Genetic estimate of gene flow We paired each extinction treatment tree with its nearest neighbor (the distance between paired trees ranged from 1 to 5 m) and calculated

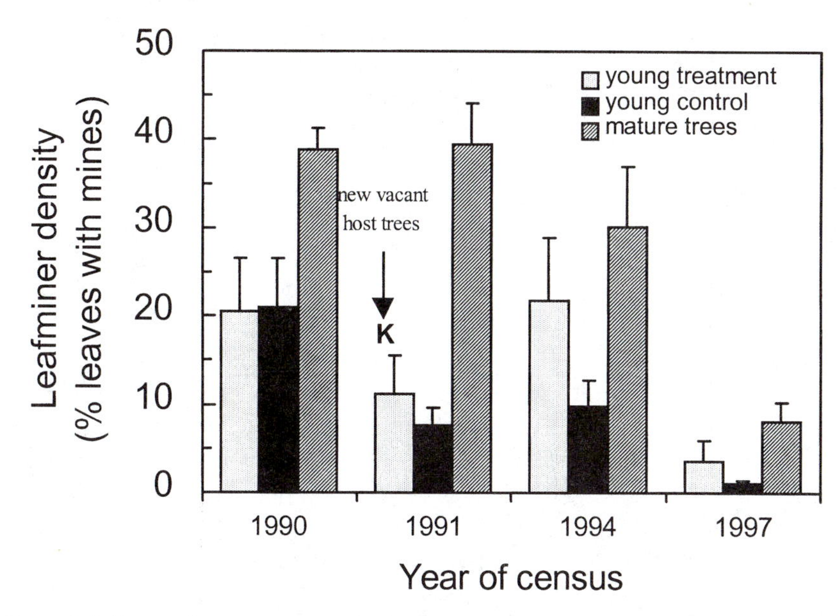

Figure 5.3 Pre- and post-treatment leafminer densities on young control and extinction treatment trees in 1990 (pretreatment), and in three post-treatment years. Mature deme formation experiment trees (Mopper et al. 1995) are also included in the comparison. Repeated-measures ANOVAs of arcsine transformed density data detected no significant group effect on leafminer densities (table 5.1), except in 1991, when there were significantly higher leafminer densities on mature trees than compared to young control and extinction trees ($P < .05$, Tukey's Studentized Range Test). K indicates the density of leafminers colonizing the six vacant treatment trees. After Mopper et al. (2000).

pairwise F_{ST} values based on allele frequencies (Wright 1965, Weir and Cockerham 1984). In 1994 we collected larvae for genetic analysis from the 10-year-old and 40-year-old trees and conducted cellulose acetate protein electrophoresis (Richardson et al. 1986, Hebert and Beaton 1989). We screened leafminers for activity at 14 loci and resolved 3 that were sufficiently polymorphic for comparison: malate dehydrogenase (*Mdh,* EC 1.1.1.37), phosphoglucose isomerase (*Pgi,* EC 5.3.1.9), and phosphoglucomutase (*Pgm,* EC 5.4.2.2).

Wright's island model (1977) describes migration as occurring randomly among all local populations and is estimated by

$$M = (1 - F_{ST}) / 4 F_{ST}$$

where M is the number of leafminers migrating between demes (notation after Slatkin 1993). Alternatively, migration occurs only between neighboring populations in the stepping stone model (Kimura and Weiss 1964), which may better predict leafminer gene flow because of the narrow circular configuration of the *Q. geminata* population (fig. 5.1b). We calculated stepping-stone gene flow as in Benzie and Williams's (1997) study of the giant clam, *Tridacna maxima*,

$$M = [(1 - F_{ST})(-\ln 2\mu)] / 4\pi F_{ST}$$

The rate of mutation (μ) for electrophoretic variants is estimated as 10^{-7} (Nei 1987).

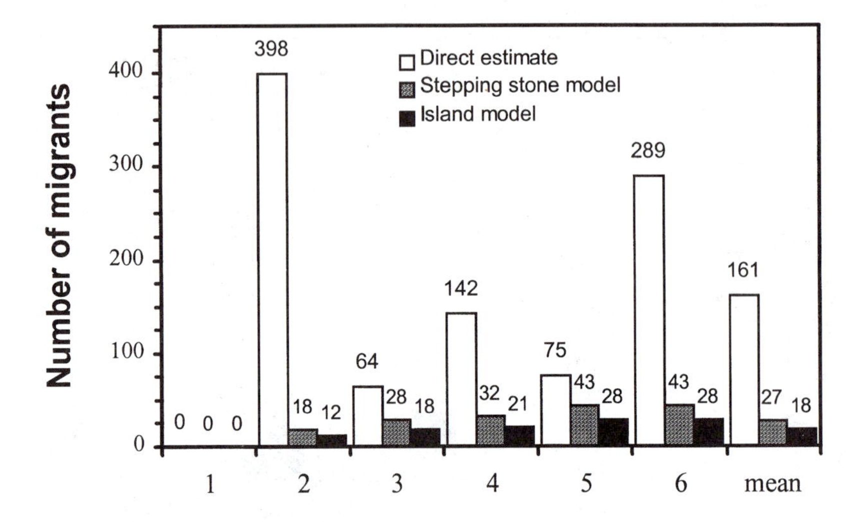

Figure 5.4 Direct and indirect estimates of leafminer gene flow (ANOVA, $F_{2,15} = 4.93$, $P = .023$; direct estimates versus island $P = .04$; stepping stone $P = .06$, Tukey pairwise comparisons). Numerical values for migrant numbers appear above the bars. After Mopper et al. (2000).

The island and stepping stone models predicted similar rates of gene flow and differed significantly from the demographic estimate (fig. 5.4). Although model predictions were much lower in magnitude (12–43 migrants/tree/generation) than the direct estimate (64–398 migrants/tree/generation), they were probably sufficiently large to homogenize the demes in the absence of local selection.

5.3.2 Adaptive population structure

If the Lost Lake leafminer population exhibits locally adaptive structure, then long-term residents of oak hosts should outperform migrants. That is, 10th-generation leafminers on young control trees should outperform the more recent 4th-generation leafminers on young extinction-treatment trees. Likewise, in the reciprocal transfer experiment, 40th-generation leafminers transferred back to natal trees should outperform 1st-generation leafminers transferred to novel trees. We predicted that plant-related mortality should be lower for long-term residents than for migrants if leafminers become locally adapted to their host plant.

In addition to comparing the mortality of migrants versus residents within the young and mature trees, we further predicted that selection should be stronger against migrants on mature trees than on young trees because mature-tree residents should be better adapted than young-tree residents. We calculated relative fitness and selection coefficients within each group (Strauss and Karban 1998). The within-group fitness (ω) of the migrants relative to the residents is

$$\omega = \frac{(\bar{X} \text{ migrants surviving plant-related mortality})}{(\bar{X} \text{ residents surviving plant-related mortality})}$$

and the selection coefficient (s) for the migrants is

$$s = 1 - \omega$$

Selection coefficients of O indicate fitness equality between migrants and residents (i.e., no local adaptation), a positive s indicates that residents outperform migrants.

Although 10th-generation leafminers on young control trees suffered less plant-related mortality than did 4th-generation leafminers on young extinction-treatment trees (Mopper et al. 1999), plant-related mortality differed by a much wider margin for migrants versus residents of mature trees (fig. 5.5). The selection coefficients ($1 - \omega$) indicate that selection against migrants increases with the age of the leafminer population.

5.3.3 Meta-analysis: Testing temporal variation in adaptive structure

An alternative method of comparing the magnitude of selection against recent and resident leafminers is meta-analysis. Meta-analysis is a relatively new method of comparing the results of separate experiments (Gurevitch and Hedges 1993). This powerful technique is rapidly increasing in popularity among ecologists (Rosenberg et al. 1997) and can be used to compare different experiments within a study as we have (see also Hechtel and Juliano 1997) or to compare independent studies conducted in different systems (Gurevitch et al. 1992, Van Zandt and Mopper 1998). Meta-analysis calculates an effect size for each study based on the sample size, average response, and variance (Hedges and Olkin 1985, Rosenberg et al. 1997). Hedges's d is calculated as

$$d = \frac{\bar{Y}_r - \bar{Y}_m}{s}$$

where $\bar{Y}_r$ is the mean plant-related mortality of residents (i.e., insects inhabiting young control trees in experiment 1, and insects transferred back to mature natal trees in experiment 2), and $\bar{Y}_m$ is the mean plant-related mortality of migrants (i.e., insects colonizing young extinction trees in experiment 1 or insects transferred to mature novel trees in experiment 2). Therefore, d is the difference between residents and migrants in standard deviation units. For this analysis, a positive value of d indicates that insects suffered less plant-related mortality on natal host trees than on recently colonized novel trees. The pooled standard deviation (s) in the equation above is estimated as

$$s = \sqrt{\frac{(N_r - 1)(s_r)^2 + (N_m - 1)(s_m)^2}{N_r + N_m - 2}}$$

where N_r and N_m are the total sample sizes of the resident and migrant groups, respectively, and s_r and s_m are the standard deviations of the resident and migrant groups.

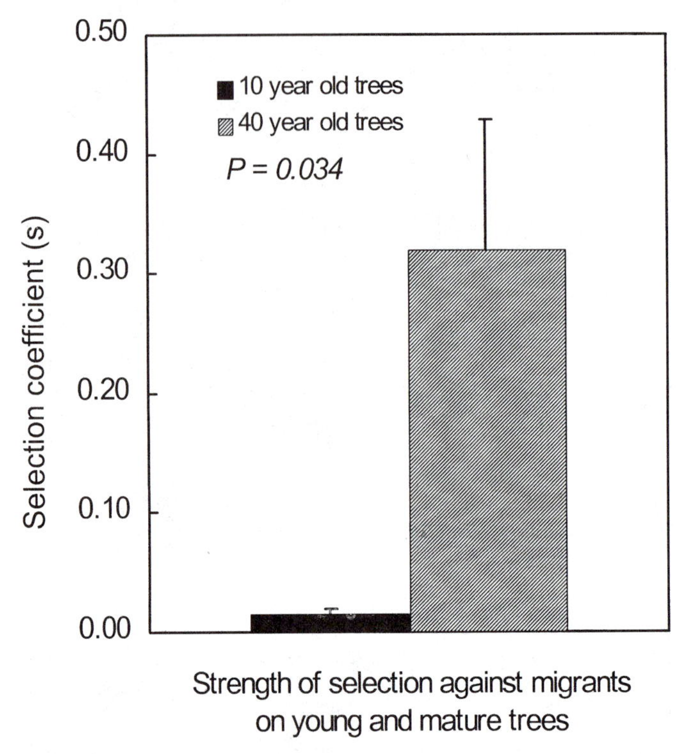

Figure 5.5 Selection against novel immigrants relative to the natal residents was significantly greater on 40-year-old than on 10-year-old trees (t-test for unequal variances = 2.8, one-tailed $P = .034$). After Mopper et al. (2000).

The meta-analysis indicated an effect size of 0.543 ± .021 for experiment 1, in which we compared plant-related mortality of 4th- versus 10th-generation leafminers (Mopper et al. 1999). The effect size was much greater in experiment 2, 1.053 ± 0.115, in which we compared plant-related mortality of 1st versus 40th-generation leafminers. These are moderate to high effect sizes (Cohen 1988) and consistent in direction with the selection coefficients calculated for mature and young demes (fig. 5.6). The lack of overlap in s indicates the dissimilarity of responses in different experiments; that neither group overlaps with zero indicates a statistically significant effect of leafminer residence on plant-related mortality in both groups.

5.3.4 *Neutral genetic structure*

We calculated several parameters of population subdivision for 4th-generation leafminers ($n = 6$ young trees), 10th-generation leafminers ($n = 6$ young trees), and 40th-generation leafminers ($n = 4$ mature trees). One 4th- and 40th-generation subpopulation was excluded because pupal sample size was insufficient for genetic comparison. To evaluate population genetic structure, we used Wright's F_{ST}, which quantifies the standardized variance in allele frequencies among local populations and reflects genetic differentiation and population sub-

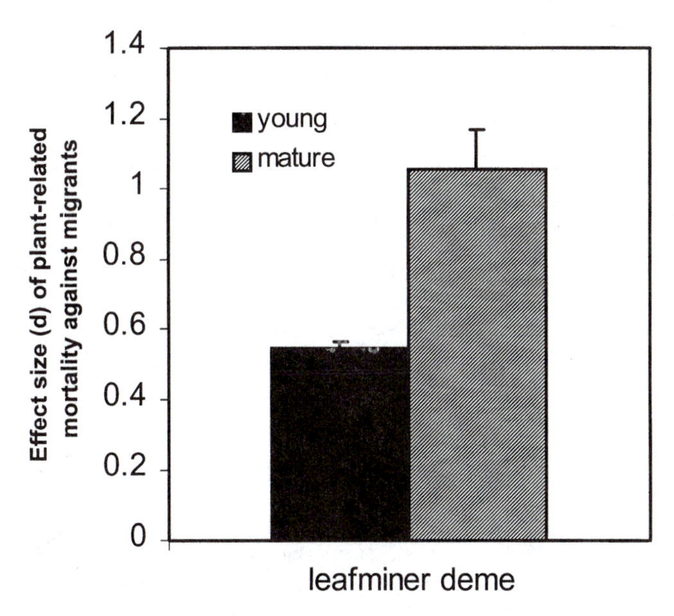

Figure 5.6 The meta-analysis comparing plant-related mortality of 4th- versus 10th-generation leafminers (Mopper et al. 2000). The lack of overlap observed in effect size indicates the dissimilarity of responses in different experiments; that neither overlap with zero indicates a statistically significant effect of leafminer residence on plant-related mortality.

division (Wright 1951, 1965, Nei 1977). An F_{ST} of zero indicates no genetic differentiation among putative demes, whereas an F_{ST} of 1 indicates complete genetic isolation (Roderick 1996, Alstad 1998). In addition to F_{ST} statistics, we also computed θ to estimate population genetic structure. Theta is a bias-corrected analog of F_{ST} that is not influenced by numbers of alleles per locus or by numbers of individuals and subpopulations sampled (Weir and Cockerham 1984).

Both F_{ST} and θ indicate significant structure of *Mdh* and *Pgi* loci among fourth-generation leafminers colonizing the extinction treatment trees (table 5.1). Both F_{ST} and θ indicated significant structure in the *Pgi* variant among 10th-generation demes on young control trees. Neither F_{ST} nor θ indicated genetic structure among leafminers inhabiting mature trees. This rapid decline in putative neutral structure across generations (fig. 5.7) is consistent with moderate to high gene flow (fig. 5.4).

5.4 Summary

5.4.1 Dispersal, gene flow, and neutral population structure

The neutral structure of incipient demes declines with successive generations (table 5.1 and fig. 5.7). This is caused by gene flow (fig. 5.4), which reduces interdemic genetic variance and increases intrademic genetic variation as demes progress from colonization to equilib-

Table 5.1 F and θ estimates of population structure for 4th-, 10th-, and 40th-generation leafminers

	Locus	θ	SE	F_{ST}	$P \pm$ SE
4th generation	*Mdh*	0.017	0.010	0.024	.028 ± 0.002*
(n = 241)	*Pgi*	0.019	0.013	0.031	.033 ± 0.002*
	Pgm	0.002	0.008	0.015	.294 ± 0.006
	All loci	0.017	0.004	0.027	
	CI	0.002–0.019			
10th generation	*Mdh*	0.000	0.003	0.016	.277 ± 0.066
(n = 245)	*Pgi*	0.008*	0.005	0.020*	.030 ± 0.002*
	Pgm	-0.001	0.000	0.010	.944 ± 0.003
	All loci	0.002	0.003	0.017	
	CI	-0.001–0.008			
40th generation	*Mdh*	-0.004	0.005	0.006	.151 ± 0.005
(n = 370)	*Pgi*	0.010	0.015	0.017	.114 ± 0.005
	Pgm	0.012	0.014	0.019	.154 ± 0.005
	All loci	0.009	0.002	0.015	
	CI	-0.004–0.012			

F-statistics were determined with BIOSYS-PC (Swofford and Selander 1989); the Genetic Data Analysis program (Lewis and Zaykin 1996) calculated θ, jackknifed standard errors, and bootstrapped confidence intervals; chi-square estimates of population structure (not shown) and P-values and were bootstrapped 5000 times using the Restriction Enzyme Analysis Program (Roff and Bentzen 1989). After Mopper et al. (2000).

*Significant at $P < .05$.

rium (Wright 1943, Kimura and Maruyama 1971, Whitlock and McCauley 1990, Slatkin 1994). Several insect species are known to exhibit ephemeral neutral genetic structure during the colonization process. These include forked fungus beetles inhabiting decaying logs (Whitlock 1992), fungus beetles colonizing flower heads (Ingvarsson et al. 1997), and aquatic beetles in freshwater ponds (Nürnberger and Harrison 1995). Our observations of the *S. quadricustatella* population are consistent with theoretical predictions and empirical observations of age-structured populations because neutral structure is eliminated by gene flow within 40 generations. At Lost Lake, colonization sampling drift produces significant genetic variance in recently established leafminer demes, but gene flow continues to sample alleles from the population such that aging demes exhibit reduced genetic variance in selectively neutral loci.

Theoretical models that estimate gene flow are based on numerous unrealistic assumptions. For example, the island model assumes that subpopulations are equally populated, randomly mating, at equilibrium, and that genetic variants are selectively neutral and do not mutate. Waples (1998) provides some caveats with respect to these models and also discusses some limitations of F_{ST} statistics. We remedied many of these concerns by collecting sufficiently large numbers of leafminers from each tree for genetic analysis, using bootstrapping resampling techniques, bias-correcting neutral structure estimates with θ, and averaging across multiple loci.

To test the neutrality of the genetic markers, we correlated the frequency of the most common allele with survival and density of leafminers on the 12 young trees (fig. 5.8A–C). There were no significant correlations between allele frequencies and estimated fitness com-

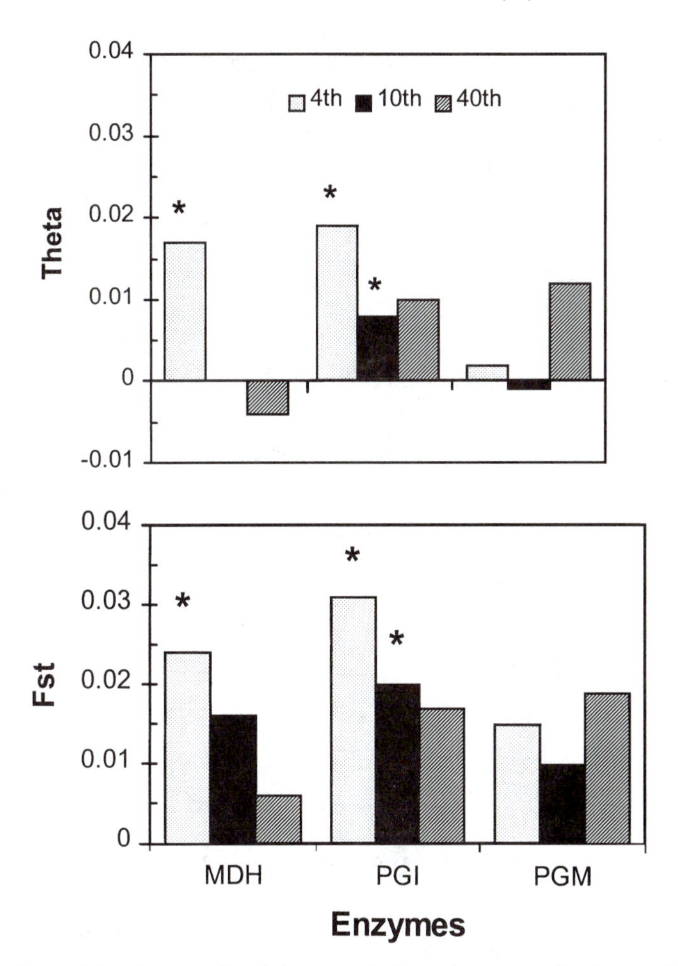

Figure 5.7 F_{ST} and θ estimates of leafminer population substructure for three polymorphic proteins sampled from 4th-, 10th-, and 40th-generation leafminers. Significance levels at $P < .05$ are indicated by asterisks. After Mopper et al. (2000).

ponents in any comparison (cf. Alstad 1998). Although this does not guarantee neutrality, it suggests that *Mdh*, *Pgi*, and *Pgm* are neutral with respect to fitness in this system.

Some evidence for population equilibrium in neutral alleles is provided by a comparison of genetic and geographic distances using a Mantel spatial analysis (Fortin and Gurevitch 1993). We determined Nei's genetic similarity for leafminers inhabiting the six young control and three mature trees with PC-BIOSYS-1 (Swofford and Selander 1989), and also measured the distance (m) between the host trees. The absence of correlation between geographic and genetic distance (table 5.2), and general loss of neutral structure over time (table 5.1 and fig. 5.7), suggests that resident demes are at genetic equilibrium with respect to the Lost Lake metapopulation, for neutral genetic variants. However, the evidence for adaptive structure indicates that multiple equilibria exist in the population.

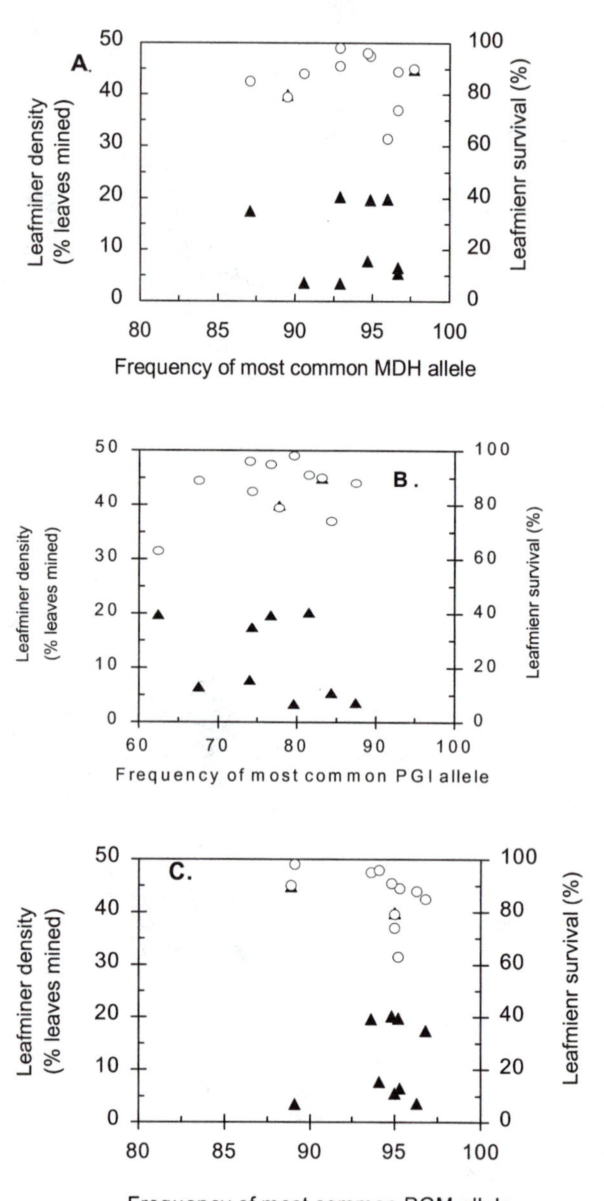

Figure 5.8 Density (triangles) and survival (circles) of leafminers plotted against the frequency of the most common *Mdh* (A), *Pgi* (B), and *Pgm* (C) alleles (all data from 1994). There were no significant correlations between allele frequencies and leafminer performance at any locus (R^2 *Mdh* survival = .01, R^2 *Mdh* density = .00; R^2 *Pgi* survival = .13; R^2 *Pgi* density = .00; R^2 *Pgm* survival = .195, R^2 *Pgm* density = .08; $P >> .10$ for all comparisons). After Mopper et al. (2000).

Table 5.2 Genetic (above diagonal) and geographic (below diagonal) distance matrices between leafminers inhabiting 10-year-old control (41,42,44,47,48,51) and 40-year-old (13,103,104) *Q. geminata* trees

Tree	41	42	44	47	48	51	13	103	104
41		0	0	0	0.002	0	0	0	0
42	130		0.004	0	0.002	0.003	0	0.003	0.003
44	364	320		0.006	0.011	0.002	0.001	0.003	0.008
47	156	26	294		0	0.001	0	0	0
48	230	101	219	75		0.003	0.004	0.004	0
51	254	124	196	98	23		0.001	0	0.001
13	62	68	388	94	169	192		0.001	0.004
103	237	107	214	81	6	18	175		0.001
104	238	108	212	82	7	16	176	2	

Geographic distances are the shortest circumferential distance in meters between trees surrounding Lost Lake. Genetic distances were calculated using Nei's (1978) unbiased estimates (BIOSYS-1, 1.7). See fig. 5.1b for location of trees. Correlation between geographic and genetic distance is nonsignificant ($r = .252$, $P = .84$, Mantel spatial analysis). After Mopper et al. (2000).

5.4.2 *Local selective mechanisms*

As neutral structure declines, adaptive structure rises (fig. 5.9). To our knowledge, this is the only study to report simultaneous temporal variation in neutral and adaptive demic structure in a natural insect population.

Local adaptation in the oak leafminer and other dispersive insect species (Komatsu and Akimoto 1995, Stiling and Rossi 1998) supports theoretical arguments that adaptive evolution is possible and even likely under conditions of strong selection, despite considerable gene flow (Ehrlich and Raven 1969, Endler 1977, Wright 1977, Slatkin 1987). In metapopulations, K, the number colonizing vacant patches, and M, the number migrating between inhabited patches, can differ considerably, and their influence on population structure will be affected by the strength of selection (Slatkin 1977, Wade and McCauley 1988). Our extinction experiment suggests that K may surpass M in the Lost Lake leafminer population, due perhaps to declining tree resources as densities rise and competitive pressures increase from locally adapted residents.

Many oak traits could act as selective mechanisms, including secondary chemical defenses, structural defenses, and leaf phenology. *S. quadricustatella* leafminers are tiny internal feeders encased in plant tissue. Nonetheless, chemical defenses such as oak tannins may be ineffective against such oak herbivores (Faeth 1995). By selectively consuming specific tissue layers, leafminers may avoid secondary compounds that are concentrated in specialized vacuoles, cuticle, and epidermal layers (Cornell 1989, Connor and Taverner 1997). The lower leaf surface of *Q. geminata* is densely covered with trichomes, but this pubescence may protect rather than deter *S. quadricustatella* leafminers. Seen through a dissecting microscope, eggs are nestled within the pubescence, and just-hatched larvae wander among the trichomes as through a forest (S. Mopper, personal observation).

Plant phenology may be the most likely source of local selection on *S. quadricustatella* leafminers because oviposition and feeding must be tightly synchronized with it (Faeth et al. 1981a,b, Crawley and Akhteruzzaman 1988, Hunter 1990, 1992, Auerbach 1991, Connor et al.

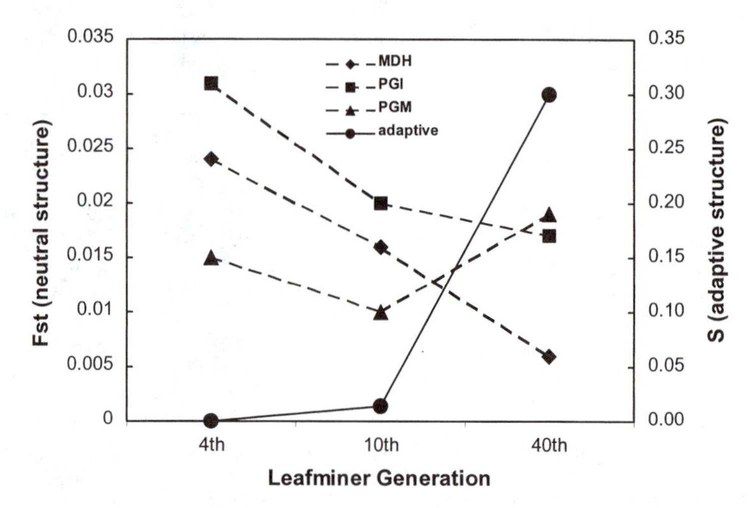

Figure 5.9 Leafminers exhibit contrasting temporal patterns in adaptive and neutral population genetic structure. F_{ST} and s are based on genetic and experimental data, except for the fourth-generation s, which we arbitrarily assigned as 0.0, assuming that differential fitness between fourth-generation leafminers and recent colonists is probably minimal at this early stage of demic evolution.

1994, Komatsu and Akimoto 1995, Van Dongen et al. 1997). Individual *Q. geminata* vary in the timing of new leaf growth and leaf abscission (Stiling and Simberloff 1989, Stiling et al. 1991, Mopper and Simberloff 1995). Bud break coincides with leafminer oviposition, when tiny first instars must excavate mines before leaf tissue toughens, and unlike free-feeding insects, *S. quadricustatella* leafminers must remain in the same leaf from egg to final instar. Prematurely abscised *Q. geminata* leaves deteriorate rapidly, and the miners within suffer significantly greater predation and desiccation than leafminers in attached leaves (Stiling and Simberloff 1989, Stiling et al. 1991, Mopper and Simberloff 1995). For leafmining insects, phenology could be a powerful mechanism promoting adaptive population structure.

5.4.3 Shifting balance evolution: A conceptual model of the S. quadricustatella *metapopulation*

The shifting balance theory was formulated for structured populations under the influence of random variation and directed selection with "continual shifting differentiation among local races" (Wright 1929: 287). This can be expressed conceptually by multiple fitness peaks in a field of genotypic frequencies (Wright 1977). There are three fundamental components of shifting balance evolution. First, demes are created via random drift within a framework of equilibrium gene frequencies that are produced by interactions between mutations, gene flow, and selection. Second, there is relatively rapid intrademic selection until a new fitness peak is reached, about which the deme drifts at random. And third, there is strong interdemic selection, such that adaptively superior demes spread through the population via dispersal, ultimately raising neighboring demes to the same fitness peak. The third point differs from our observations in the *S. quadricustatella* population because, theoretically, each *Q. geminata* host plant could select for a unique *S. quadricustatella* fitness peak.

Our conceptual model of the Lost Lake *S. quadricustatella* leafminer metapopulation is consistent with the shifting balance theory. We propose that vacant host plants are colonized rapidly, and, over time, the fitness of new migrants relative to residents diminishes, residents outcompete migrants for plant resources, and the tree becomes dominated by locally adapted genotypes. Mature demes eventually go extinct, either gradually as hosts senesce or abruptly from catastrophic events (Mopper 1998). Colonization, adaptation, and extinction create a metapopulation of semi-isolated leafminer demes in which gene flow is a creative rather than a constraining force that interacts with genetic drift to provide the raw material for natural selection to act upon (Wright 1929). In the leafminer system, shifting balance may produce a genetically heterogeneous population able to survive conditions imposed by catastrophes such as forest fires and tropical storms, frequent events in north Florida. Wright did not frame his theory in the context of local parasite–host interactions, but it seems that evolutionary resilience in the face of unstable environments may be most prevalent in these dynamic and heterogeneous systems.

Acknowledgments This research was supported by the National Science Foundation, the State of Louisiana, and the Lafayette Parish Medical Society. We thank Tim Mousseau for inviting us to participate in this volume. Thanks also to Mark Hunter, Tim Mousseau, Joe Neigel, and two anonymous reviewers for their insightful comments. We greatly appreciate Lisa Schile's editing, Lance Gorham's mapping, and Paul Leberg's help with numerous aspects of this study.

References

Akimoto, S. 1990. Local adaptation and host race formation of a gall-forming aphid in relation to environmental heterogeneity. Oecologia 83:162–170.

Alstad, D. N. 1998. Population structure and the conundrum of local adaptation. Pp. 3–18 *in* S. Mopper and S. Y. Strauss, eds., Genetic Structure and Local Adaptation in Natural Insect Populations: Effects of Ecology, Life History, and Behavior. Chapman and Hall, New York.

Alstad, D. N., and K. W. Corbin. 1990. Scale insect allozyme differentiation within and between host trees. Evol. Ecol. 4:43–56.

Auerbach, M. 1991. Relative impact of interactions within and between trophic levels during an insect outbreak. Ecology 72:1599–1608.

Barton, N. H., J. S. Jones, and J. Mallet. 1988. No barriers to speciation. Nature 336:13–14.

Berg, E. E., and J. L. Hamrick. 1995. Fine-scale genetic structure of a turkey oak forest. Evolution 49:110–120.

Benzie, J. A. H., and S. T. Williams. 1997. Genetic structure of giant clam (*Tridacna maxima*) populations in the west Pacific is not consistent with dispersal by present-day ocean currents. Evolution 51:768–783.

Boecklen, W. J., and S. Mopper. 1998. Local adaptation in specialist herbivores: theory and evidence. Pp. 64–90 *in* S. Mopper and S. Y. Strauss, eds., Genetic Structure and Local Adaptation in Natural Insect Populations: Effects of Ecology, Life History, and Behavior. Chapman and Hall, New York.

Bossart J. L., and D. P. Prowell. 1998. Genetic estimates of population structure and gene flow: limitations, lessons and new directions. Trends Ecol. Evol. 13:202–206.

Bossart, J. L., and J. M. Scriber. 1995. Maintenance of ecologically significant genetic variation in the tiger swallow butterfly through differential selection and gene flow. Evolution 49:1163–1171.

Bush, G. 1969. Sympatric host race formation and speciation in frugivorous flies of the genus *Rhagoletis* (Diptera: Tephritidae). Evolution 23:237–251.

Bush, G. 1994. Sympatric speciation in animals: new wine in old bottles. Trends Ecol. Evol. 9:285–288.

Carroll, S. P., and C. Boyd. 1992. Host race radiation in the soapberry bug: natural history with the history. Evolution 46:1052–1069.

Carroll, S. P., H. Dingle, and S. P. Klassen. 1997. Genetic differentiation of fitness-associated traits among rapidly evolving populations of the soapberry bug. Evolution 51:1182–1188.

Cobb, N. S., and T. G. Whitham. 1993. Herbivore deme formation on individual trees: a test case. Oecologia 94:496–502.

Cobb, N. S., and T. G. Whitham. 1998. Prevention of deme formation by the pinyon needle scale: problems of specializing in a dynamic system. Pp. 37–63 *in* S. Mopper and S. Y. Strauss, eds., Genetic Structure and Local Adaptation in Natural Insect Populations. Chapman and Hall, New York.

Cohen, J. 1988. Statistical Power Analysis for the Behavioral Sciences. Lawrence Erlbaum Associates, Hillsdale, NJ.

Connor, E. F., R. H. Adams-Manson, T. Carr, and M. W. Beck. 1994. The effects of host plant phenology on the demography and population dynamics of the leaf-mining moth, *Cameraria hamadryadella* (Lepidoptera: Gracillariidae). Ecol. Entomol. 19:111–120.

Connor, E. F., and M. P. Taverner. 1997. The evolution and adaptive significance of the leaf-mining habit. Oikos 79:6–25.

Cornell, H. V. 1989. Endophage-ectophage ratios and plant defense. Evol. Ecol. 3:64–76.

Costa, J. T. III, and K. G. Ross. 1993. Seasonal decline in intracolony genetic relatedness in eastern tent caterpillars: implications for social evolution. Behav. Ecol. Sociobiol. 32: 47–54.

Craig, T. P., J. D. Horner, and J. K. Itami. 1997. Hybridization studies of the host races of *Eurosta solidaginis*: implications for sympatric speciation. Evolution 51:1552–1560.

Crawley, M. J., and M. Akhteruzzaman. 1988. Individual variation in the phenology of oak trees and its consequences for herbivorous insects. Funct. Ecol. 2:409–415.

Edmunds, G. F., and D. N. Alstad. 1978. Coevolution in insect herbivores and conifers. Science 199: 941–945.

Ehrlich, P. R., and P. H. Raven. 1969. Differentiation of populations. Science 165:1228–1232.

Endler, J. A. 1977. Geographic variation, speciation, and clines. Princeton University Press, Princeton, NJ.

Faeth, S. H. 1995. Quantitative defense theory and patterns of feeding by oak insects. Oecologia 68:34–40.

Faeth, S. H., E. F. Connor, and D. Simberloff. 1981a. Early leaf abscission: a neglected source of mortality for folivores. Am. Nat. 117:409–415.

Faeth, S. H., S. Mopper, and D. S. Simberloff. 1981b. Abundance and diversity of leaf mining insects on three oak species: effects of host plant phenology and nitrogen content of leaves. Oikos 32:238–251.

Feder, J. L., S. H. Berlocher, and S. B. Opp. 1998. Sympatric host race formation and speciation in *Rhagoletis* (Diptera: Tephritidae): a tale of two species for Charles D. Pp. 408–434 *in* S. Mopper and S. Y. Strauss, eds., Genetic Structure and Local Adaptation in Natural Insect Populations: Effects of Ecology, Life History, and Behavior. Chapman and Hall, New York.

Feder, J. L., C. A. Chilcote, and G. L. Bush. 1990. The geographic pattern of genetic differentiation between host associated populations of *Rhagoletis pomonella* (Diptera: Tephritidae) in the eastern United States and Canada. Evolution 44:570–594.

Fortin, M., and J. Gurevitch. 1993. Mantel tests: spatial structure in field experiments. Pp. 342–359 *in* S. M. Scheiner and J. Gurevitch, eds., Design and Analysis of Ecological Experiments. Chapman and Hall, New York.

Futuyma, D. J., and S. Peterson. 1985. Genetic variation in the use of host resources by insects. Annu. Rev. Entomol. 30:217–238.

Gurevitch, J., and L. V. Hedges. 1993. Meta-analysis: combining the Results of Independent Experiments. Pp. 378–398 *in* S. M. Scheiner and J. Gurevitch, eds., Design and Analysis of Ecological Experiments. Chapman and Hall, New York.

Gurevitch, J., L. L. Morrow, A. Wallace, and J. S. Walsh. 1992. A meta-analysis of competition in field experiments. Am. Nat. 140:539–572.

Hanks, L. M., and R. F. Denno. 1998. Dispersal and adaptive deme formation in sedentary coccoid insects. Pp. 239–262 *in* S. Mopper and S. Y. Strauss, eds., Genetic Structure and Local Adaptation in Natural Insect Populations: Effects of Ecology, Life History, and Behavior. Chapman and Hall, New York.

Hanski, I., and M. Gilpin. 1997. Metapopulation Biology: Eiology, Genetics, and Evolution. Academic Press, San Diego, CA.

Hassell, M. P., H. N. Comins, and R. M. May. 1991. Spatial structure and chaos in insect population dynamics. Nature 353:255–258.

Hastings, A., and S. Harrison. 1994. Metapopulation dynamics and genetics. Annu. Rev. Ecol. Syst. 25:167–188.

Hebert, P. D. N., and M. J. Beaton. 1989. Methodologies for Allozyme Analysis Using Cellulose Acetate Electrophoresis. Helena Laboratories, Beaumont, TX.

Hechtel, L. J., and S. A. Juliano. 1997. Effects of a predator on prey metamorphosis: plastic responses by prey or selective mortality? Ecology 78:838–851.

Hedges, L. V., and I. Olkin. 1985. Statistical methods for meta-analysis. Academic Press, Orlando, FL.

Holt, R. D., and R. Gomulkiewicz. 1997. How does immigration influence local adaptation? A reexamination of a familiar paradigm. Am. Nat. 149:563–572.

Hunter, M. D. 1990. Differential susceptibility to variable plant phenology and its role in competition between two insect herbivores on oak. Ecol. Entomol. 15:401–408.

Hunter, M. D. 1992. A variable insect-plant interaction: the relationship between tree budburst phenology and population levels of insect herbivores among trees. Ecol. Entomol. 16:91–95.

Ingvarsson, P. K. 1997. The effect of delayed population growth on the genetic differentiation of local populations subject to frequent extinctions and recolonizations. Evolution 51: 29–35.

Ingvarsson, P. K., K. Olsson, and L. Ericson. 1997. Extinction-recolonization dynamics in the mycophagous beetle *Phalacrus substriatus*. Evolution 51:187–195.

Kimura, M., and Maruyama, T. 1971. Patterns of neutral variation in a geographically structured population. Genet. Res. 18:125–131.

Kimura, M., and G. Weiss. 1964. The stepping stone model of population structure and the decrease of genetic correlation with distance. Genetics 49:561–576.

Komatsu, T., and S. Akimoto. 1995. Genetic differentiation as a result of adaptation to the phenologies of individual host trees in the galling aphid *Kaltenbachiella japonica*. Ecol. Entomol. 20:33–42.

Levin, D. A. 1995. Metapopulations: an arena for local speciation. J. Evol. Biol. 8:635–644.

McCauley, D. E. 1989. Extinction, colonization, and population structure: a study of a milkweed beetle. Am. Nat. 134:365–376.

McCauley, D. E., and W. F. Eanes. 1987. Hierarchical population structure analysis of the milkweed beetle, *Tetraopes tetraophthalmus* (Forster). Heredity 58:193–201.

McCauley, D. E., and P. Goff. 1998. Intrademic genetic structure and natural selection in insects. Pp. 181–204 *in* S. Mopper and S. Y. Strauss, eds., Genetic Structure and Local Adaptation in Natural Insect Populations: Effects of Ecology, Life History, and Behavior. Chapman and Hall, New York.

McPheron, B. A., D. Courtney Smith, and S. H. Berlocher. 1988. Genetic differences between host races of *Rhagoletis pomonella*. Nature 336:64–66.

Michalakis, Y., A. W. Sheppard, V. Noel, and I. Olivieri. 1993. Population structure of a herbivorous insect and its host plant on a microgeographic scale. Evolution 47:1611–1616.

Mitter, C., D. J. Futuyma, J. C. Scheider, and J. D. Hare. 1979. Genetic variation and host plant relations in a parthenogenetic moth. Evolution 33:777–790.

Mitton, J. B. 1994. Molecular approaches to population biology. Annu. Rev. Ecol. Syst. 25:45–69.

Mopper, S. 1996. Adaptive genetic structure in phytophagous insect populations. Trends Ecol. Evol. 11: 235–238.

Mopper, S. 1998. Local adaptation and stochastic events in an oak leafminer population. Pp. 139–155 *in* S. Mopper and S. Y. Strauss, eds., Genetic Structure and Local Adaptation in Natural Insect Populations: Effects of Ecology, Life History, and Behavior. Chapman and Hall, New York.

Mopper, S., M. Beck, D. Simberloff, and P. Stiling. 1995. Local adaptation and agents of selection in a mobile insect. Evolution 49: 810–815.

Mopper, S., S.H. Faeth, W. J. Boecklen, and D. Simberloff. 1984. Host-specific variation in leaf miner population dynamics: effects on density, natural enemies and behavior of *Stilbosis quadricustatella* (Lepidoptera: Cosmopterigidae). Ecol. Entomol. 9:169–177.

Mopper, S., J. B. Mitton, T. G. Whitham, N. S. Cobb, and K. M. Christensen. 1991. Genetic differentiation and heterozygosity in pinyon pine associated with herbivory and environmental stress. Evolution 45:989–999.

Mopper, S., and D. Simberloff. 1995. Differential herbivory in an oak population: the role of plant phenology and insect performance. Ecology 76: 1233–1241.

Mopper, S., and S. Y. Strauss. 1998. Genetic Structure and Local Adaptation in Natural Insect Populations: Effects of Ecology, Life History, and Behavior. Chapman and Hall, New York.

Mopper, S., P. Stiling, K. Landau, D. Simberloff, and P. Van Zandt. 2000. Spatiotemporal variation in leafminer population structure and adaptation to individual oak trees. Ecology, in press.

Nei, M. 1977. F-statistics and analysis of gene diversity in subdivided populations. Ann. Hum. Genet. 41:225–233.

Nei, M. 1987. Molecular Evolutionary Genetics. Columbia University Press, New York.

Neigel, J. E. 1997. A comparison of alternative strategies for estimating gene flow from genetic markers. Annu. Rev. Ecol. Syst. 28:105–128.

Nürnberger, B., and R. G. Harrison. 1995. Spatial population structure in the whirligig beetle *Dinetus assimilis*: evolutionary inferences based on mitochondrial DNA and field data. Evolution 49:266–275.

Olivieri, I., Y. Michalakis, and P. Gouyon. 1995. Metapopulation genetics and the evolution of dispersal. Am. Nat. 146:202–228.

Pashley, D. P. 1988. Quantitative genetics, development, and physiological adaptation in host strains of fall armyworm. Evolution 42:93–102.

Peterson, M. A., and R. F. Denno. 1998. Life-history strategies and the genetic structure of phytophagous insect populations. Pp. 263–324 *in* S. Mopper and S. Y. Strauss, eds.,

Genetic Structure and Local Adaptation in Natural Insect Populations: Effects of Ecology, Life History, and Behavior. Chapman and Hall, New York.

Price, P. W. 1980. Evolutionary Biology of Parasites. Princeton University Press, Princeton, NJ.

Richardson, B. J., P. R. Baverstock, and M. Adams. 1986. Allozyme Electrophoresis: A Handbook for Animal Systematics and Population Genetic Studies. Academy Press, Sydney, Australia.

Roderick, G. K. 1996. Geographic structure of insect populations: gene flow, phylogeography, and their uses. Annu. Rev. Entomol. 41:325–352.

Roff, D. A., and P. Bentzen. 1989. The statistical analysis of mitochondrial DNA polymorphisms—Chi-2 and the problem of small samples. Molecular Biology and Evolution 6: 539–545.

Roininen, H., J. Vuorinen, J. Tahvanainen, and R. Julkunen-Tiitto. 1993. Host preference and allozyme differentiation in shoot galling sawfly, *Euura atra*. Evolution 47:300–308.

Rosenberg, M. S., D. C. Adams, and J. Gurevitch. 1997. MetaWin: Statistical Software for Meta-analysis with Resampling Tests, Version 1.0. Sinauer Associates, Sunderland, MA.

Schmitt, J., and S. E. Gamble. 1990. The effect of distance from the parental site on offspring performance and inbreeding depression in *Impatiens capensis*: a test of the local adaptation hypothesis. Evolution 44:2022–2030.

Simberloff, D., and P. Stiling. 1987. Larval dispersion and survivorship in a leaf-mining moth. Ecology 68:1647–1657.

Singer, M. C., C. D. Thomas, and C. Parmesan. 1993. Rapid human-induced evolution of insect-host association. Nature 366:681–683.

Slatkin, M. 1977. Gene flow and genetic drift in a species subject to frequent local extinctions. Theor. Popul. Biol. 12:253–262.

Slatkin, M. 1987. Gene flow and the geographic structure of natural populations. Science 236:787–792.

Slatkin, M. 1993. Isolation by distance in equilibrium and non-equilibrium populations. Evolution 47:264–279.

Slatkin, M. 1994. Gene flow and population structure. Pp. 3–17 *in* L. Real, ed., Ecological Genetics. Princeton University Press, Princeton, NJ.

Slatkin, M., and N. H. Barton. 1989. A comparison of three indirect methods for estimating average levels of gene flow. Evolution 43:1349–1368.

Sork, V. L., K. A. Stowe, and C. Hochwender. 1993. Evidence for local adaptation in closely adjacent subpopulations of northern red oak (*Quercus rubra* L.) expressed as resistance to leaf herbivores. Evolution 142:928–936.

Stiling, P., and A. M. Rossi. 1998. Deme formation in a dispersive gall-forming midge. Pp. 22–36 *in* S. Mopper and S. Y. Strauss, eds., Genetic Structure and Local Adaptation in Natural Insect Populations: Effects of Ecology, Life History, and Behavior. Chapman and Hall, New York.

Stiling, P., and D. Simberloff. 1989. Leaf abscission: induced defense against pests or response to damage? Oikos 55:43–49.

Stiling, P., D. Simberloff, and B. Brodbeck. 1991. Variation in rates of leaf abscission between plants may affect the distribution patterns of sessile insects. Oecologia 88:367–370.

Strauss, S. Y., and R. Karban. 1998. The strength of selection: intraspecific variation in host-plant quality and the fitness of herbivores. Pp. 156–177 *in* S. Mopper and S. Y. Strauss, eds., Genetic Structure and Local Adaptation in Natural Insect Populations: Effects of Ecology, Life History, and Behavior. Chapman and Hall, New York.

Sturgeon K. B., and J. B. Mitton. 1986. Allozyme and morphological differentiation of mountain pine beetles *Dendroctonus ponderosae* Hopkins (Coleoptera: Scolytidae) associated with host tree. Evolution 40:290–302.

Swofford, D. L., and R. B. Selander. 1989. Biosys-1. A computer program for the analysis of allelic variation in populations genetics and biochemical systematics. Illinois Natural History Survey, Urbana.

Tavormina, S. J. 1982. Sympatric genetic divergence in the leaf-mining insect *Liriomyza brassicae* (Diptera: Agromyzidae). Evolution 36:523–534.

Thomas, C. D., and M. C. Singer. 1998. Scale-dependent evolution of specialization in a checkerspot butterfly: from individuals to metapopulations and ecotypes. Pp. 343–374 *in* S. Mopper and S. Y. Strauss, eds., Genetic Structure and Local Adaptation in Natural Insect Populations. Chapman and Hall, New York.

Unruh, T. R., and R. F. Luck. 1987. Deme formation in scale insects: a test with the pinyon needle scale and a review of other evidence. Ecol. Entomol. 12:439–449.

Van Dongen, S., T. Backeljau, E. Matthysen, and A. Dhondt. 1997. Synchronization of hatching date with budburst of individual host trees (*Quercus robur*) in the winter moth (*Operophtera brumata*) and its fitness consequences. J. Anim. Ecol. 66:113–121.

Van Zandt, P.A., and S. Mopper. 1998. A meta-analysis of adaptive deme formation in phytophagous insect populations. Am. Nat. 152:597–606.

Via, S. 1990. Ecological genetics and host adaptation in herbivorous insects: the experimental study of evolution in natural and agricultural systems. Annu. Rev. Entomol. 35:421–446.

Via, S. 1991. Specialized host plant performance of pea aphid clones is not altered by experience. Ecology 72:1420–1427.

Wade, M. J., and D. E. McCauley. 1988. Extinction and recolonization: their effects on the genetic differentiation of local populations. Evolution 42:995–1005.

Waples, R. S. 1998. Separating the wheat from the chaff: patterns of genetic differentiation in high gene flow species. J. Hered. 89:438–450.

Weir, B., and C. C. Cockerham. 1984. Estimating F-statistics for the analysis of population structure. Evolution 38:1358–1370.

Whitlock, M. C. 1992. Nonequilibrium population structure in forked fungus beetles: extinction, colonization, and the genetic variance among populations. Am. Nat. 139:952–970.

Whitlock, M. C., and D. E. McCauley. 1990. Some population genetic consequences of colony formation and extinction: genetic correlations within founding groups. Evolution 44:1717–1724.

Wright, S. 1929. Evolution in a Mendelian population. Anat. Rec. 44:287.

Wright, S. 1943. Isolation by distance. Genetics 28:114–138.

Wright, S. 1951. The genetical structure of populations. Ann. Eugenics 15:323–354.

Wright, S. 1965. The interpretation of population structure by F-statistics with special regard to systems of mating. Evolution 19:395–420.

Wright, S. 1977. Evolution and the Genetics of Populations, vol. 3. Experimental Results and Evolutionary Deductions. University of Chicago Press, Chicago.

Zangerl, A. R., and M. R. Berenbaum. 1990. Furanocoumarin induction in wild parsnip: genetics and population variation. Ecology 71:1933–1940.

6

Reaching New Adaptive Peaks

Evolution of Alternative Bill Forms in an African Finch

THOMAS B. SMITH AND DEREK J. GIRMAN

We examine the adaptive significance of bill size polymorphism in small- and large-billed morphs of the African finch *Pyrenestes ostrinus* and the extremely large-billed race *P. o. maximus*. Using information on the morphology, fitness, feeding ecology, and the genetic basis of the bill forms, we examine the evolution of the polymorphism. Results suggest that small, large, and mega-billed forms are distinct morphs within a single species and that the large and the mega-billed forms arose by a mutation of large effect. Genetic correlations among bill characters are generally higher than in other seed-eating species. The simple genetic basis of the polymorphism suggests that single mutations have allowed populations to cross adaptive valleys in which larger morphs occupy separate adaptive peaks through differences in feeding performance. The significance of resource polymorphisms under simple genetic control to speciation is discussed.

6.1 Genes of Small and Large Effect

A pervasive view in evolutionary biology is that most adaptations result from allelic substitutions of small effect at many loci (Fisher 1958, Mayr 1963). This "micromutational" view that evolutionary change proceeds primarily through the accumulation of many small mutations was recently challenged by Orr and Coyne (1992). Examining both the theory and data from available studies, they concluded that although some adaptations are the result of the action of many genes, numerous adaptations, perhaps half of those known, result from one or two genes of large effect. Although the idea that large mutations may be responsible for some adaptations is not new (Haldane 1932, Wright 1968, Templeton 1981, Charlesworth 1990), there are few cases where we understand the ecological conditions under which mutations of large effect arise and are maintained (Orr and Coyne 1992).

Are there ecological conditions that might favor adaptations arising from single genes? Despite considerable recent progress toward understanding adaptation (Rose and Lauder

1996), our ability to answer this question is limited by the scarcity of adaptations in which both the genetic basis and the ecological circumstances are well understood. Nevertheless, one can imagine certain ecological situations where adaptations determined by single genes of large effect might be favored. For example, one circumstance might be found in environments where the majority of available niches are filled, but where resource "islands" nevertheless occur. In environments with particularly high species richness, interspecific competition may create adaptive valleys that are difficult or impossible to cross through incremental changes. Under these circumstance, new adaptations might be more likely to arise through mutations of large effect because they would lead to sufficiently large enough phenotypic changes capable of leaping across valleys.

A corollary of the importance of major genes in adaptation is their role in speciation. If some new adaptations arise by mutations of large effect, how often are these associated with reproductive isolation and speciation? Numerous theoretical models have shown that polymorphisms may lead to speciation (Levene 1953, Maynard Smith 1966, Maynard Smith and Hoekstra 1980, Hoekstra et al. 1985, Johnson et al. 1996), especially when traits important in reproductive isolation are correlated with the traits important in resource use (Rice and Hostert 1993). Work on diverse taxa suggest that resource polymorphisms in particular, in which discrete intraspecific morphs show differential resource use and varying degrees of reproductive isolation, may represent important intermediate stages in speciation (Smith and Skulason 1996, Robinson and Schluter, this volume). Examples of resource polymorphism are especially common in fish, where resource-based morphs may differ dramatically in morphology and life-history characteristics yet co-exist within the same freshwater system (Smith and Skulason 1996). For example, in the arctic char (*Salvelinus alpinus*), a circumpolar salmonid, many lakes have from two to four sympatric resource morphs. Morphs differ in shape, size, and life-history characteristics (Snorrason et al. 1994). In some cases resource segregation is clear and stable, such as between benthic and limnetic habitats (Sandlund et al. 1987, Malmquist et al. 1992), whereas in other cases, habitat and food segregation are less dramatic or even seasonal (Hindar and Jonsson 1982, Sparholdt 1985, Riget et al. 1986, Walker et al. 1988). Often different morphs or trophic forms show varying degrees of resource use and reproductive isolation, resulting in many being identified as distinct species rather than as morphs (Hindar 1994, Schluter and Nagel 1995). There are few resource polymorphisms where the genetic mechanisms are understood, but several seem to involve cases where the polymorphism is controlled by only a few alternate alleles at a locus or a major locus with modifiers (Hori 1993, Smith 1993a, Bush 1994). Thus, if nearly half of all adaptations are determined by single genes of large effect, as Coyne and Orr (1992) contend, examining resource polymorphisms may be helpful in understanding adaptive genetic variation and speciation.

Avian bill size and shape are particularly useful characters in the study of adaptation. First, bill morphology can be easily measured and quantified (Grant 1986). Second, there is generally a close relationship between bill size and shape and feeding ecology (Benkman 1988, 1991, Grant and Grant 1993, Peterson 1993). Finally, bill characters tend to be highly heritable (Boag and van Noordwijk 1987, Benkman 1988, Benkman and Lindholm 1991, Grant and Grant 1993, Grant and Grant, this volume Peterson 1993).

In this chapter we review what is known about adaptive genetic variation in the polymorphic finch *Pyrenestes ostrinus*. We focus particular attention on a recently described form with an extremely large bill. Specifically, we examine (1) the morphological basis of the polymorphism, (2) how ecological factors such as seed quality may affect differences in performance and fitness, (3) the genetic basis of the polymorphism, (4) the relationships among bill types using genetic markers from the mitochondrial DNA control region, and (5) how such polymorphisms might lead to speciation.

6.2 Characteristics of the Bill Polymorphism in *Pyrenestes*

The African finch, *P. ostrinus*, or black-bellied seedcracker, is remarkable in birds in showing a non–sex-linked polymorphism in bill size (Smith 1987, 1993a). Small and large bill morphs occur together and interbreed across their range, which extends across most of central Africa (Chapin 1924, Smith 1987). Based on breeding experiments, the polymorphism is determined by a single locus, and the allele for large bill is dominant (Smith 1993a). Because of the discontinuous nature of bill size, seedcracker taxonomy has been a source of considerable debate. Until recently bill forms were classified as distinct races rather than as morphs, despite the fact that all forms co-occurred (Chapin 1924, 1932, Smith 1990b). In addition to the small and large morph found in forested zones of Africa, recent research shows the existence of a third extremely large-billed form (mega-billed) which occurs in forest–savanna transition zones (Smith 1993b, 1997). First described by Chapin (1924) as the distinct subspecies *Pyrenestes o. maximus*, this mega-billed form feeds on extremely hard seeded sedges, co-occurs with both small and large-billed morphs in a portion of its range, and is thought to interbreed with small-billed birds (Smith 1993b).

6.2.1 Small- and large-billed morphs

Seedcrackers are commonly found along water courses in rainforest and heavily wooded savanna where sedge seeds (*Cyperaceae*), their principal food, is abundant. Finches typically reach their highest densities in seasonally flooded forests where semiclosed canopies offer ideal growing conditions for sedges. Morphs differ dramatically in their ecology, particularly in diets across the year and feeding efficiencies on sedge seeds in the genus *Scleria* (Smith 1990b, 1991). In rainforest regions of south-central Cameroon, the major rainy season extends from September to the middle of November. During this period seeds are maximally abundant, and both morphs specialize on the soft-seeded sedge *S. goossensii*. It is also during this food-abundant period that finches breed. During food-scarce times, January–May, when all seeds are scarce, small morphs become generalized in their food habits, feeding on a variety of soft-seeded sedges and grass seeds. In contrast, large morphs switch to hard seeds, which they obtain from dried stalks and the leaf litter, as a result of competition from soft-seed specialists (Smith 1990c, 1991). Based on feeding performance trials, large morphs were found to crack the hard seeds of *S. verrucosa* significantly more rapidly than small morphs, with just the reverse true for small morphs feeding on soft seeds of *S. goossensii*. Despite these differences, large morphs still crack soft seeds in less time than hard seeds, explaining why they feed primarily on *S. goossensii* when these seeds are in great abundance (Smith 1990c). Seeds are most scarce following the major dry season. During this food-scarce period natural selection is most intense.

Depending on the character in question, morphological variation in the small-billed and large-billed finch may be discontinuous between morphs and continuous within morphs (fig. 6.1). Consequently, the nature of morphologic variation can be examined using different approaches. For example, bill size can be treated as a continuous character, and selection analyses appropriate for addressing continuous characters can be applied (Smith 1990a). Alternatively, morphs defined on the basis of lower bill width can be treated as a simple bivariate character with two forms, small and large, allowing pedigree analyses appropriate for determining the genetic basis of discrete characters (Smith 1993a). Our research on *Pyrenestes* has used both approaches. Analyses of selection on continuously distributed morphological characters show that bill morphology is maintained in part by disruptive selection acting on bill size, as a consequence of differences in feeding performance on important sedge seeds. Such within-morph analyses of feeding performance, fitness, and morphology are similar to those carried out on Darwin's finches (Grant and Grant, this volume).

6.2.2 Genetic correlations

Although bill polymorphism is under simple genetic control (Smith 1993a), it is still useful to examine the heritabilities and genetic correlations of the various characters. Within each morph, bill characters such as lower mandible width are normally distributed and allow for analyses more typical of characters under polygenetic inheritance. Genetic correlations and variances may evolve by natural selection, and their strengths may reveal the degree of functional relationships between traits (Lande 1980). Depending on the strength of the genetic correlation, selection may also lead to peak shifts (Lande and Kirkpatrick 1988, Price et al. 1993). In *Pyrenestes* the most important bill character with respect to feeding performance is lower bill width (table 6.1; Smith 1987). However, the upper and lower mandible must operate together to be effective in feeding; thus the nature of the genetic correlations among traits can reveal information about the adaptive nature and evolutionary constraints of characters (Schluter and Smith 1986a). Heritabilities estimated from mid-parent regressions are generally high and significant for most characters except for tarsus length and bill depth (table 6.2). The genetic correlation between characters, r_A, was estimated from 12 families and is given by:

$$r_A = \frac{\mathrm{Cov}_{xy}}{(\mathrm{Cov}_{xx}\,\mathrm{Cov}_{yy})^{1/2}}$$

where Cov_{yy} and Cov_{xx} are the offspring–parent covariance of each character, and y and x, separately, and Cov_{xy} is the cross-covariance between characters (Falconer 1981). Standard errors were estimated following Falconer (1981). The mean genetic correlation across all morphological characters was 0.42, nearly twice as high as that found in the song sparrow (*Melospiza melodia*), which averaged 0.28, but only slightly higher than the Galápagos finch, *Geospiza conirostris*, which averaged 0.37 (Schluter and Smith 1986b, Grant and Grant 1989). However, the average genetic correlation among three bill characters (bill length, bill depth, and bill width) were 1.5 times higher in *Pyrenestes* than *G. conirostris* (0.54 versus 0.36, respectively). The higher genetic correlations among bill characters is consistent with pedigree analyses that suggest that the bill polymorphism is controlled by a single gene. If many genes were involved, one might expect genetic correlations to be lower and similar to those found in song sparrows and Galápagos finches, in which many genes are believed to control bill size (Boag 1983, Schluter and Smith 1986a).

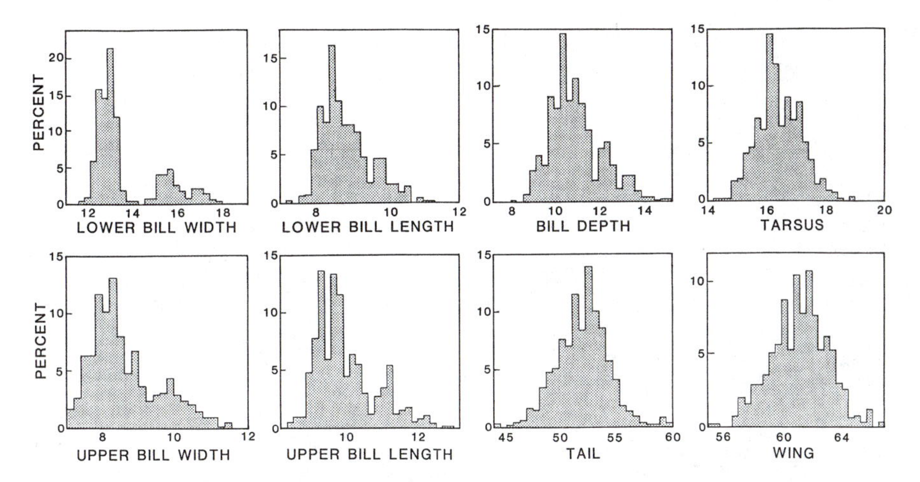

Figure 6.1 Distributions of eight mensural characters for adult male *P. ostrinus* (small and large bill morphs combined) captured at Ndibi, Cameroon. Upper and lower bill refer to upper and lower mandibles; units are in millimeters (see Smith 1990b).

6.2.3 Mega-billed form

In contrast to small and large morphs, which predominate in rainforest areas, the evolutionary status of the mega-billed form and the nature of the adaptive landscape created by seeds of different quality are uncertain. For instance, is the mega-billed form another morph, freely interbreeding with large and small-billed finches, a distinct species, or a hybrid? Did the existence of seeds of differing hardness create distinct adaptive peaks that could only be reached by mutations of large effect, as appears to have been the case in the large morph, or could

Table 6.1 Character sizes of adult male *Pyrenestes ostrinus* captured on the main study area in south-central Cameroon

	Small		Large		Mega	
Character	*n*	Mean ± SE	*n*	Mean ± SE	*n*	Mean ± SE
Weight (g)	624	18.6 ± 0.05	260	20.9 ± 0.08	35	29.0 ± 0.62
Wing (mm)	618	60.1 ± 0.007	256	60.2 ± 0.01	35	69.7 ± 0.47
Tarsus	624	10.6 ± 0.004	258	16.3 ± 0.009	35	19.4 ± 0.36
Lower mandible width	629	10.3 ± 0.001	260	15.9 ± 0.005	35	19.4 ± 0.16
Upper mandible width	622	8.2 ± 0.002	258	10.2 ± 0.006	35	11.6 ± 0.13
Lower mandible length	623	8.5 ± 0.001	259	9.8 ± 0.003	35	11.8 ± 0.14
Upper mandible length	624	9.7 ± 0.002	259	11.3 ± 0.004	35	13.0 ± 0.1
Bill depth	606	10.0 ± 0.003	253	12.6 ± 0.006	34	15.0 ± 0.15
Upper mandible depth	621	3.7 ± 0.001	259	4.9 ± 0.003	35	5.8 ± 0.06

"Small" and "large" refer to each morph; "mega" to individuals with extremely large bills captured at all northern sites in Cameroon except Bétaré Oya. From Smith (1997). Character differences among bill types are all significant (ANOVA, $P < .001$).

Table 6.2 Heritabilities ($h^2 \pm$ SE), genetic correlations (above diagonal), and standard errors (below diagonal) estimated from breeding experiments of small and large morphs (Smith 1993)

Character	h^2	R	1	2	3	4	5	6
1. Wing	0.73 ± 0.32*	77		0.50	0.22	0.50	0.35	0.45
2. Tarsus	0.42 ± 0.25	48	0.27		0.50*	-0.25	-0.35	0.22*
3. Lower mandible width	1.13 ± 0.31**	99	0.23	0.21		0.63*	0.47*	0.50*
4. Bill depth	0.56 ± 0.35	85	0.28	0.40	0.18		0.44	0.50*
5. Lower mandible length	1.06 ± 0.46*	96	0.27	0.31	0.19	0.30		0.39
6. Upper mandible length	1.19 ± 0.26***	98	0.17	0.24	0.13	0.19	0.31	

$^*P<.05$; $^{**}P<.01$, $^{***}P<.001$.

populations have crossed adaptive valleys gradually by directional selection? What ecological circumstances would favor one possibility over another? Central to addressing these questions is an understanding of the genetic relatedness of the mega-billed form to the large and small-billed morphs and the nature of the adaptive landscape formed by seeds of differing quality.

6.3 Morphologic Variation among Bill Forms

Character means for the small and large morph and the mega-billed form from across Cameroon are shown in table 6.1 (Smith 1997). Large differences in body size (as indexed by wing and weight) and bill size are apparent. Large morphs are on average 12% greater in body weight than small morphs, and the mega form is 30% greater than the large morph. In contrast, the increase in bill width from small to large morph is 54%, whereas the increase from large to mega form is only 23%. When bill characters are compared (fig. 6.2), each bill form can be clearly differentiated, with mega forms larger in both wing length and lower mandible width. In other words, mega forms, although exhibiting very large bills, have proportionally larger bodies relative to either the small or large morph (Smith 1997). Examining multivariate variation in size and shape using principal components (PC) analysis, one sees a similar pattern. A plot of PC1 versus PC2 clearly separates each of the bill types along the PC1, a size axis, while revealing considerable overlap along PC2, a shape axis (fig. 6.3).

6.4 Molecular Genetic Comparisons

Research to date has shown that large and small bill morphs interbreed freely and constitute the same species. However, it is unclear whether the mega-billed form is a third bill morph or a separate species. Of four nests discovered during 1991, three consisted of homotypic mega-billed pairs, and one consisted of a small-billed male paired with a mega-billed female (Smith

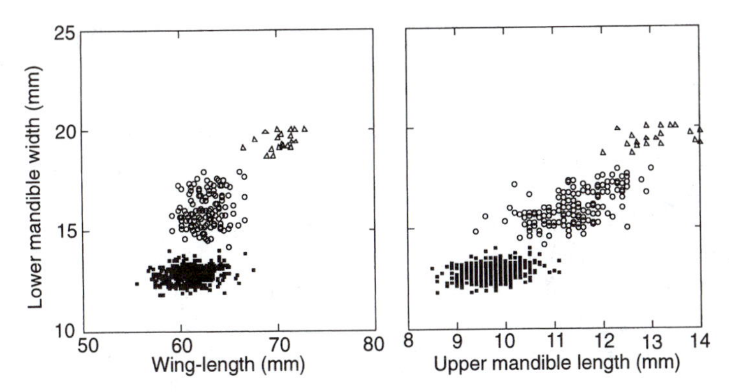

Figure 6.2 Bivariate plots of lower mandible width on wing length and culmen for adult males. Squares represent small morphs, circles represent large morphs, and triangles represent mega-billed forms (see Smith 1997).

1993b, 1997). Given the few nesting pairs, it was necessary to examine other sources of evidence regarding the relationship of mega-billed forms to large- and small-billed forms.

Using samples from all bill types, we amplified a portion of the mitochondrial control region and found 11 mtDNA genotypes among 39 seedcrackers sampled (fig. 6.4). Of these genotypes, D and H were shared by all three bill morphs (Girman and Smith, in preparation). Genotype I was shared by mega- and small-billed individuals, and genotypes J and B were shared by large- and small-billed individuals. An unweighted parsimony analysis, with the genotype from *S. haematima* as an outgroup, the closest related taxa to *Pyrenestes* (Sibley and

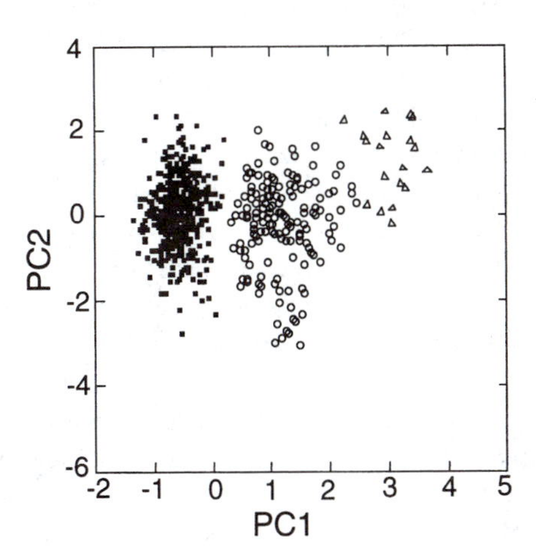

Figure 6.3 Principle component plot of five bill characters for adult males; PC1 and PC2 are primarily size and shape axes, respectively (Smith 1990b, 1997). Squares represent small morphs, circles represent large morphs, and triangles represent mega-billed forms.

Ahlquist 1990), suggests that the genotypes within *P. ostrinus* are relatively undifferentiated from each other, although they are supported as a monophyletic group distinct from *P. sanguineus* in 70% of 1000 bootstrap replicates (fig. 6.4; Girman and Smith, in preparation).

In sum, the molecular genetic results suggest that the mega-billed form is a third bill morph of *P. ostrinus* (Girman and Smith in preparation). There are no genotypes that form a distinct monophyletic grouping of mega-billed individuals. In fact, several mtDNA genotypes are shared between mega-billed and other morphs (fig. 6.4). In addition, the levels of sequence divergence found among all *P. ostrinus* morphs were typical of within-species differences. For example, compared to sequence divergence values between any *P. ostrinus* genotype and any genotype from *P. sanguineus*, divergence values among *P. ostrinus* individuals were small (Girman and Smith, in preparation). Thus, there is no evidence that mega-billed forms are a separate species from small- and large-billed morphs.

6.5 Adaptive Significance of Mega-billed Form

The extremely large bill of the mega-billed form is only found in ecotone regions between rainforest and savanna where the extremely hard-seeded sedge (*Scleria racemosa*) occurs, generally in gallery forests along small rivers and streams (Smith 1990b, 1993b), and appears to be an adaptation for feeding on these seeds (Smith 1997). Although similar morphologically to the hard-seeded sedge *S. verrucosa* found in rainforest to the south, *S. racemosa* seeds require, on average, twice the force to crack (fig. 6.5). Results of feeding trials of large morphs on the hard-seeded sedge *S. verrucosa* have shown that many seeds are rejected after unsuccessful attempts, indicating that even a proportion of the seeds on which large-billed birds depend during the dry season are too hard to crack successfully (Smith 1990c). Because the maximum force a bill can exert in cracking a seed is a function of bill width and depth (Lederer 1975), it is doubful that large morphs would be able to crack the extremely hard seeds of *S. racemosa*.

Body size increases considerably more rapidly than bill size in the mega-billed form than it does in either the small- or large-billed morphs (fig. 6.2; Smith 1997). Although body size (as indexed by wing length) and bill size are correlated, it is useful to examine how changes in body size may influence the ability to distinguish morphs. This is possible by controlling for body size by log transforming measurements and regressing lower mandible on wing length. The resulting plot of residuals shows that bill morphs and the mega-billed form may be distinguished even after controlling for differences in body size (Smith 1990b). These results are consistent with the small, nonsignificant genetic correlation between the two traits, suggesting both are likely under independent genetic control. Previous work also suggests that selective forces operating on wing length are different from those working on bill morphology (Smith 1990a). An examination of natural selection found evidence of directional selection for increasing wing length in adults and juveniles of each morph, whereas selection on bill characters more frequently showed disruptive selection (Smith 1990a). However, the reasons for directional selection were unclear.

One possible ecological explanation for the large scaling difference between bill size and wing length in the mega morph is aerodynamic consequences imposed by the impact of aerial predators in open habitats. Aerodynamic efficiency is hypothesized to decrease as bills become

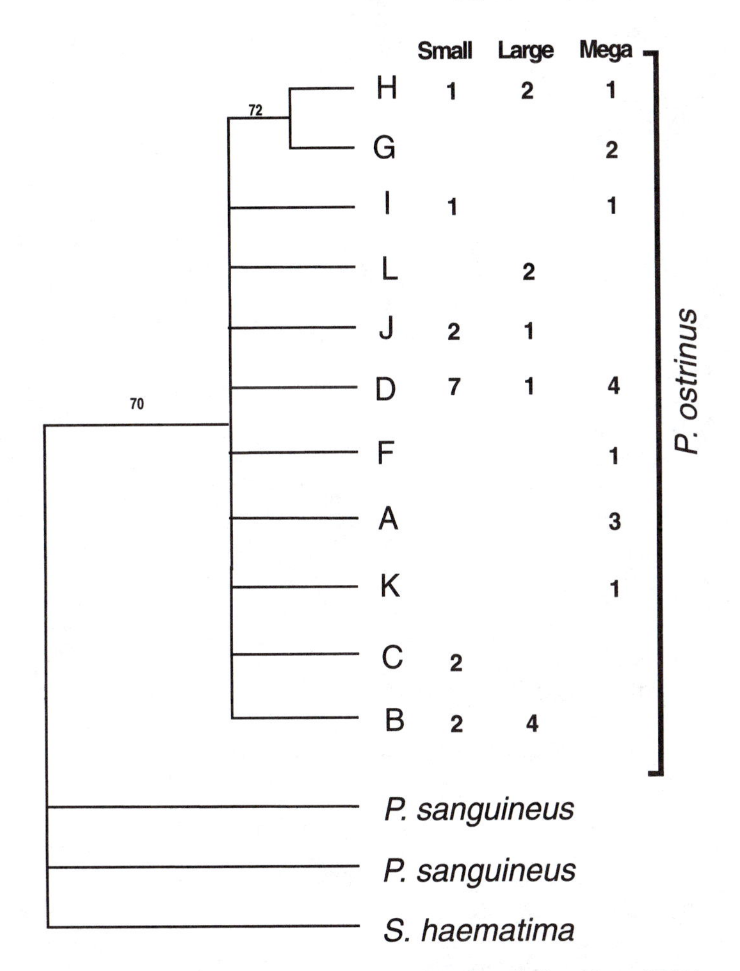

Figure 6.4 An unweighted parsimony tree depicting the relationships of DNA sequences from 367 base pairs of the mitochondrial control region. Eleven haplotypes from *P. ostrinus* (four ecotone sites and one forest site), two from *P. sanguinus*, and one from *Sphermaphaga haematima* are shown. Bootstrap values are indicated for branches supported by 50% of 1000 bootstrap replications. The number of small, large, and mega-billed individuals are given for each *P. ostrinus* genotype (Girman and Smith, in preparation).

larger; thus finches subject to pressure from avian predators are believed to evolve relatively larger body sizes to compensate for the increased bill size (Schluter 1988b, Benkman 1991). Benkman (1991) hypothesized that continental ground-feeding finches exhibit comparatively larger body weights than scrub and forest-dwelling finches because of this increased predation pressure. Although there are few comparative data on the relative vulnerability to predators in northern and southern Cameroon, there is a marked difference in the extent of forest cover (Letouzey 1968, Smith 1997). Forest patches in the forest/savanna ecotone in northern Cameroon are separated by open savanna, requiring mega-bill finches to fly between patches

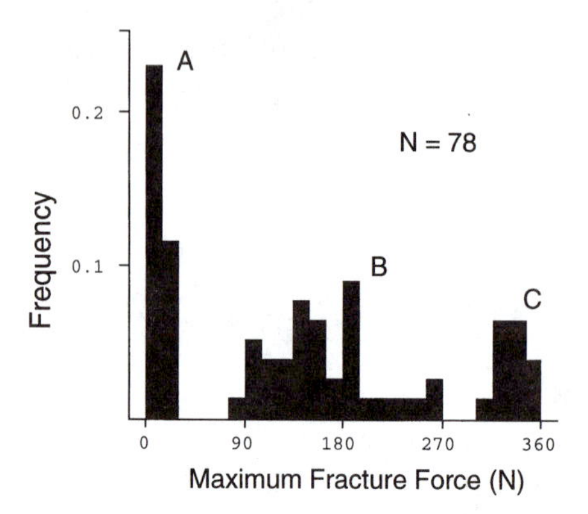

Figure 6.5 Histogram of maximum fracture forces for (A) soft-seeded sedge (*S. goossensii*), (B) hard-seeded sedge (*S. verrucosa*) collected on the main site in the forest, and (C) the sedge *S. racemosa* collected at the Meiganga site in northern Cameroon. Mean (± SE) maximum fracture forces to crack seeds are 12.3 ± 2.0, 146.9 ± 2.0, and 299.5 ± 11.9, respectively. Mean seed diameters are 2.9, 3.2, and 4.3 mm (Smith 1990b). Except for *S. racemosa*, there are no known seeds on which finches feed that exceed the hardness of *S. verrucosa*.

where they would be more at risk to aerial predators than if they were in contiguous forest. Furthermore, other nonfinch species of small birds also show a similar increase in wing length in response to more open habitats, consistent with the predation hypothesis (Smith et al. 1997, 1999).

6.6 Evolution of Bill Forms and Significance of Alternative Adaptive Phenotypes

The simple genetic basis of the bill polymorphism suggests that a mutation of large effect has allowed morphs to cross adaptive valleys formed by sympatric granivorous species feeding on specific seeds. The separate adaptive peaks are reinforced by the absence of seeds of intermediate hardness between *S. verrucosa*, on which the large morph specializes, and the extremely hard sedge *S. racemosa*, on which the mega morph specializes. When morphology (lower mandible length), performance (on respective seeds), and fitness (survival of juveniles over a 7-year period) are plotted for small and large morphs, two peaks are distinguishable (fig. 6.6). Estimates of performance and fitness are unavailable for the mega bill; however, the close relationship between mean bill size of bill types and mean seed hardness of the different sedge seeds suggests a similar association.

The close relationship between bill width and preferred seeds in the mega-bill morph suggests that it either evolved gradually, with intermediate bill forms subsequently going extinct, or that a mutation arose enabling the population to reach a third adaptive peak in a single step, as hypothesized in the case of the large morph (Smith 1993a, 1993b). Several lines of evidence support the latter hypothesis. First, based on museum specimens, although there are slight shifts in mean bill width across regions of the Democratic Republic of Congo (Smith

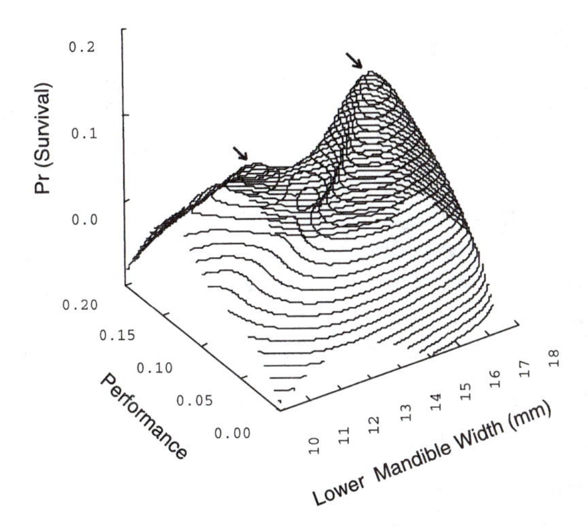

Figure 6.6 Three-dimensional plot of fitness, lower mandible width, and respective performances on soft- (*Scleria goossensii*) and hard- (*S. verrucosa*) seeded sedges. The two sets of contours represent fitness peaks for small- and large-billed morphs, arrows indicate respective fitness peaks. Fitnesses were estimated from juveniles of either sex measured and released between 1983 and 1986 and compared to those recaptured during an 8-month period, 1989–1990 (Smith 1993a). Fitness surfaces were estimated for lower mandible width using the cubic spline technique (Schluter 1988a). Performance (inverse of mean cracking time) was estimated from feeding trials using wild-caught small and large-billed birds of either sex (Smith 1987). A performance surface (based on feeding trials involving 30 wild-caught individuals of each morph) was estimated separately for each morph using the spline technique to determine the curve of best fit.

1990b), in no region in Central Africa is bill size intermediate between mega- and large-billed morphs (Smith 1990b). Second, as previously mentioned, there are no known seeds of intermediate hardness between *S. verrucosa* and *S. racemosa*. Finally, Africa has one of the highest diversities of seed-eating birds on any continent (Keast 1972), and even if seeds of intermediate hardness existed, crossing adaptive valleys incrementally via directional selection, in which many genes are involved with varying pleiotropic effects, would be difficult. In Cameroon alone, there are more than 100 granivorous species and subspecies of birds (Louette 1981), many of which are potential competitors with *Pyrenestes*. Studies examining interspecific competition are scarce for central African bird species. However, we found evidence for intense interspecific competition between small morphs and other small-billed granivorous species during food-scarce times (Smith 1991), suggesting that incremental shifts to new adaptive peaks might be impeded by competition. Using approaches developed for Darwin's finches (Schluter et al. 1985), future research hopes to estimate the magnitude and location of adaptive peaks.

While reaching a new adaptive peak may have occurred through a mutation of large effect, the nature and dynamics of ecologically based natural selection on bill size within each morph is similar to that reported in Darwin's finches (Grant and Grant, this volume).

In Darwin's finches, as in *P. ostrinus*, the availability of important seeds, feeding performance, and morphology interact along with other ecological and evolutionary processes to shape morphologic variation within and between species (Grant and Grant 1993, 1994, this volume).

Are there other bird species where morphs or ecotypes have adapted to different resource peaks? Although the genetic basis of bill morphology is unknown, a similar example to *Pyrenestes* may also occur in North American red crossbills (*Loxia curvirostra*), where an adaptive landscape is envisaged by virtue of five bill types, each exhibiting different feeding efficiencies on species of conifer cones (Benkman 1993). Other possible examples, such as the small seed-eating finches in the neotropical genus *Oryzoborus* (Chapin 1924) await investigation.

6.7 Resource Polymorphisms and Speciation

Resource polymorphisms, although generally rare, may provide important stepping stones for speciation (Bush 1994, Skulason and Smith 1995, Smith and Skulason 1996). Ecological segregation arising as a function of correlations between phenotype and resource use has been documented in several taxa (Echelle and Kornfield 1984, Sigurjonsdottir and Gunnarsson 1989, Skulason et al. 1989, Meyer 1993, Robinson et al. 1993, 1996, Schluter and McPhail 1993, Schluter 1996). For instance, within species of freshwater fish, population segregation and genetic divergence has repeatedly occurred within the same freshwater system, sometimes within the same lake (Foote et al. 1989, Taylor and Bentzen 1993, Hindar 1994, Schluter 1996) in response to differences in foraging habitat. In some cases, gene flow is unimpeded, while in others, sympatric morphs may appear partially or completely reproductively isolated (Hindar 1994). A recent survey of laboratory studies suggests that niche-specific adaptation, typical of resource polymorphisms, may be a key element in speciation (Rice and Hostert 1993). One example of where a niche-specific adaptation has resulted in speciation, apparently under sympatric conditions, is found in some cichlid fishes from Cameroon, West Africa (Trewavas et al. 1972): mtDNA analysis of cichlid species flocks endemic to two crater lakes strongly suggests that each lake contains a monophyletic group of species that has originated sympatrically (Schliewen et al. 1994). Species within lakes are more closely related to each other than they are to riverine species or to species from adjacent lakes. Furthermore, there are no geographic features of the lakes that could have provided geographic isolation. Although species were reproductively isolated and no polymorphism was demonstrated, resource morphs, or ecotypes, were likely an intermediate step given that numerous lakes in East and West Africa contain polymorphic populations of cichlids (Meyer et al. 1990, Meyer 1993).

Discrete intraspecific variation has been hypothesized as the initial step in speciation and has been emphasized in theoretical models of sympatric speciation (Bush 1994, West-Eberhard 1986). A central component to some of the early models of sympatric speciation (Levene 1953, Maynard Smith 1966, Bush 1994) is that a stable polymorphism evolves first, with reproductive isolation occurring later by morphs reproducing separately in the two niches. The close relationship between morphology, fitness, and performance among the three seed-cracker morphs raises the possibility of reproductive isolation via habitat selection. In contrast to southern Cameroon, where small- and large-billed morphs breed when food is abundant, show high diet overlap, and no habitat selection (Smith 1990c, 1991), habitats in the ecotone of northern Cameroon are patchy, with the diversity of sedges often differing among patches (T. Smith unpublished data). Under these circumstances positive assortative mating could arise in response to seed quality and feeding performance if morphs bred in patches where they feed. For example, in some habitat patches in northern Cameroon only the extremely hard-seeded sedge (*S. racemosa*) is present and only the mega-billed morph occurs.

Most models for the evolution of prezygotic isolation in response to disruptive selection in the presence of gene flow fall under the heading of the double variation model (fig. 6.7A). These have been considered unlikely to result in speciation because, in order to make progress toward reproductive isolation, a nonrandom association between the selected traits and the ones causing prezygotic isolation must occur (Rice and Hostert 1993). Under the double variation model, most simulations suggest that homogenizing effects of recombination tend to overpower even strong selection, and prezygotic isolation is not expected to evolve (Felsenstein 1981). However, this problem is bypassed in the single-variation model (fig. 6.7B; Rice and Hostert 1993). This is because the trait under selection is identical to the one important in reproductive isolation (Rice and Hostert 1993). In the case of mega-bill morphs occurring in patches containing only the hardest seeds, positive assortative mating might occur for purely ecological reasons. Only the morph capable of using resources in such patches would breed in them. Under the single variation model, assortative mating has been shown to be a powerful generator of reproductive isolation via pleiotropy (Rice and Hostert 1993). Current research examines evidence for the single variation model, the genetic basis of the mega-billed form, and the role of these transitional habitats or ecotones generally in promoting speciation (Smith et al. 1997, 1999).

6.8 Conclusions

Understanding the influence of ecology on adaptive genetic variation in the wild is in its infancy. Additional studies will be needed to determine the genetic basis of adaptation and the extent to which ecological conditions may favor adaptations controlled by one or two genes of large effect or many genes of small effect. In the case of the bill polymorphisms in *Pyrenestes*, the ecological conditions, particularly the discrete hardness of sedge seeds, has clearly favored adaptation via major genes. Ultimately, the precise genetic mechanisms producing new adaptations are the byproduct of processes at individual and population levels and will be determined by mutation rates, pleiotropic effects, drift, and other processes (Kirkpatrick 1996). However, the ecological context in which populations exist needs careful consideration when examining the nature of adaptive genetic variation. Whether resources are continuously or discontinuously distributed and the intensity of inter- and intraspecific competition may ultimately determine the genetic basis of adaptation.

Alternative adaptive phenotypes that characterize resource polymorphisms may represent important building blocks in speciation by linking resource use and assortative mating under certain conditions. This has been shown to circumvent the problems associated with linkage disequilibrium and may generate rapid reproductive isolation via pleiotropy and/or genetic hitchhiking (Rice and Hostert 1993). Determining the likelihood of reproductive isolation will be largely influenced by the ecological context and the genetic architecture of the adaptation. In the case of the mega-billed morph, the fact that only this form can use the hardest sedge may preclude other bill forms from using patches containing only these sedges.

A. Double-variation Models (isolation via link. disequilibrium)

Disruptively selected trait

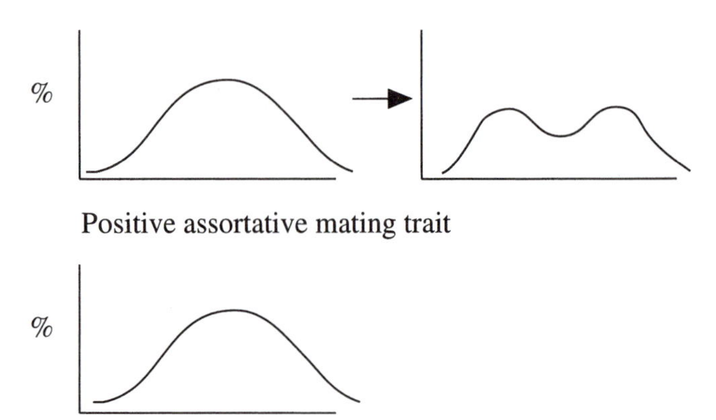

B. Single-variation Models (isolation via pleiotropy)

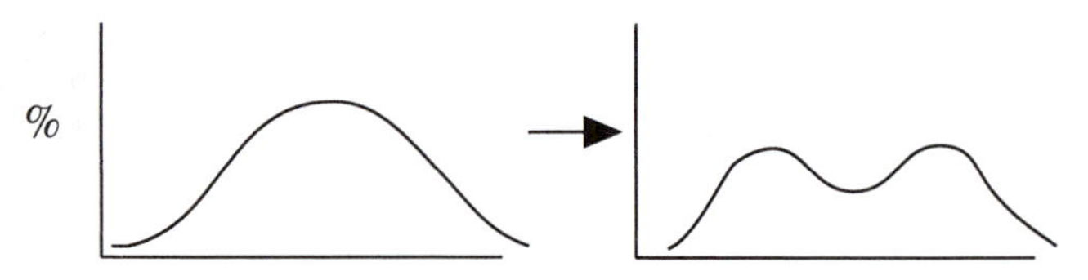

Figure 6.7 Two models of speciation in the presence of gene flow (modified from Rice and Hostert 1993). In the double variation model, reproductive isolation is unlikely to occur because linkage of traits important in reproductive isolation and selected traits are broken down by recombination. Under the single variation method, the trait under selection is the same trait important in prezygotic isolation. Under this model reproductive isolation evolves through pleiotropy and/or genetic hitchhiking (sampling error-induced linkage disequilibrium between alleles affecting positive assortative mating and alleles affecting divergently selected characters) (Rice and Hostert 1993). Speciation may still occur even with some gene flow if the trait under selection is also important in reproductive isolation. One could envision such a circumstance if bill morphs tend to reproduce where they feed and the hard-seeded sedges occur in patches where soft sedges do not occur (Smith and Skulason 1996).

Under these conditions, depending on the genetic basis, and the degree of assortative mating, reproductive divergence could occur rapidly via a variety of mechanisms (Johnson et al. 1996).

Acknowledgments We thank the Government of the Republic of Cameroon for permission to conduct the research, R. Zebell and the Conservation Genetics Lab at San Francisco State University for technical assistance, and B. Larison, C. Schneider, P. Grant, J. Endler, and E. Routman for helpful comments on various versions of the manuscript. The work was supported by grants from San Francisco State University, National Institutes of Health (RIMI), The National Geographic Society, and National Science Foundation, BR-88-17336 and DEB-96-28978.

References

Benkman, C. W. 1988. On the advantage of crossed mandibles: an experimental approach. Ibis 130: 288–293.
Benkman, C. W. 1991. Predation, seed size partitioning and the evolution of body size in seed-eating finches. Evolutionary Ecology 5: 118–127.
Benkman, C. W. 1993. Adaptation to single resources and the evolution of crossbill (*Loxia*) diversity. Ecol. Monogr. 63: 305–325.
Benkman, C. W., and A. K. Lindholm. 1991. The advantages and evolution of a morphological novelty. Nature 349: 519–520.
Boag, P. T. 1983. The heritability of external morphology in Darwin's ground finches. Evolution 37: 877–894.
Boag, P. T., and A. van Noordwijk. 1987. Quantitative genetics. In F. Cooke and P. A. Buckley, eds., Avian Genetics: A Population and Ecological Approach (pp. 45–78). New York: Academic Press.
Bush, G. L. 1994. Sympatric speciation in animals: old wine in new bottles. Trends Ecol. Evol. 9: 285–288.
Chapin, J. P. 1924. Size-variation in *Pyrenestes*, a genus of weaver-finch. Am. Mus. Nat. Hist. Bull. 49: 415–441.
Chapin, J. P. 1932. The birds of the Belgium Congo, vol. 65. Bull. Am. Mus. Nat. Hist.
Charlesworth, B. 1990. The evolutionary genetics of adaptation. In M. H. Nitecki, ed., Evolutionary Innovations (pp. 47–70). Chicago: University of Chicago Press.
Echelle, A. A., and I. Kornfield. 1984. Evolution of Fish Species Flocks. Orono: University of Maine Press.
Falconer, D. S. 1981. Introduction to Quantitative Genetics, 2nd ed. London: Longman.
Felsenstein, J. 1981. Skepticism towards Santa Rosalia, or why are there so few kinds of animals? Evolution 35: 124–138.
Fisher, R. A. 1958. The genetical theory of natural selection, 2nd rev. ed. New York: Dover.
Foote, C. J., C. C. Wood, and R. E. Withler. 1989. Biochemical genetic comparison of sockeye and kokanee, the anadromous and non-anadromous forms of *Oncorhynchus nerka*. Can. J. Fish Aquat. Sci. 46: 149–158.
Grant, B. R., and P. R. Grant. 1989. Evolutionary Dynamics of a Natural Population: The Large Cactus Finch of the Galapagos. Chicago: University of Chicago Press.
Grant, B. R., and P. R. Grant. 1993. Evolution of Darwin's finches caused by a rare climatic event. Proc. R. Soc. Lond. 251: 111–117.
Grant, P. R. 1986. Ecology and evolution of Darwin's Finches. Princeton, NJ: Princeton University Press.
Grant, P. R., and B. R. Grant. 1994. Phenotypic and genetic effects of hybridization in Darwin's finches. Evolution 48: 297–316.
Haldane, J. B. S. 1932. The causes of evolution. Ithaca, NY: Cornell University Press.
Hindar, K. 1994. Alternative life histories and genetic conservation. In V. Loeschke, J. Tomink, and S. K. Jain, eds., Conservation Genetics (pp. 323–336). Basel: Birkhauser.
Hindar, K., and B. Jonsson. 1982. Habitat and food segregation of dwarf and normal arctic charr (*Salvelinus alpinus*) from Vangsvatnet Lake, western Norway. Can. J. Fish. Aqatic. Sci. 39: 1030–1045.
Hoekstra, R. F., R. Bijlsma, and A. J. Dolman. 1985. Polymorphism from environmental heterogeneity: models are only robust if the heterozygote is close in fitness to the favored homozygote in each environment. Genetic Research 45: 299–314.
Hori, M. 1993. Frequency-dependent natural selection in the handedness of scale-eating cichlid fish. Science 260: 216–219.

Johnson, P. A., F. C. Hppensteadt, J. J. Smith, and G. L. Bush. 1996. Conditions for sympatric speciation: a diploid model incorporating habitat fidelity and non-habitat assortative mating. Evol. Ecol. 10: 187–205.

Keast, A. 1972. Faunal elements and evolutionary patterns: some comparisons between the continental avifaunas of Africa, South America, and Australia. Paper presented at the Proceedings of the XVth International Ornithological Congress, Leiden.

Kirkpatrick, M. 1996. Genes and adaptation: a pocket guide to the theory. In M. R. Rose & G. V. Lauder, eds., Adaptation (pp. 125–146). New York: Academic Press.

Lande, R. 1980. The genetic covariance between characters maintained by pleitopic mutations. Genetics 94: 203–215.

Lande, R., and M. Kirkpatrick. 1988. Ecological speciation by sexual selection. J. Theor. Biol. 133: 85–98.

Lederer, J. R. 1975. Bill size, food size and jaw forces of insectivorous birds. Auk 92: 385–387.

Letouzey, R. 1968. Etude phytogeographique du Cameroun. Paris: Lechevalier.

Levene, H. 1953. Genetic equilibrium when more than one ecological niche is available. Am. Nat. 87: 331–333.

Louette, M. 1981. The birds of Cameroon. An annotated checklist. Verhandelingen van de Koniklijke Academie voor Wetenschappen, Letteren Schone Kunsten van Belgie. Klasse de Wetenschappen 43: 1–295.

Malmquist, H. J., S. S. Snorrason, S. Skulason, B. Jonsson, O. T. Sandlund, and P. M. Jonasson. 1992. Diet differentiation in polymorphic arctic charr in Thingvallavatn, Iceland. J. Anim. Ecol. 61: 21–35.

Maynard Smith, J. 1966. Sympatric speciation. Am. Nat. 100: 637–650.

Maynard Smith, J., and R. Hoekstra. 1980. Polymorphism in a varied environment: how robust are the models? Genet. Res. 35: 45–57.

Mayr, E. 1963. Animal species and evolution. Cambridge, MA: Harvard University Press.

Meyer, A. 1993. Phylogenetic relationships and evolutionary processes in East African cichlid fishes. Trends Ecol. Evol. 8: 279–284.

Meyer, A., T. D. Kocher, P. Basasibwaki, and A. C. Wilson. 1990. Monophyletic origin of Lake Victoria cichlid fishes suggested by mitochondrial DNA sequences. Nature 347: 550–553.

Orr, H. A., and J. A. Coyne. 1992. The genetics of adaptation: a reassessment. Am. Nat. 140: 725–742.

Peterson, A. T. 1993. Adaptive geographical variation in bill shape of scrub jays (*Aphelocoma coerulescens*). Am. Nat. 142: 508–527.

Price, T., M. Turelli, and M. Slatkin. 1993. Peak shifts produced by correlated respose to selection. Evolution 47: 280–290.

Rice, R. R., and E. E. Hostert. 1993. Laboratory experiments on speciation: what have we learned in 40 years? Evolution 47: 1637–1653.

Riget, A., K. H. Nygaard, and B. Christensen. 1986. Population structure, ecological segregation, and reproduction in a population of arctic char (*Salvelinus alpinus*). Can. J. Fish. Aquat. Sci. 43: 985–992.

Robinson, B. W., D. S. Wilson, A. S. Margosian, and P. T. Lotito. 1993. Ecological and morphological differentiation of pumpkinseed sunfish in lakes without bluegill sunfish. Evol. Ecol. 7: 451–464.

Robinson, B. W., D. S. Wilson, and G. O. Shea. 1996. Trade-offs of ecological specialization: An intraspecific comparison of pumpkinseed sunfish phenotypes. Ecology 77: 170–178.

Rose, M. R., and G. V. Lauder. 1996. Post-Spandrel adaptationism. In M. R. Rose and G. V. Lauder eds., pp. 1–8 Adaptation. New York: Academic Press.

Sandlund, O. T., B. Jonsson, H. J. Malmquist, R. Gydemo, T. Lindem, S. Skulason, S. S. Snorrason, and P. M. Jonasson. 1987. Habitat use of arctic charr *Salvelinus alpinus* in Thingvallavatn, Iceland. Environ. Biol. Fish. 20: 263–274.

Schliewen, U. K., D. Tautz, and S. Paabo. 1994. Sympatric speciation suggested by monophyly of crater lake cichlids. Nature 368: 629–632.

Schluter, D. 1988a. Estimating the form of natural selection on a quantitative trait. Evolution 42: 849–861.

Schluter, D. 1988b. The evolution of finch communities on islands and continents: Kenya vs. Galapagos. Ecol. Monogr. 58: 229–249.

Schluter, D. 1996. Ecological speciation in postglacial fishes. Phil. Trans. R. Soc. London 351: 807–814.

Schluter, D., and J. D. McPhail. 1993. Character displacement and replicate adaptive radiation. Trends Ecol. Evol. 8: 197–200.

Schluter, D., and L. M. Nagel. 1995. Parallel speciation by natural selection. Am. Nat. 146: 292–301.

Schluter, D., T. D. Price, and P. R. Grant. 1985. Ecological character displacement in Darwin's Finches. Science 227: 1056–1059.

Schluter, D., and J. N. M. Smith. 1986a. Genetic and phenotypic correlations in a natural population of song sparrows. Biol. J. Linn. Soc. 29: 23–36.

Schluter, D., and J. N. M. Smith. 1986b. Natural selection on beak and body size in the song sparrow. Evolution 40: 221–231.

Sibley, C. G., and J. E. Ahlquist. 1990. Phylogeny and Classification of Birds: A Study in Molecular Evolution. New Haven, CT: Yale University Press.

Sigurjonsdottir, H., and K. Gunnarsson. 1989. Alternative mating tactics of arctic charr, *Salvelinus alpinus*, in Thingvallavatn, Iceland. Environ. Biol. Fish. 26: 159–176.

Skulason, S., and T. B. Smith. 1995. Resource polymorphisms in vertebrates. Trends Ecol. Evol. 10: 366–370.

Skulason, S., S. S. Snorrason, D. L. G. Noakes, M. M. Ferguson, and H. J. Malmquist. 1989. Segregation in spawning and early life history among polymorphic Arctic charr, *Salvelinus alpinus*, in Thingvallavatn, Iceland. J. Fish Biol. 35: 225–232.

Smith, T. B. 1987. Bill size polymorphism and intraspecific niche utilization in an African finch. Nature 329: 717–719.

Smith, T. B. 1990a. Natural selection on bill characters in the two bill morphs of the African finch *Pyrenestes ostrinus*. Evolution 44: 832–842.

Smith, T. B. 1990b. Patterns of morphological and geographic variation in trophic bill morphs of the African finch *Pyrenestes*. Biol. J. Linn. Soc. 41: 381–414.

Smith, T. B. 1990c. Resource use by bill morphs of an African finch: evidence for intraspecific competition. Ecology 71: 1246–1257.

Smith, T. B. 1991. Inter- and intra-specific diet overlap during lean times between *Quelea erythrops* and bill morphs of *Pyrenestes ostrinus*. Oikos 60: 76–82.

Smith, T. B. 1993a. Disruptive selection and the genetic basis of bill size polymorphism in the African finch, *Pyrenestes*. Nature 363: 618–620.

Smith, T. B. 1993b. Ecological and evolutionary significance of a third bill form in the polymorphic finch *Pyrenestes ostrinus.* Paper presented at the Birds and the African Environment: Proceedings of the 8th Pan African Ornithological Congress, Bujumbura, Burundi.

Smith, T. B. 1997. Adaptive significance of the mega-billed form in the polymorphic black-bellied seedcracker *Pyrnestes ostrinus*. Ibis 139: 382–387.

Smith, T. B., and S. Skulason. 1996. Evolutionary significance of resource polymorphisms in fish, amphibians and birds. Annu. Rev. Ecol. Syst. 27: 111–133.

Smith, T. B., R. K. Wayne, D. J. Girman, and M. W. Bruford. 1997. A role for ecotones in generating rainforest biodiversity. Science 276: 1855–1857.

Smith, T. B., R. K. Wayne, D. J. Girman, and M. W. Bruford. 1999. Evaluating the divergence-with-gene-flow model in natural populations: the importance of ecotones in rainforest speciation. In C. Mortiz and E. Bermingham, eds., Rainforests Past and Future. Chicago: University of Chicago Press.

Snorrason, S. S., S. Skulason, B. Johsson, H. J. Malquist, P. M. Jonasson, O. T. Sandlund, and T. Lindem 1994. Trophic specialization in arctic charr Salvelinus alpinus (Pisces: Salmonidae): morphological divergence and ontogenetic niche shifts. Biol. J. Linn. Soc. 52: 1–18.

Sparholdt, H. 1985. The population, survival, growth, reproduction and food of arctic charr, *Salvelinus alpinus* (L.), in four unexploited lakes in Greenland. J. Fish. Biol. 26: 313–330.

Taylor, E. B., and P. Bentzen. 1993. Evidence for multiple origins and sympatric divergence of trophic ecotypes of smelt (*Osmerus*) in northeastern North America. Evolution 47: 813–832.

Templeton, A. R. 1981. Mechanisms of speciation-a population genetic approach. Annu. Rev. Ecol. Syst. 12: 23–48.

Trewavas, E., J. Green, and S. A. Corbert. 1972. Ecological studies on crater lakes in West Cameroon, fishes of Barombi Mbo. J. Zool. 167: 41–95.

Walker, A. F., R. B. Greer, and A. S. Gardner. 1988. Two ecologically distinct forms of artic charr *Salvelinus alpinus* (L.) in Lock Rannoch. Biol. Conserv. 43: 43–61.

West-Eberhard, M. J. 1986. Alternative adaptations, speciation, and phylogeny (a review). Proc. Natl. Acad. Sci. USA 83: 1388–1392.

Wright, S. 1968. Evolution and the genetics of populations, vol. 1. Chicago: University of Chicago Press.

7

Geographic Variation in Flower Size in Wild Radish

The Potential Role of Pollinators in Population Differentiation

SUSAN J. MAZER AND DANIEL E. MEADE

In spite of a long history of interest in the evolution of floral traits, and the intuitively appealing view that pollinator preferences play an important role in the evolution of flower size and shape, few studies have demonstrated the selective forces leading to genetic differentiation in floral traits within or among species. In this chapter we review several key studies that have been conducted in wild species to evaluate the processes responsible for floral trait evolution. Then we focus on previous and current work on wild radish (*Raphanus sativus* L.: Brassicaceae) to examine the ability of pollinators to impose selection on floral traits. We suggest that, as a model system for the evolution of flower size in particular, wild radish holds much promise as a species in which to test alternative hypotheses concerning the selective forces that mold this trait. Within wild radish populations, common garden experiments and paternal sib analyses reveal significant heritable variation in flower size (the area of displayed petals). Moreover, California populations of wild radish have genetically differentiated with respect to this trait. Previous work on pollinator preferences and their fitness consequences in wild radish suggests that pollinators discriminate among plants on the basis of flower size (generally favoring large flowers), but that in spite of this preference, plants visited primarily by honeybees may experience selection favoring small flowers. Our recent survey of the insect visitors in 32 California populations of wild radish detected a strong association between mean flower size and the abundance and identities of their pollinators. In contrast to the expectation that high honeybee densities should select for small-flowered genotypes, however, populations visited by medium to high densities of honeybees have larger-flowered plants relative to populations where honeybees are rare. To determine whether selection on flower size may depend on the pollinator community, we experimentally exposed a range of floral phenotypes either to ambient pollinators (including honeybees) or to small native bees and syrphid flies. Selection gradient analysis using maternal lifetime seed yield (the total biomass of seeds produced) as a fitness estimate determined that under ambient conditions (honeybees included), there was no direct selection on flower size; however, when large pollinators were excluded, selection favored large-flowered genotypes. This result suggests that the observed association among populations between honeybee density and mean flower size cannot be explained by the selective process observed in this or in previous experiments. While we cannot yet rule out genetic drift as the cause of population differentiation, it is likely that floral evolution in wild radish is influenced by selection imposed by abiotic and biotic factors, such as water availability, that merit further investigation.

7.1 Evidence for the Roles of Pollinators in Floral Trait Evolution

The mutualistic relationships between plants and their pollinators have been of interest to biologists since the beginning of the study of natural history. Darwin viewed these intimate relationships as evidence for a strong evolutionary influence of pollinators on floral morphology, and since his time, evolutionists have examined pollinator behavior with the aim of understanding the mechanisms of floral trait evolution (Darwin 1877, Müller 1883, Leppik 1957, Faegri and van der Pijl 1979, Feinsinger 1983, Waser 1983, Willemstein 1987). Due to the ability of pollinators to learn which floral and inflorescence traits are most strongly associated with reliable rewards within plant populations, pollen vectors clearly hold the potential for being important agents of selection on floral phenotype and floral display (Waser 1983, Harder 1990, Seeley et al. 1991, Hill et al. 1997). We know that pollinators are sensitive to floral traits such as nectar production and flower size (e.g., Schemske and Ågren 1995), that pollinator visitation rates can have a strong effect on components of plant reproduction (particularly on male fitness), and that pollinators vary widely in their efficiency of delivering and exporting pollen. Together, these observations tell us that pollinator abundance and preferences may play a strong role in determining the direction and magnitude of adaptive evolutionary change in floral traits. Indeed, there are numerous species in which pollinators appear to generate direct selective forces on floral traits within populations.

Species with flower color polymorphisms were the first to be studied in detail. Pollinator preferences for particular flower colors are well known (Paige and Whitham 1985, Stanton 1987, Pohtio and Teräs 1995, Lunau 1996, Picker and Midgley 1996, Meléndez-Ackerman et al. 1997), but they do not necessarily result in the evolution of flower color due to their effects on male fitness. In *Ipomoea purpurea*, for example, which has a Mendelian flower color polymorphism, Schoen and Clegg (1985) found that, in spite of previous work showing that white morphs were discriminated against by this species' principle pollinators (bumblebees) (Brown and Clegg 1984), white-flowered genotypes had higher male reproductive success relative to purple or pink ones. This kind of discrepancy between floral attractiveness and male reproductive success underscores the fact that pollinator visitation rates are not always reliably correlated with realized male fitness. Accordingly, this observation has motivated investigators either to measure the amount of pollen exported from flowers with different phenotypes or to use genetic markers to determine the relationship between floral traits and paternal success. In one intensive set of studies using electrophoretic markers, Stanton et al. (1986, 1989) reported that in experimental populations of *Raphanus raphanistrum*, the primary insect visitor, *Pieris rapae*, discriminated against white flowers relative to yellow ones and that the difference in attractiveness between yellow- and white-flowered genotypes results in lower male fitness for the latter. More recently, Campbell et al. (1997) and Meléndez-Ackerman et al. (1997) have used selection gradient analysis in an *Ipomopsis aggregata* × *I. tenuituba* hybrid zone to detect directional selection on flower color mediated by hummingbird preferences.

Additional floral traits that appear to experience selection mediated by pollinator preferences include nectar production, scent, nectary–stigma distance, nectary spur length, floral symmetry, floral color change, and flower size. In *Mirabilis multiflora*, Hodges (1995) found that hawkmoths visit more flowers per individual on plants producing relatively high quantities of nectar per flower, resulting in both increased pollen deposition and pollen removal as a function of nectar production. While this pattern of visitation would appear to favor increased nectar production, rates of self-pollination also increased with visitation rate. Because *M. multiflora* is self-incompatible, this increase in self-pollination with nectar production actually resulted in a predicted pattern of stabilizing selection on nectar production. Similarly, in *Polemonium viscosum*, Cresswell and Galen (1991) found that both the number of flowers probed per visit and the time spent by bumblebees probing a flower increased strongly with nectar reward, although this did not result in an increase in the number of pollen grains received by these flowers. The observation by Galen and Stanton (1989) that the amount of time spent probing a flower was positively correlated with the amount of pollen removed, however, suggests that selection on nectar production in this species may operate through its effects on male reproductive success rather than through female fitness. In *Ipomopsis aggregata*, Mitchell (1993, 1994) likewise found that hummingbird pollinators responded to increased levels of nectar by probing a higher proportion of flowers per plant, and he estimated that plants with naturally high levels of nectar production exported more pollen per flower than those with low nectar production.

Strong evidence that floral scents evolve in response to the preferences of local pollinators is provided by the detailed studies of *Polemonium viscosum* (Galen et al. 1987, Galen and Newport 1988). Populations where bumblebees are abundant produce the large-flowered, sweet-scented forms that these pollinators favor, while populations where flies predominate produce the skunk-scented genotypes that flies prefer. Analagously, populations of *Leavenworthia crassa* that rely on pollinators for reproduction produce much more strongly scented flowers than do highly self-pollinating populations that do not rely on insect pollen vectors (Lloyd 1965, Lyons and Antonovics 1991).

Examples of selection on relatively subtle traits also point to a role for pollinators in floral evolution. For example, in *Lobelia cardinalis*, Johnston (1991) detected directional selection favoring higher nectar–stigma distances in a population in which seed production was relatively pollen limited. Numerous orchids show extremes of nectary spur length that require highly specific hawkmoth pollinators for pollen transfer, and studies in which the spur length of such orchids is manipulated confirm that fitness is maximized when the spurs are longer than the tongues of the moths that visit them, which in turn may experience selection favoring longer tongues to gain access to nectar (Nilsson 1988, 1992). Evidence that pollinators prefer symmetrical flowers over asymmetrical ones has been reported by Møller (1995) and by Møller and Eriksson (1995) for 10 species in Spain, Denmark, and Sweden. Finally, the phenomenon of floral color change has also been interpreted as a trait that has evolved due to its effect on pollinator behavior. When pollinated flowers change color but remain turgid, they retain their ability to attract pollinators to plants while directing them away from flowers that have already been pollinated (Gori 1989, Weiss 1991, 1995, Weiss and Lamont 1997). In sum, these detailed studies of single populations indicate the strong potential for pollinators to exert an effective evolutionary force on a wide range of floral traits.

7.2 The Role of Pollinators in Flower Size Evolution

Perhaps because flower size is so readily measured, this trait has received the most attention by those seeking to detect evolutionary change mediated by pollinators. One common observation that strongly suggests that pollinator preferences and behavior are key factors in determining the trajectory of flower size evolution is that species that require no pollinators for reproduction often produce much smaller flowers than do closely related species that are dependent on pollen vectors for seed set. For example, individuals in a highly self-pollinating population ($t = 0.03$) of *Leavenworthia crassa* produce much smaller flowers than those in a population with a higher outcrossing rate ($t = 0.33$) (Lyons and Antonovics 1991). Similarly, flower size in self-pollinating populations of *Arenaria uniflora* is smaller than in their outcrossing progenitors (Hill et al. 1992). At a higher taxonomic level, flower size in the cleistogamous self-pollinating *Mosla chinensis* is much smaller than in its pollinator-dependent sibling species, *M. hangchouensis* (Zhou et al. 1996). Likewise, populations of the self-pollinating *Mimulus micranthus* produce smaller flowers than those of the outcrossing *M. guttatus* (Carr and Fenster 1994). The results of these comparative studies are consistent with several rigorous studies conducted within populations that illustrate the power of pollinators to effect evolutionary change in flower size.

Flower size has been found to be under selection by pollinators even in the absence of variation in mating system. Galen's work on the self-incompatible, pollen limited, perennial *Polemonium viscosum* (Polemoniaceae) in Colorado provides perhaps the most complete study of this process (Galen and Kevan 1983, Galen et al. 1987, Galen and Newport 1988, Galen 1989, 1992, 1996, Galen and Stanton 1989). In this species, there are two morphs (with variation between them): a sweet-scented morph with relatively wide corolla lobes and long corolla tubes, and a skunk-scented morph with smaller flowers. The large-flowered morph predominates in upper alpine tundra sites where pollen-gathering bumblebees are the principal pollinator, while the smaller-flowered morph is favored in timberline populations where pollen-collecting flies are the most frequent visitors. Typically, in the sweet morph, the amount and purity of pollen received by flowers increases with corolla width, as does seed production, while in the skunk morph flower size is neutral with respect to the amount of pollen received and is negatively associated with pollen purity (Galen et al. 1987).

These differences between floral morphs suggest the presence of disruptive selection on flower size with respect to maternal reproductive success, a process that may generate and maintain divergence in flower size and associated traits in this species. Detailed studies of the pollinators of these two forms indicate that selection by bumblebees favors relatively large (and sweetly scented) flowers in two ways. First, in populations where bumblebees predominate, large flowers receive a higher proportion of intraspecific pollen and a higher quantity of outcross pollen than do small flowers (Galen et al. 1987, Galen and Newport 1988). These advantages translate into a higher seed set (Galen 1989). Second, in larger-flowered plants, more pollen is removed per visit than in small-flowered plants, potentially resulting in increased male fitness (Galen and Stanton 1989). Large-flowered phenotypes are therefore favored by bumblebees with respect to components of both male and female fitness, apparently accounting for the evolutionary increase in flower size observed in populations serviced primarily by bumblebees relative to those pollinated largely by flies.

A second set of studies revealing the efficacy of pollinators in creating selection differentials on the basis of flower size is provided by Campbell's investigation of the self-incompatible, pollen-limited monocarpic perennial *Ipomopsis aggregata* (Polemoniaceae), at the Rocky Mountain Biological Laboratory in Colorado. In a field study using fluorescent dye particles to estimate pollen removal and delivery by hummingbirds, Campbell (1989) found that selection on male reproductive success favored phenotypes with wider corolla tubes, while selection on female reproductive success favored phenotypes with highly exserted stigmas. Further work found that phenotypic selection on corolla width was mediated more by its direct effects on the amount of pollen exported per visit than by its effects on hummingbird visitation rates (Campbell et al. 1991). Similar to *Polemonium viscosum, I. aggregata* exhibits a positive correlation between corolla width and pollen production, providing an additional advantage in male function to large-flowered phenotypes (Campbell et al. 1996). Both Galen's and Campbell's work show clearly that selection on flower size can operate through its effects on male fitness, and Galen's work suggests that selection may also operate through its effects on female reproductive success.

In contrast to the view that flower size contributes more to male than to female reproductive success (see Bell 1985), a 3-year study of the self-incompatible annual *Raphanus raphanistrum* (Brassicaceae) indicated that large-flowered phenotypes showed no advantage in male reproductive success in any year, although selection favored large flowers through differences in female fitness in one year (Conner et al. 1996a, 1996b). Visitation rates to *R. raphanistrum* by syrphid flies have been observed to increase with flower size, potentially accounting for the relationship between flower size and female success, but the effect of flower size on visitation rates by small bees is less clear (Conner and Rush 1996, Conner 1997). The rate of evolution of flower size in this species is therefore likely to depend on the relative abundances of different pollinators as well as on the degree to which seed production is pollen limited. Similarly detailed work on the role of pollinators in the evolution of flower size has been conducted by Stanton and colleagues on the closely related *Raphanus sativus* (e.g., Stanton and Preston 1988, Young and Stanton 1990, Stanton et al. 1991). This work is reviewed below in the context of our current investigation.

The detailed studies cited above show clearly that, within populations, pollinators can exert a strong selective force on flower size. These studies do not, however, resolve the issue of whether pollinator preferences can account for genetic divergence among populations in flower size. That is, do flower size differences among populations reflect an adaptive evolutionary response to population- or habitat-specific pollinator abundances or preference? In fact, few studies have aimed to determine whether geographic variation in floral traits may be due to pollinator-mediated selection, although recent work by Galen (1992, 1996), Campbell et al. (1997), and Gilbert et al. (1996) has begun to address this question.

Over the last 4 years, we have taken advantage of the large number of ecological and genetic studies of wild radish (*Raphanus sativus* L.) in California to ask whether floral traits in this species — particularly variation in flower size — have differentiated genetically among populations. We have begun to detect geographic patterns of flower size variation in the field, to assess its genetic basis, and to explore whether these patterns might represent the adaptive response to local pollinator assemblages, which also vary geographically. We have addressed questions concerning the evolution of flower size in wild radish from the following perspective: If there are high levels of heritable variation in flower size in wild populations, and if flower size influences reproductive success due to pollinator preferences or efficiencies, then if we observe genetically based flower size variation among populations, it is reasonable to ask whether such variation represents adaptive responses to local pollinator behavior or abundances? In this chapter, then, we first review some of what is known about the first two conditions. Then we discuss our recent observations concerning variation among wild radish populations and whether pollinators appear to be responsible for it.

7.3 Introduction to Wild Radish

Raphanus sativus has attracted the attention of many evolutionary ecologists because it has a number of convenient characteristics (Stanton 1984, 1985, Marshall and Ellstrand 1986, 1988, Mazer et al. 1986, Mazer 1987a, 1987b, 1989, Stanton et al. 1987, Lewis et al. 1988, Nakamura and Stanton 1989). Wild radish is an annual plant that colonizes disturbed habitats and can be found throughout California in coastal and inland roadsides, airport fields, feral agricultural fields, and cleared woodlands. It is easy to cultivate and to hand-pollinate, and its self-incompatibility means that variation in fitness cannot be influenced by inbreeding depression due to self-fertilization (although biparental inbreeding can occur). Wild radish has large flowers and seeds that can be easily measured, counted, and weighed to obtain measures of lifetime maternal seed production and seed yield (the total mass of seeds produced). In addition, several studies of quantitative trait variation have revealed high levels of heritable variation in a variety of floral traits and reproductive components (Mazer 1987b, Mazer and Schick 1991a, 1991b), suggesting that evolution of these traits by natural selection may be an ongoing process in wild populations.

7.4 Sources of Variation in Flower Size in Wild Radish

Previous work has documented both genetic and environmental sources of phenotypic variation in flower size in *Raphanus sativus* (Young and Stanton 1990, Mazer and Schick 1991a, 1991b, Stanton and Young 1994, Young et al. 1994, Mazer and Wolfe, unpublished data). In this chapter, we refer to "flower size" as the corolla area that is displayed by a flower and visible to an approaching pollinator. This value can be estimated as the length × the width of the visible portion of one of the four similar-sized petals produced by each wild radish flower (e.g., Mazer and Schick 1991a, 1991b, Mazer and Wolfe, unpublished data), or as the maximum width of the corolla multiplied by the maximum width of a petal blade (Meade 1996).

We know that, within natural populations of wild radish, flower size varies enormously among individuals and among genetic lineages. For example, a large breeding experiment conducted to evaluate sources of variation in floral and fitness-related traits in wild radish detected both heritable and density-dependent variation in flower size (Mazer and Wolfe, unpublished data). In this experiment, a nested breeding design was conducted in the greenhouse in which groups of four maternal plants raised from field-collected seed (from a natural population in a disturbed coastal grassland on University of California, Santa Barbara property) were each pollinated by 1 of 19 distinct pollen donors. Nineteen paternal sibships were thereby produced, from which a total of 2700 progeny were raised in a common garden (adjacent to the natural population) in 9 blocks representing 3 different planting densities. Among many other traits, flower size and fecundity were measured on all surviving plants, and the statistical effects of population density (high, medium, and low density, corresponding to planting densities of ~400, 100, and 36 plants/m^2, respectively), maternal parent, and paternal parent on variation in flower size and fitness were estimated. In this kind of analysis, significant paternal effects on offspring phenotype can be interpreted as evidence for heritable variation in the observed traits, assuming no strong effects of dominance or epistasis on flower size.

The analysis of variance for this common garden experiment revealed both heritable and environmentally induced variation in flower size. Among individuals, mean flower size (measured as the visible petal length × width) varied from 20.2 to 191.3 mm^3 (fig. 7.1). Part of this variation was clearly due to local population density, as mean flower size increased as plants were cultivated at decreasing population density (mean petal area ± 1 SE, high density = 80.6 mm^2 ± 0.98; medium density = 85.2 mm^2 ± 0.97; low density = 92.0 mm^2 ± 1.03). Second, we found that among the paternal sibships, petal area means varied significantly within and across densities, indicating a heritable component to this trait (fig. 7.2; $P < .0001$ in all cases; statistical analysis not shown). The observation that genetic variation in flower size was expressed even across population densities suggests that in field populations, where environmental heterogeneity is expected to be high, genetic variation in flower size is also likely to be expressed. In sum, the presence of genotypic variation in flower size indicates that it is a trait that should be open to natural selection in wild populations.

In two ways, this common garden experiment revealed evidence for genotype × environment interactions. First, the between-sire component of variance in flower size varied among densities, with the highest value (58.36) being expressed at high density (medium density between-sire variance component = 28.12; low density = 46.12). This suggests that phenotypic variation among genotypes might be most strongly expressed when conditions are relatively unfavorable and intraspecific competition is highest, providing the greatest opportunity for selection (or pollinators) to discriminate among genotypes. The relatively high between-sire variance component observed in the low-density population suggests that phenotypic variation among genotypes may also be high when conditions are highly favorable.

Second, the relative flower size of the 19 paternal sibships changed among densities [ANOVAs (not shown) detected significant sire × density interaction effects on flower size]. For example, no single sibship was among the three largest-flowered sibships at all three densities; sibships 1, 4, and 5 were the three largest-flowered sibships at low density, but each of these sibships was among the top three in only one of the other two densities. At the other end of the distribution, although sibship 17 was among the three smallest-flowered sibships at all densities, sibships 7, 9, 13, and 20 were among the three smallest-flowered sibships in only one of the three densities. This kind of environment-dependent phenotype has been suggested to contribute to the maintenance of genetic variation in fitness-related traits within populations, potentially constraining the rate of their evolution in heterogeneous environments (Falconer and Mackay 1996).

Finally, we glimpsed the possibility that flower size might be under selection by pollinators, as we found that in all population densities large-flowered plants had significantly higher maternal yield (the total biomass of seeds produced per individual) than small-flowered plants (fig. 7.3). Although the slopes of the ln-ln regression of yield on petal area do not vary significantly among densities, the greatest increase in yield with increasing flower size appears to occur where intraspecific competition among wild radish plants is the lowest. This suggests that the strength of phenotypic selection on flower size would be greatest when conditions are relatively favorable. Although it is possible that pollinators played a role in generating these fitness functions, these data could also simply mean that large-flowered genotypes were, overall, the most vigorous genotypes in our sample.

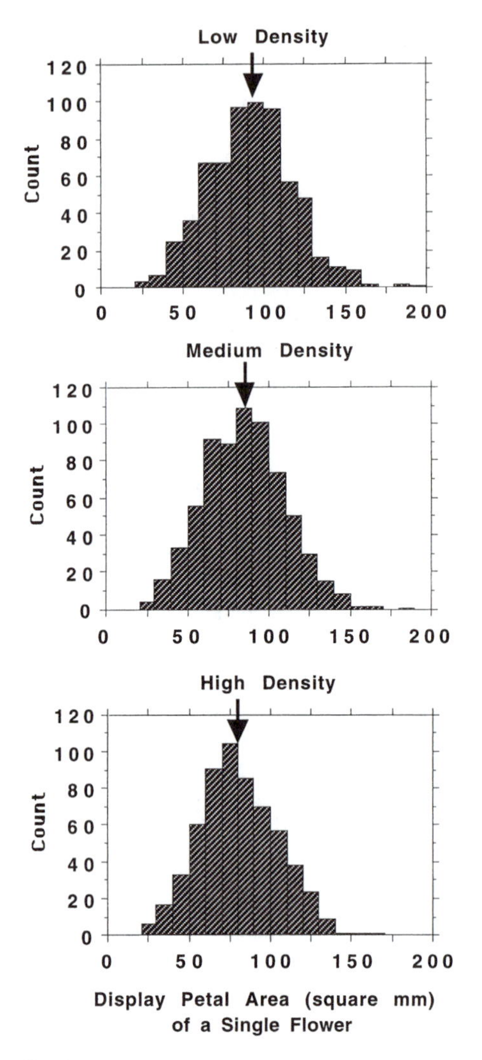

Figure 7.1 Frequency distribution of mean petal size among wild radish individuals cultivated in three density treatments. Although density has a significant effect on mean flower size ($F_{2,1919} = 32.6$; $P < .0001$), within densities much of the variation is genetically based. Vertical arrows indicate the position of the population mean in each density.

7.5 Evidence that Pollinators Mediate Natural Selection on Flower Size within Populations of Wild Radish

Although our common garden data do not provide any direct evidence for a role of pollinators in the evolution of flower size, there is considerable evidence from other studies that pollinators mediate natural selection on flower size within wild radish populations. Young and Stanton (1990), for example, found that honeybees prefer large flowers over small ones in controlled observations and that, in general, more pollen is removed from larger flowers,

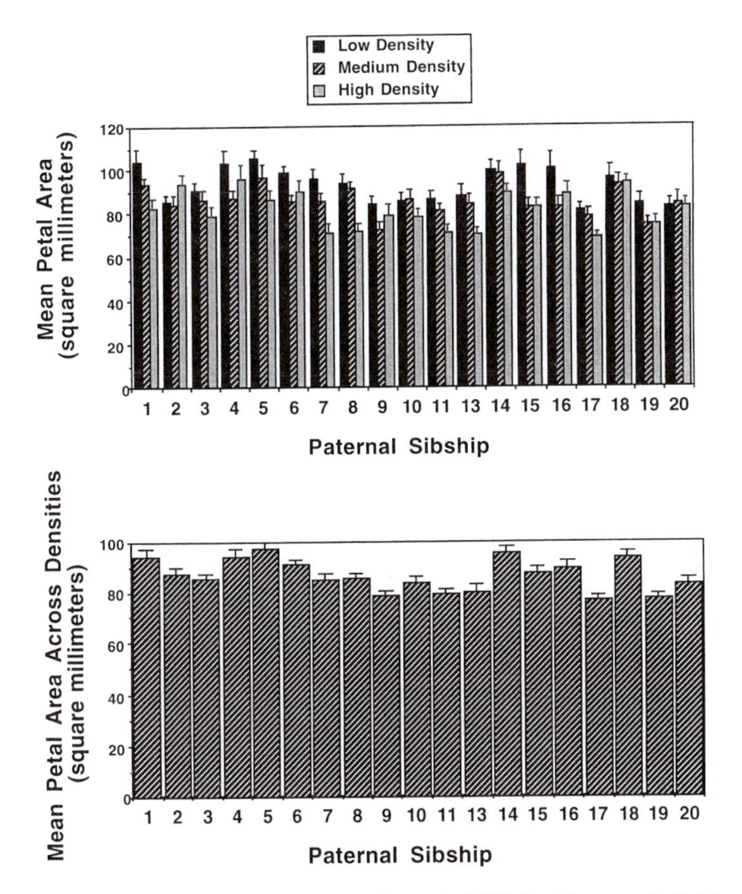

Figure 7.2 Bar graphs showing the mean petal area (+1 SE) for 19 paternal sibships of wild radish in three planting density treatments and across densities. Mean petal area differed significantly among paternal sibships both within and across densities.

although this does not always translate into higher siring success for large-flowered than for small-flowered plants (Stanton et al. 1991). It has also been found in experimental populations that pollinator visitation increases with flower size, that large flowers produce more pollen than small ones, and that there is sometimes no correlation between flower size and female reproductive success measured on a per-flower or whole-plant basis (Stanton and Preston 1988, Young and Stanton 1990, Stanton et al. 1991).

Details concerning pollen production and its reliable transport to stigmas of other plants are important where female reproductive success tends to be limited by resources rather than by pollinator visitation. Under these conditions, if pollinators account for the differential fitness of wild radish genotypes, it is likely to be through their effects on male fitness. To predict the selective forces imposed on flower size by particular pollinator taxa, the effects of their visitation on male fitness must take into account their post-visit behavior. Stanton et al. (1991) found, for example, that honeybee visitation rates are not a good predictor of male reproductive success in wild radish, as increased honeybee visitation actually diminishes

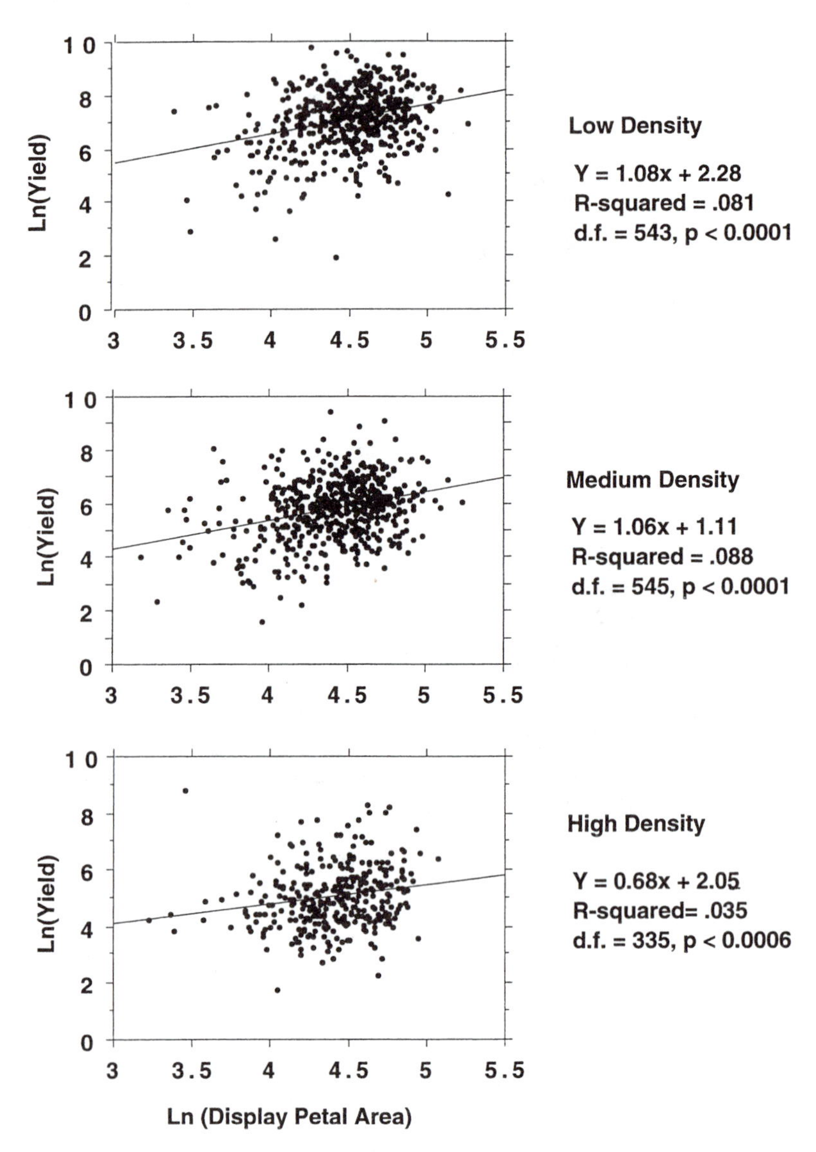

Figure 7.3 Regressions of ln-transformed maternal seed yield on ln-transformed display petal area (one flower per plant) for plants cultivated in three density treatments. For each plant, seed yield was measured as total lifetime seed production multiplied by mean individual seed mass.

male reproductive success. Large flowers are preferred by honeybees and have the potential to contribute disproportionately to plant fitness through the export of pollen to other plants, but there is an inverse relationship between honeybee visitation rates and expected male fitness (particularly when honeybee visitation rates are high; Young and Stanton 1990). In other words, for wild radish, honeybees appear to be "ugly" pollinators (see Thomson and

Thomson 1992), removing large amounts of pollen, but possibly reducing male function either by collecting the pollen rather than depositing it, by displacing more effective small pollinators, or by spreading self-incompatible pollen among the flowers of the same plant rather than by exporting it to other plants. Contrary to what one might infer from honeybee preferences alone, high densities of honeybees in the pollinator fauna of wild radish would be expected to impose selection against large flower size. It is important to emphasize, however, that these studies controlled for flower number, thereby evaluating the effects of flower size on male fitness on a per-flower basis. Lifetime paternal reproductive success was not measured. If, for example, there are strong trade-offs between flower size and lifetime flower production, then disadvantages to small-flowered plants observed on a per-flower basis might not apply to measures of whole-plant reproductive success.

It appears that native bees generate a different relationship between flower size and reproductive success than honeybees. Stanton et al. (1991) observed that, in contrast to honeybee performance, increases in the number of visits by small, native bees do result in increased male success, again measured on a per-flower basis. Stanton et al. (1991) proposed that when small, native bees visit wild radish, there is a higher probability that a pollen grain will be successfully transported to a receptive conspecific stigma than when a honeybee visits. There is no evidence that small, native bees prefer large flowers, but perhaps simply because larger flowers produce more pollen, visits by small bees result in higher fitness for large than for small flowers. Large, native bees (at least *Agapostemma* spp.) do appear to have a preference for large flowers (Stanton et al. 1991), although this does not translate into higher rates of paternity. Similarly, syrphid flies appear to have a slight preference for large flowers in *R. sativus* (Stanton and Preston 1988) (and they have a strong preference for large flowers in *R. raphanistrum*; Conner and Rush 1996).

A full picture of the nature of selection on flower size requires a multivariate approach that considers the associations between flower size, pollen production per flower, pollinator visitation rates, the quantity of pollen received, the quantity of pollen exported and successfully deposited on the stigmas of other plants, lifetime flower production, and lifetime fruit or seed production. A step in this direction was made by Stanton et al. (1991), who conducted a path analysis that illustrates the complexity of the relationships between some of these factors (fig. 7.4). It is evident, for example, that petal size can influence paternal success in a positive way due to its positive effects on pollen production per flower and on the number of pollinator visits received by some (but not by all) visitors. For example, only for some small, native bees does an increase in visitation rate result in an increase in seed paternity.

It should be emphasized that, even though Stanton et al. (1991) provide strong evidence that visitation by small bees favors large-flowered phenotypes, whereas visitation by honeybees selects against them, the relationship between petal size and paternal success that they observed does not provide a complete picture of the nature of selection on petal size. Evidence that large flowers are visited more frequently or export more pollen than small flowers must be balanced by the possibility that large-flowered plants produce fewer flowers. Such a negative correlation between mean flower size and rates of flower production could compensate for the apparent male advantage to large-flowered genotypes (such a trade-off was observed by Stanton et al. 1991). It is also critical to note that the relationship between petal size and female reproductive success does not necessarily mirror that between petal size and paternal success. In most cases, seed production in wild radish appears to be resource limited rather than pollen limited, such that there is no selection on flower size through

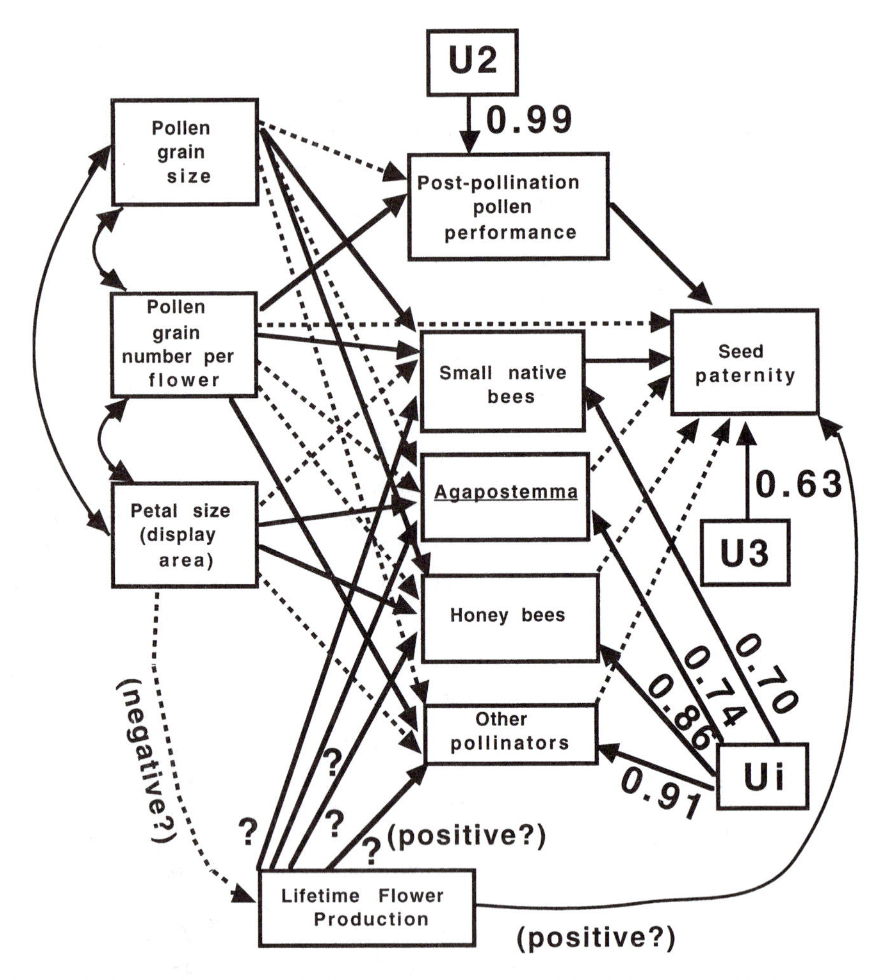

Figure 7.4 Path diagram illustrating qualitative relationships between floral traits, pollinator visitation rates, pollen performance, and paternal reproductive success observed in *Raphanus sativus* (modified from Stanton et al. 1991). Double-headed arrows indicate correlated traits between which the causal relationship is unknown. Single-headed arrows indicate the inferred direction of a causal relationship between traits. Solid lines indicate positive correlations; dashed lines indicate negative associations between traits. For each endogenous variable, the residual variance is indicated by the arrows emerging from Ui (for individual pollinator categories), U2, and U3. Lifetime flower production has been included as an additional whole-plant trait that may complicate the nature of selection on petal size. We propose that a negative relationship between mean flower size and flower production could confound the relationship between flower size and paternal reproductive success. If there is a trade-off between flower size and lifetime flower production, then any positive effects of petal size on reproduction may be offset by lower flower production.

female function due to pollinator behavior. However, here, too, if a reduction in flower size results in a compensatory increase in flower and fruit production, petal size might experience negative directional selection, at least to a point. And, although resource-limited seed production has been observed under experimental conditions, seed production in natural populations with a depauperate pollinator fauna may in fact be pollen limited. In such cases, petal size may be under selection by pollinators through its effects on both male and female reproductive success.

7.6 Scaling up: A Predicted Relationship between Pollinators and Flower Size across Populations

Combining the observations that flower size is heritable and that pollinators are often sensitive to flower size variation, and ignoring a possible trade-off between flower size and lifetime flower production, we can make a preliminary prediction concerning selection on flower size in *R. sativus* populations exposed to different pollinators. If, in contrast to honeybees, high visitation rates of small, native bees increase male fitness (fig. 7.4; cf. Stanton et al. 1991), then populations in which honeybees are absent or rare but native bees are abundant should experience selection favoring larger flowers through male function. Although syrphid flies are also significant visitors to *R. sativus*, they are relatively slow moving (remaining on flowers about 12 times as long as honeybees; D. E. Meade, personal observation) and carry relatively few pollen grains between flowers (D. E. Meade, personal observation). Although syrphid flies have been observed to respond positively to flower size in the closely related *R. raphanistrum* (Conner and Rush 1996), they do not appear to be highly effective pollinators of *R. sativus*. We believe that, to the extent that syrphid flies exert a selective force on flower size in *R. sativus*, their effects should be similar to those of small, native bees.

If selection on flower size through male function is not overwhelmed by selection on flower size through female function or through an indirect effect of flower size on flower production, and if genetic variation in flower size among field populations is the result of variation among them in pollinator abundance and preferences, then large-flowered populations should be associated with low honeybee densities but high native pollinator densities, and relatively small-flowered populations should be associated with high honeybee densities. Our recent work has aimed to determine whether geographic variation in flower size in field populations might be the result of selection imposed by pollinator preferences. To begin with, we aimed to determine whether geographic patterns are consistent with the above expectation.

To date, we have taken two approaches to the study of geographic variation in wild radish flower size. First we sought indirect evidence of coevolution between pollinator assemblages and flower size by seeking a geographic association between flower size and pollinator identity. Second, we conducted an experiment in which we manipulated pollinator access to wild radish to see whether the relationship between flower size and lifetime female fitness might depend on the composition of the pollinator assemblage. This kind of experiment had not previously been conducted on wild radish. Evaluating the relationship between flower size and male fitness was beyond the scope of this study, but we accept the primary result of previous work: that high honeybee densities select against large-flowered phenotypes. We are still in the early stages of evaluating the causes of differentiation in flower size in wild radish, and below we discuss alternative processes (in addition to pollinator-mediated selection) that may account for it.

7.7 Geographic Variation in and Genetic Differentiation of Flower Size

At the onset of this population survey, flower size variation among multiple populations of wild radish had not yet been documented. To determine whether wild radish populations in California differ in flower size and, if so, whether this has a genetic basis, was a two-step process. First, 32 field populations were surveyed from San Francisco to near the Mexican border over 3 years, sampling 3 flowers from each of 30 individuals near the beginning of the flowering season (sampled plants had produced no more than 20 flowers). In this survey, flower size was estimated in square millimeters as the product of the maximum corolla width (when the flower was observed face-on) and the maximum width of a single petal. To determine whether there were any clear regional trends in flower size, populations were classified as "inland" or "coastal," with "inland" populations defined as those where a significant geographic feature (e.g., a mountain range) separated them from coastal influences (e.g., fog, onshore wind, and rain).

To determine whether flower size differences observed in the field persisted in a common environment, we conducted a common garden experiment using seeds produced in four relatively large-flowered and four relatively small-flowered populations. Representing each of these populations were 100 seeds collected from 30 field-grown individuals. In the garden, we used seeds representing the same seed mass range from each of the populations to reduce any early maternal effects that might influence seedling growth, plant vigor, and, ultimately, flower size (details in Meade 1996).

We found significant variation among field populations in flower size. Among 28 populations that were sampled in 1994 (fig. 7.5), for example, mean flower size varied from 159.6 to 231.3 mm^2 (fig. 7.6). Moreover, these differences were consistent from year to year. The populations with the smallest petals in 1994 were also those with the smallest petals in 1995 (fig. 7.7). In addition, we found that variation among the populations persisted in the common garden (fig. 7.8). The populations with the smallest petals in the field produced progeny that developed into small-petaled adults in the common garden, while the largest-flowered field populations produced relatively large-flowered progeny. These data suggest strongly that the flower size differences observed in the field reflect genetic differentiation among populations. We also found that the geographic variation in flower size that we observed was associated in part with general geographic region. Populations near the coast in both northern (north of Point Conception) and southern California had larger flowers than inland populations (fig. 7.9).

7.8 Geographic Variation in Pollinator Communities

Over 3 years, nearly 2500 insects were sampled visiting wild radish (table 7.1). The largest proportion of visitors were honeybees, followed by syrphid flies and, collectively, by small, native bees. To determine whether pollinator abundances might be associated with flower

Figure 7.5 Map of California with 28 locations indicating field populations of wild radish sampled in 1994.

Table 7.1 Percentages of different insect visitors to wild radish plants sampled in field populations, 1993–1995

Insect type	Percent in sample
Honeybees	48.0
Small syrphid flies (< 10 mm)	24.5
Medium-sized syrphid flies (10–15 mm)	14.0
Halictid bees	5.5
Bumblebees	4.0
Lepidopterans	3.5
Colletid bees	1.5
Anthophorid bees	<1.0
Andrenid bees	<1.0
Megachilid bees	<1.0
Bee flies	<1.0
Large hover flies (> 15 mm)	<1.0

A total of 2482 insects were collected in sweep samples of inflorescences in each population at the beginning of the flowering period and at peak flowering.

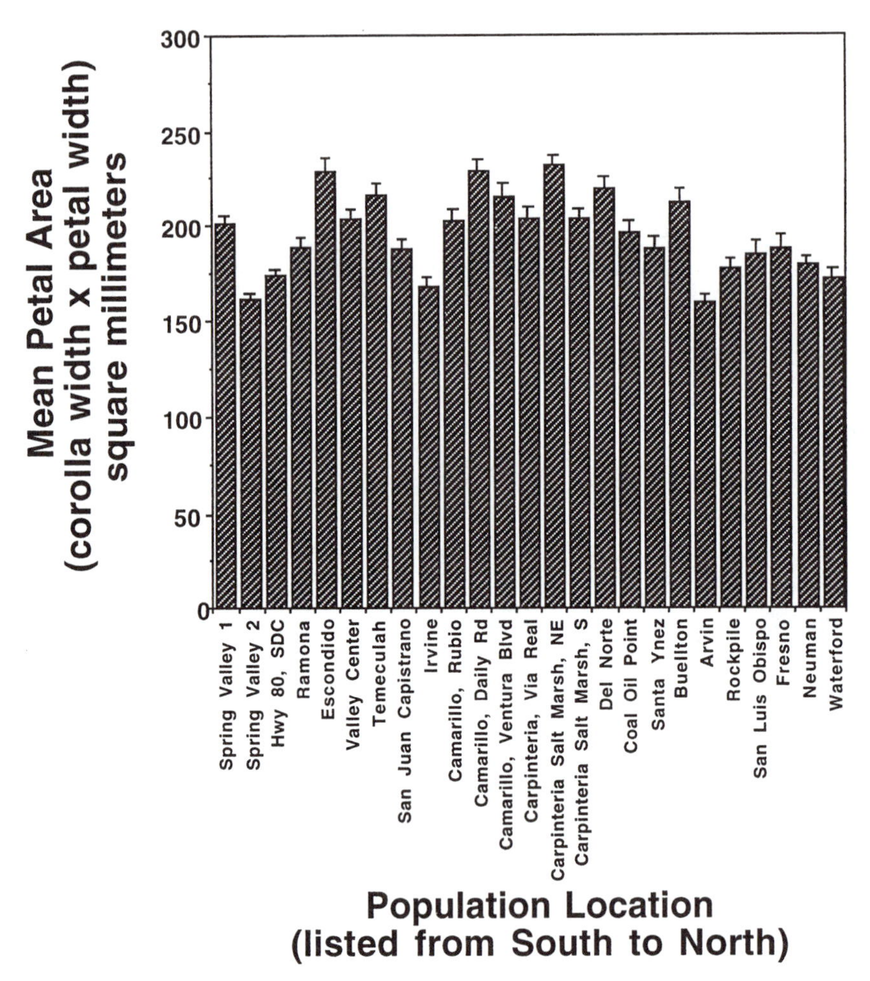

Figure 7.6 Bar graphs showing the mean (with 1 SE) of petal area for locations sampled in 1994. Mean petal area differed significantly among populations.

size, we used a variety of statistical procedures to detect patterns of association between the abundances of honeybees, all bees, or nonbee pollinators and mean population flower size (Meade 1996). With the exception of honeybees, no differences were detected among populations in the abundances of other insect visitors as a function of mean population flower size.

The strongest nonrandom association appeared between honeybee abundances and flower size when honeybee abundances were estimated in each population according to their frequency of capture in 5-min sweep samples that targeted flowering branches. These sweep samples were all conducted on sunny days between 1000 and 1500 h, and most populations were sampled twice a year at the beginning of the flowering season and at peak bloom. Populations with low densities of honeybees were those in which 5-min sweep samples captured a mean of less than one honeybee per sample. Medium density populations were those in

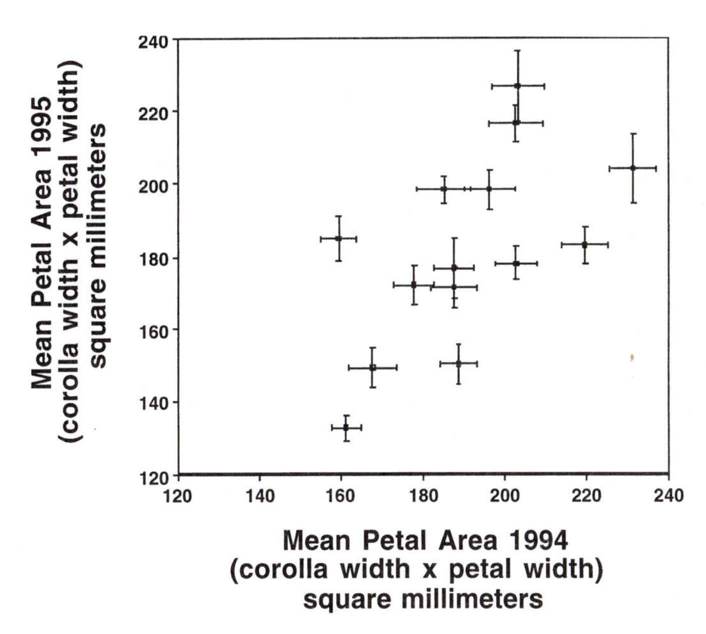

Figure 7.7 Scatterplot of the mean petal area (± 1 SE) of populations sampled in both 1994 and in 1995.

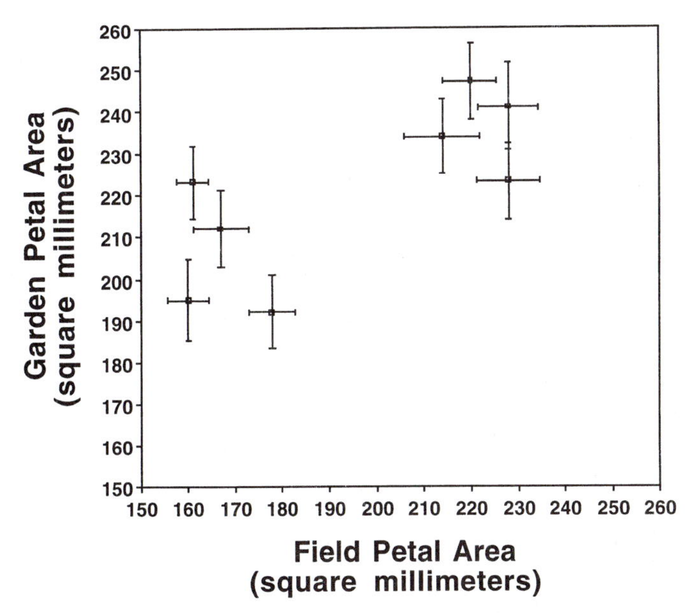

Figure 7.8 Scatterplot of the mean petal area (± 1 SE) of plants sampled from eight field populations of wild radish and of progeny of the same field populations grown in a common garden.

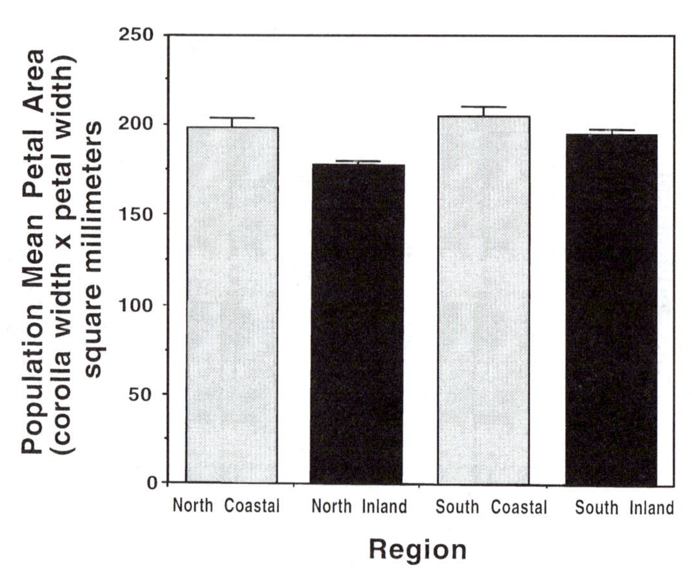

Figure 7.9 Comparison among geographic regions in California of petal area mean (with one standard error) per population of wild radish. Northern populations are those north of Point Conception. Inland populations are those separated from coastal influences by major geographic barriers.

which 5-min sweep samples captured a mean of one to five honeybees, and high density populations were those in which sweep samples captured a mean of five or more honeybees.

In these field populations, honeybee density was strongly associated with mean population flower size. Using population means for flower size, the two-way ANOVA of the effects of year and honeybee density on flower size detected no strong year-to-year variation in flower size, but did detect significant variation in flower size among populations classified as having low, medium, and high honeybee densities (table 7.2). Additional evidence that honeybee densities remained constant from year to year within populations is the high correlation detected among 13 populations between the mean number of honeybees per sweep-sample collected in 1994 and 1995 ($y = 0.39x + 1.28$; $R^2 = .74$; $P < .001$). As year had no effect on mean flower size, we conducted a two-way ANOVA to evaluate the effects of region and hon-

Table 7.2 Summary of two-way ANOVA to detect variation among populations in flower size associated with sampling year and honey bee abundances

Source	df	MS	*F*	*P*
Year	2	627.7	0.98	.384
Honeybee density	2	2092.4	3.27	.049
Year × density	4	536.7	0.84	.509
Residual	39	639.4		

The outcome variable in this ANOVA was the mean flower size of each population in each year.

Table 7.3 Summary of two-way ANOVA to detect variation among populations in flower size associated with region (inland versus coast) and honeybee abundances

Source	df	MS	*F*	*P*
Region	1	4352.4	8.95	.005
Honeybee density	2	1629.4	3.35	.045
Honeybee density × region	4	1204.4	2.48	.0961
Residual	42	486.0		

The outcome variable in this ANOVA was the mean flower size of each population in each year.

eybee density on flower size. This ANOVA reveals that, in spite of a significant difference in flower size between coastal and inland populations, flower size was significantly associated with honeybee density (table 7.3).

The quantitative relationship between honeybee density and flower size among populations can be seen in figures 7.10 and 7.11. In populations sampled in 1993, 1994, and 1995, the largest-flowered populations were always associated with medium or high honeybee densities (fig. 7.10). In inland populations, there was no significant association between honeybee density and mean flower size (fig. 7.11). In coastal populations, however, flower size tended to increase with honeybee density. The absence of a significant density × region interaction

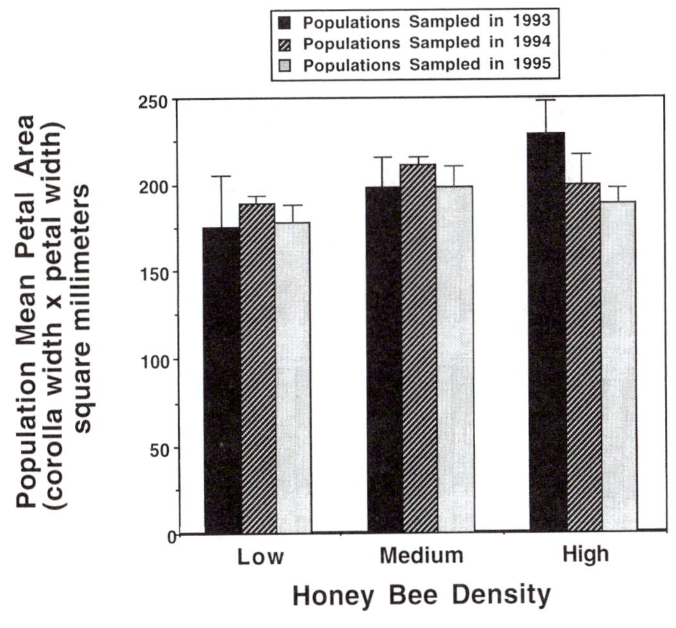

Figure 7.10 Bar graph showing yearly means for the mean petal areas (with 1 SE) of field populations experiencing three levels of honeybee density.

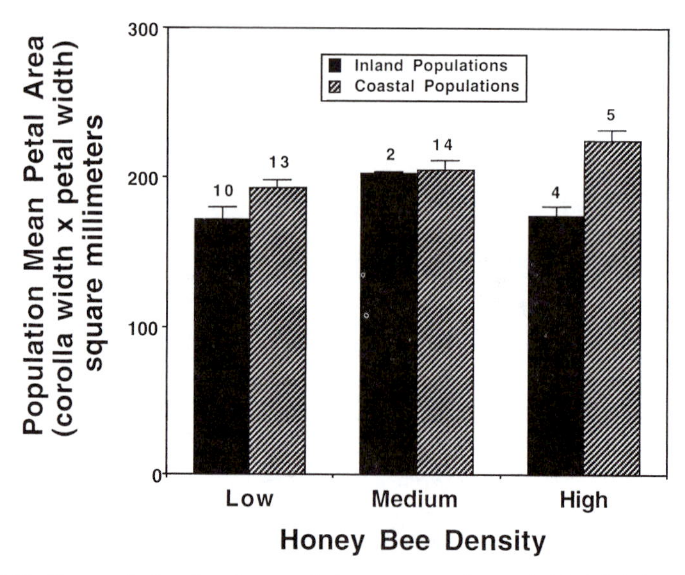

Figure 7.11 Bar graph showing the population mean petal area (with 1 SE) for populations experiencing three levels of honeybee density, and monitored in inland versus coastal populations. The numbers above the bars indicate the number of populations representing each region and honeybee density.

term (table 7.3) may be due to small sample sizes in some combinations. The observation that the inland populations with the highest honeybee densities were composed of relatively small-flowered plants could be due to the effects of a large number of honeybee colonies being placed near a few (small-flowered) inland populations for agricultural purposes.

The correspondence between mean flower size and honeybee abundance could be interpreted as indirect evidence that pollinators impose a selective force on flower size, causing the genetic differentiation observed among populations. However, our particular result, that large-flowered populations are associated with medium and high honeybee densities, is contrary to what we predicted from the effectiveness of honeybees as pollinators of large- versus small-flowered plants. Based on Stanton et al.'s (1991) results, which show that honeybees prefer large flowers over small flowers but that their visitation is associated with a reduction in the male fitness of large-flowered plants, we expected that high honeybee densities would be associated with smaller flowers than would low honeybee densities. That is, if honeybees displace native pollinators from large-flowered plants (or are otherwise "ugly" pollinators), then in wild radish populations with high honeybee densities, selection should favor small-flowered plants.

7.9 Natural Selection on Flower Size in Controlled Populations Exposed to and Protected from Honeybees

To determine whether the absence of honeybees results in selection favoring small-flowered genotypes (as the field data suggest), we conducted an experiment in which we controlled

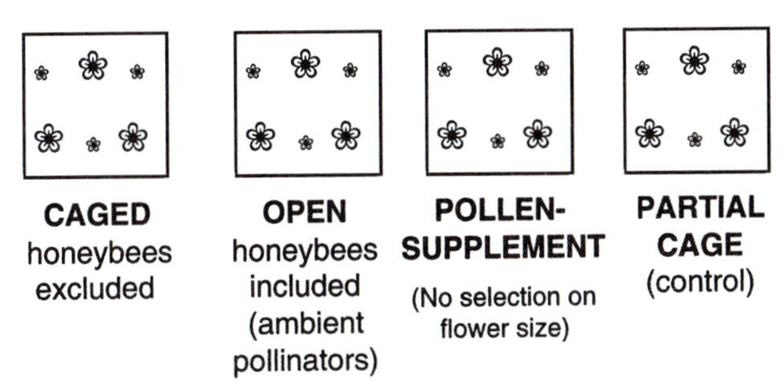

- **Each plot contained 3 small- and 3 large-flowered plants**
- **4 replicates per treatment = 24 plants per treatment**
- **The relationship between flower size and lifetime seed yield determined for each treatment**

Figure 7.12 Experimental treatments to simulate populations exposed to different pollinator assemblages. Selection gradient analyses indicated that large-flowered genotypes were favored with respect to female fitness only where honeybees were excluded (the caged plots).

access to flowers by honeybees and native bees. We then examined the effects of visits by honeybees versus native pollinators on female reproductive output using selection gradient analysis (Lande and Arnold 1983). This allowed us to determine the direction and magnitude of selection on flower size through female function, independently of variation in other floral traits (corolla length, stigma–nectary distance, and total flower production) that were also recorded in our experimental plants. We expected that, if honeybees are in fact responsible for the geographic pattern we saw in the field, when honeybee abundances are high, large-flowered phenotypes should be favored.

Within a coastal population of wild radish at the Carpinteria Salt Marsh (CSM) (a reserve in Santa Barbara County managed by the University of California at Santa Barbara), we established experimental plots of wild radish in which different pollinator assemblages were exposed to plants of different flower sizes in March–April 1993 (details in Meade 1996). We created 4 treatments, each including 4 plots with 3 small-flowered and 3 large-flowered genotypes, providing a total of 24 plants per treatment in which to examine selection on flower size (fig. 7.12). These small- and large-flowered genotypes were derived from the common garden experiment described above (Mazer and Wolfe, unpublished data), in which genotypes of varying flower size were detected in the coastal grassland population at the University of California at Santa Barbara (about 25 km away).

First, we caged 1-m^3 plots (n = 4) using nylon netting that excluded honeybees but allowed access by native bees and flies smaller than honeybees. This treatment was meant to simulate populations where honeybee densities are low. Second, we created open plots (n = 4) which allowed access to all ambient pollinators, including honeybees. In the pollen supplementation treatment (n = 4 plots), most flowers were hand-pollinated with a mixture of wild radish pollen. This treatment allowed us to search for fitness effects of floral traits when pollinators could not be imposing selection on flower size (at least with respect to maternal fitness); no such effects were found. The partial caging treatment (n = 4 plots) provided caging material above and on two sides of the target plants. This treatment tested for any adverse effects on fitness or any selective forces on flower size due to the proximity of the cage itself; again, no such effects were detected.

Pollinator visitation rates were observed for all treatments during a 7-week period. Every other day, each plot was observed for 1 min, six times a day, during peak pollinator activity (0900–1300 h). Each time a pollinator was observed within or entered a plot, it was tallied as a "visit" if the pollinator visited at least one flower. Because visits by different pollinators often occurred simultaneously, it was not possible to keep track of all individual flowers visited.Therefore, each visit indicates that at least one flower (and probably many more) was visited in a plot. Our treatments had the desired effects on pollinator assemblage composition (table 7.4). The caged treatment had significantly lower visitation by honeybees than the open treatment. In the open treatment, honeybees seemed to have no effect on visitation by native bees and syrphid flies.

For all 96 plants included in this study (4 treatments × 4 plots/treatment × 6 plants/plot), a suite of traits was measured, including mean petal display area, total flower production, nectary–stigma distance, and corolla tube length. Using total seed yield (female fecundity × mean individual seed mass) as our estimate of female fitness, we conducted a selection gradient analysis to detect directional selection on flower size while controlling for variation in the other traits (table 7.5). The only instance in which we detected significant selection on flower size (favoring larger flowers) was in the caged treatment, in which visits by honeybees were comparatively rare. There was also no direct selection on total flower production in the caged treatment, meaning that maternal reproductive success was independent of total flower production but did depend on flower size. Even when the survivorship of seeds produced by these plants was included as a component of their fitness, the results were the same (Meade 1996). All in all, in the absence of honeybees, selection favored large-flowered plants with respect to female function. This result is consistent with those of Stanton et al. (1991) for male function, but is counter to what we predicted based on the relationship among field populations between honeybees and flower size, where the lowest honeybee densities were generally associated with small-flowered plants.

7.10 Contradiction between Associations between Flower Size and Pollinator Visitors in the Field and in Controlled Conditions

With few exceptions, in the field, small-flowered populations occur where honeybees are rare, and large-flowered populations occur where honeybee densities are comparatively high. This led us to predict that, when honeybees were prevented from accessing wild radish, selection on female function would favor small-flowered phenotypes. Contrary to our prediction, we found that selection favored larger flowered phenotypes when honeybees were excluded from caged plants. This result is consistent with Stanton et al.'s (1991) data on male and female fitness, which showed that native bees impose directional selection favoring large flowers. When honeybees were allowed access to plants, however, we found that flower size was neutral with respect to female fitness. This result differs from that of Stanton et al. (1991); they observed that flower size had a negative effect on female fecundity due to the negative relationship between flower size and total flower production.

There are several possible explanations for the discrepancy between the field and exper-

Table 7.4 Differences among experimental caging treatments in pollinator visitation frequencies

Pollinator	Total No. of pollinator visits	No. of visits to each treatment by each group of pollinators: Cage	Open	Pollen supplement	Partial cage	χ^{2a}
Apis mellifera	41	4	12	17	8	30.91*
Small bees	125	26	25	45	29	86.16*
Small syrphid flies	61	6	16	22	17	45.5*
Other syrphid flies	14	0	4	5	5	
Hesperid butterflies	8	0	4	2	2	
Anthophorid bees	8	2	2	1	3	

For each pollinator category, the total number of pollinator visits represents the number of times that an insect in that category was observed within or entering any plot during the 7-week census period. The number of visits to each treatment by each group of pollinators is the number of times that each pollinator type was observed within or entering a plot in the particular treatment.

[a] Significant chi-square values indicate variation among treatments in the frequency of visits by the most common pollinators; $*P < .05$. Chi-square tests were not performed when visitation frequencies were extremely low.

imental data. First, natural selection on flower size in inland populations may favor small-flowered genotypes for physiological reasons unrelated to pollinator preferences. Small-flowered genotypes may lose less water per flower through transpiration, for example, increasing floral longevity or individual life span under more arid conditions. Alternatively, honeybees may favor coastal populations for reasons unrelated to flower size. For example, Stanton et al. (1991) have found that honeybees are the most common pollinators in large populations near Davis, California, whereas native bees predominate in smaller ones. Although wild radish population size was not closely monitored in our study (all populations had more than

Table 7.5 Summary of selection gradient analyses conducted in each experimental treatment

Trait (standardized)	Caged (honeybees excluded)		Open (all visitors included)		Pollen supplement (no selection)		Partial cage (control)	
	β	*P*	β	*P*	β	*P*	β	*P*
Intercept	-0.01	.96	.14	.50	-0.22	.28	0.25	.28
Petal area	**0.78**	**.04**	.12	.65	0.34	.26	-0.23	.37
Corolla tube length	-0.38	.16	.13	.59	-0.04	.84	0.09	.76
Nectary–stigma distance	-0.89	.75	.12	.65	.37	.25	0.06	.80
Lifetime flower production	0.12	.57	-0.02	.92	0.21	.33	-0.26	.29

Fitness was estimated as the total biomass of seeds (seed yield) produced by each maternal plant ($n = 24$ plants/treatment). The standardized partial regression coefficients for floral and plant traits exposed to different pollinator assemblages are shown. Only in the caged treatment was there significant directional selection on flower size (bold-faced values). When lifetime flower production was controlled statistically, large-petaled plants were favored in the absence of honey bees. When all insect visitors had access to plants, flower size was neutral with respect to lifetime seed yield.

100 individuals), there were no dramatic differences between inland and coastal populations with respect to population size or area. Moreover, the geographic scale of the pattern we observed is so much larger than the Davis study that it seems unlikely that honeybees were visiting wild radish populations across California on the basis of population size. We cannot, however, exclude the possibility that honeybees recruited to large-flowered populations simply because they may provide more resources than small-flowered ones.

Another possibility is that in the field, selection mediated by native bees might favor small-flowered plants with respect to male and/or female fitness when honeybees are rare (as they tend to be in the inland populations), although this would contradict our (and others') experimental observations. Finally, the temporary placement of large numbers of honeybee colonies for agricultural purposes may generate associations between flower size and honeybee density that obscure evolutionary processes. As an example, two of the inland populations sampled here (Arvin and Rockpile) were located in agricultural areas where honeybee densities may have been artificially high, thereby accounting for the small mean flower size observed in inland populations with high honeybee densities (fig. 7.11).

While intriguing, the geographic relationship between flower size and honeybee density in California provides no direct evidence that the pollinator community is responsible for the observed genetic differentiation in flower size among populations. Three additional kinds of experiments and field observations would be required to support this claim: measures of selection on flower size in situ; long-term observations of genetic change in flower size from generation to generation; and the functional significance of flower size in different habitats.

7.11 Measuring Selection on Flower Size

To rule out genetic drift as the primary cause of variation among populations in flower size, we need to determine whether selection favors small-flowered phenotypes more strongly in smaller flowered than in larger flowered populations. If selection is responsible for differences between populations, the optimum flower size should differ among populations in the direction predicted by their mean flower size. Measurements of flower size and components of male and female fitness on field-grown plants would be a first step toward evaluating the direction of phenotypic selection under natural conditions. This method has been used successfully by Campbell (1996), who established that due to their effects on pollen transport, hummingbird pollinators favor large-flowered plants of *Ipomopsis aggregata* (Polemoniaceae). Similarly, Andersson (1996) found that early flowering, large-petaled plants of *Saxifraga granulata* (Saxifragaceae) raised under garden conditions had higher seed set than plants with small petals, but that this association was largely mediated by plant vigor. Herrera (1993) also used this method in a 5-year study of *Viola cazorlensis* (Violaceae) to determine the effects of floral morphology, flower and fruit production, plant size, and herbivory by ungulates on maternal reproductive success. Although cumulative seed production was significantly associated with both flower size and shape in his study, floral morphology had a much weaker effect on fitness than the traits related to plant size and herbivore resistance.

Another way to establish the relative fitness of different flower sizes within each population is to cultivate lineages representing a wide range of floral phenotypes that are then transplanted back into the parental population. If each genotype is represented by several replicates (e.g., siblings), this would allow the measurement of the direct relationship between flower size genotype and fitness. An advantage of this design is that flower size genotypes would be expressed in the context of the genetic background in which they evolved. A disadvantage, however, is that the fitness effects of genes strongly associated with flower size due to pleiotropy or linkage could not be easily distinguished from the direct effects of genes for flower size per se.

An alternative experimental design would be to conduct reciprocal transplants among individuals or genotypes from wild radish populations that differ in flower size. Monitoring home site and "alien" wild radish genotypes representing a range of flower sizes would permit the measurement of the fitness function associated with flower size for both native and foreign genotypes. Reciprocal transplant experiments, however, run the risk of introducing unusual phenotypes that enjoy a "rare advantage." For example, the appearance of particularly large alien flowers in a small-flowered population might generate a strong positive relationship between flower size and pollinator visitation that would not naturally occur. This kind of rare advantage has been observed in bumblebee preference tests on artificial flowers (Smithson and Macnair 1997a). Smithson and Macnair observed that bumblebees preferentially visited the most common color morph when the artificial flowers provided sucrose solution rewards, but disproportionately visited rare morphs when rewards were absent. In addition to such frequency-dependent fitness functions (see also Smithson and Macnair 1997b), strong genotype × environment interactions could also prevent transplanted genotypes from exhibiting fitness functions that reflect the direct effects of flower size on reproductive success.

7.12 Long-term Observations of Genetic Change in Flower Size

Long-term observations of flower size and pollinator communities from generation to generation could provide convincing evidence that evolutionary changes in flower size could be caused by consistent pollinator-mediated selection. This kind of monitoring might also show that abrupt or gradual changes in the pollinator fauna could generate short-term genetically based changes in mean population flower size.

7.13 Functional Significance of Flower Size

Finally, it is necessary to know the relationship between flower size and fitness components (e.g., longevity, total flower production, and seed yield) when pollinators are not limiting. If there is a relationship between flower size and reproductive success independently of pollinator behavior, we would expect to observe this relationship when plants are saturated with pollen. Flower size might be positively associated with male reproductive success simply because larger flowers produce more pollen, or positively correlated with maternal seed production if large flower size is a "symptom" of high vigor. In either case, selection would favor genes that directly as well as indirectly result in large-flowered plants. Conversely, flower size might be negatively associated with reproductive success in habitats where the costs of maintaining individual flowers are particularly high (see Ashman and Schoen 1997). The argument that in arid habitats relatively small-flowered wild radish genotypes are favored due to their water efficiency would depend on this kind of information.

7.14 Final Comments

The abundance of feral populations of wild radish, the high level of genetic differentiation among them in flower size and other floral traits (Meade 1996), and the high frequency of honeybees as pollinators throughout California provide a unique opportunity to examine the potential roles of these and native pollinators in directing floral evolution at both local and regional geographic scales. Although the European honeybee is the most widely introduced pollinator throughout the world, little is known about its ecological and evolutionary effects on wild and naturalized plant species. From an ecological perspective, it is possible that competition between this "ugly" pollinator and native pollinators could result in lower seed production by native plants. The fact that *Apis mellifera* pollinated *Raphanus sativus* in its native habitats in Europe could conceivably provide wild radish with a reproductive advantage as an exotic plant in the New World, where honeybees remain one of its most important pollinators. From an evolutionary point of view, now that the genetic differentiation of wild radish floral traits among California populations has been documented, we can begin to evaluate the processes that may be responsible for it. Although we cannot yet rule out genetic drift as the cause of population differentiation, it is likely that floral evolution in wild radish is influenced by abiotic factors (e.g., inland aridity) and biotic factors (e.g., pollinator abundances and composition) that merit further investigation.

Acknowledgments We thank Maureen L. Stanton for her particularly helpful comments on a previous version of the manuscript.

References

Andersson, S. 1996. Floral variation in *Saxifraga granulata*: phenotypic selection, quantitative genetics and predicted response to selection. Heredity 77: 217–223.

Ashman, T. L., and D. J. Schoen. 1997. The cost of floral longevity in *Clarkia tembloriensis*: an experimental investigation. Evolutionary Ecology 11: 289–300.

Bell, G. 1985. On the function of flowers. Proceedings of the Royal Society of London B 224: 223–265.

Brown, B. A., and M. T. Clegg. 1984. Influence of flower color polymorphism on genetic transmission in a natural population of the common morning glory, *Ipomoea purpurea*. Evolution 38: 796–803.

Campbell, D. R. 1989. Measurements of selection in a hermaphroditic plant: variation in male and female pollination success. Evolution 43: 318–334.

Campbell, D. R. 1996. Evolution of floral traits in a hermaphroditic plant: field measurements of heritabilities and genetic correlations. Evolution 50: 1442–1453.

Campbell, D. R., N. M. Waser, and E. J. Meléndez-Ackerman. 1997. Analyzing pollinator-mediated selection in a plant hybrid zone: hummingbird visitation patterns on three spatial scales. American Naturalist 149: 295–315.

Campbell, D. R., N. M. Waser, and M. V. Price. 1996. Mechanisms of hummingbird-mediated selection for flower width within *Ipomopsis aggregata*. Ecology 77: 1463–1472.

Campbell, D. R., N. M. Waser, M. V. Price, E. A. Lynch, and R. J. Mitchell. 1991. Components of phenotypic selection: pollen export and flower corolla width in *Ipomopsis aggregata*. Evolution 45: 1458–1467.

Carr, D. E., and C. B. Fenster. 1994. Levels of genetic variation and covariation for *Mimulus* (Scrophulariaceae) floral traits. Heredity 72: 606–618.

Conner, J. K. 1997. Floral evolution in wild radish: the roles of pollinators, natural selection, and genetic correlations among traits. International Journal of Plant Sciences 158: S108–S120.

Conner, J. K., and S. Rush. 1996. Effects of flower size and number on pollinator visitation to wild radish, *Raphanus raphanistrum*. Oecologia 105: 509–516.

Conner, J. K., S. Rush, and P. Jennetten. 1996a. Measurement of natural selection on floral traits in wild radish (*Raphanus raphanistrum*). I. Selection through lifetime female fitness. Evolution 50: 1127–1136.

Conner, J. K, S. Rush, S. Kercher, and P. Jennetten. 1996b. Measurements of natural selection on floral traits in wild radish (*Raphanus raphanistrum*). II. Selection through lifetime male and total fitness. Evolution 50: 1137–1146.

Cresswell, J. E., and C. Galen. 1991. Frequency-dependent selection and adaptive surfaces for floral character combinations: the pollination of *Polemonium viscosum*. American Naturalist 138: 1342–1353.

Darwin, C. 1887. The various contrivances by which orchids are fertilised by insects. 2nd ed. J. Murray, London.

Faegri, K., and L. van der Pijl. 1980. Principles of Pollination Ecology, 3rd ed. Pergamon Press, Oxford.

Falconer, D. S., and T. F. C. Mackay. 1996. Introduction to Quantitative Genetics, 4th ed. Longman, Essex, UK.

Feinsinger, P. 1983. Coevolution and pollination. Pp. 285–319 *in* D. J. Futuyma and M. Slatkins, eds., *Coevolution*, Sinauer Associates, Sunderland, MA.

Galen, C. 1989. Measuring pollinator-mediated selection on morphometric floral traits: bumblebees and the alpine sky pilot, *Polemonium viscosum*. Evolution 43: 882–890.

Galen, C. 1992. Pollen dispersal dynamics in an alpine wildflower, *Polemonium viscosum*. Evolution 46: 1043–1051.

Galen, C. 1996. Rates of floral evolution: adaptation to bumblebee pollination in an alpine wildflower, *Polemonium viscosum*. Evolution 50: 120–125.

Galen, C., and P. G. Kevan. 1983. Bumblebee foraging and floral scent dimorphism: *Bombus kirbyellus* Curtis (Hymenoptera: Apidae) and *Polemonium viscosum* Nutt (Polemoniaceae). Canadian Journal of Zoology 61: 1207–1213.

Galen, C., and M. E. A. Newport. 1988. Complementary effects of variation in flower scent and size. American Journal of Botany 75: 900–905.

Galen, C., and M. L. Stanton. 1989. Bumble bee pollination and floral morphology: factors influencing pollen dispersal in the alpine sky pilot, *Polemonium viscosum* (Polemoniaceae). American Journal of Botany 76: 419–426.

Galen, C., K. A. Zimmer, and M. E. Newport. 1987. Pollination in floral scent morphs of *Polemonium viscosum*: a mechanism for disruptive selection on flower size. Evolution 41: 599–606.

Gilbert, F. P. Willmer, F. Semida, J. Ghazoul, and S. Zalat. 1996. Spatial variation in selection in a plant-pollinator system in the wadis of Sinai, Egypt. Oecologia 108: 479–487.

Gori, D. F. 1989. Floral color change in *Lupinus argenteus* (Fabaceae): why should plants advertise the location of unrewarding flowers to pollinators? Evolution 43: 870–881.

Harder, L. D. 1990. Behavioral responses by bumble bees to variation in pollen availability. Oecologia 85: 41–47.

Herrera, C M. 1993. Selection on floral morphology and environmental determinants of fecundity in a hawk moth-pollinated violet. Ecological Monographs 63: 251–275.

Hill, J. P., E. M. Lord, and R. G. Shaw. 1992. Morphological and growth rate differences among outcrossing and self-pollinating races of *Arenaria uniflora* (Caryophyllaceae). Journal of Evolutionary Biology 5: 559–573.

Hill, P. S. M., P. H. Wells, and H. Wells. 1997. Spontaneous flower constancy and learning in honeybees as a function of colour. Animal Behaviour 54: 615–627.

Hodges, S. A. 1995. The influence of nectar production on Hawkmoth behavior, self pollination, and seed production in *Mirabilis multiflora* (Nyctaginaceae). American Journal of Botany 82: 197–204.

Johnston, M. O. 1991. Natural selection on floral traits in two species of *Lobelia* with different pollinators. Evolution 45: 1468–1479.

Lande, R., and S. J. Arnold. 1983. The measurement of selection on correlated characters. Evolution 37: 1210–1226.

Leppik, E. E. 1957. Evolutionary relationships between entomophilous plants and anthophilous insects. Evolution 11: 466–481.

Lewis, D., S. C. Verma, and M. I. Zuberi. 1988. Gametophytic-sporophytic incompatibility in the Cruciferae: *Raphanus sativus*. Heredity 61: 355–366.

Lloyd, D. G. 1965. Evolution of self-compatibility and racial differentiation in *Leavenworthia* (Cruciferae). Contributions of the Gray Herbarium 195: 3–134.

Lunau, K. 1996. Signalling functions of floral colour patterns for insect flower visitors. Zoologischer Anzeiger 235: 11–30.

Lyons, E. E., and J. Antonovics. 1991. Breeding system evolution in *Leavenworthia*: breeding system variation and reproductive success in natural populations of *Leavenworthia crassa* (Cruciferae). American Journal of Botany 78: 270–287.

Marshall, D.L., and N. C. Ellstrand. 1986. Sexual selection in *Raphanus sativus*: experimental data on nonrandom fertilization, maternal choice and consequences of multiple paternity. American Naturalist 127: 446–461.

Marshall, D.L., and N. C. Ellstrand. 1988. Effective mate choice in wild radish: evidence for selective seed abortion and its mechanisms. American Naturalist 131: 739–756.

Mazer, S. J. 1987a. Parental effects on seed development and seed yield in *Raphanus raphanistrum*: implications for natural and sexual selection. Evolution 41: 355–371.

Mazer, S. J. 1987b. The quantitative genetics of life history and fitness components in *Raphanus raphanistrum* L. (Brassicaceae): ecological and evolutionary consequences of seed-weight variation. American Naturalist 130: 891–914.

Mazer, S. J., 1989. Family mean correlations among fitness components in wild radish: controlling for maternal effects on seed weight. Canadian Journal of Botany 67: 1890–1897.

Mazer, S. J., and C. T. Schick. 1991a. Constancy of population parameters for life history and floral traits in *Raphanus sativus* L. I. Norms of reaction and the nature of genotype by environment interactions. Heredity 67: 143–156.

Mazer, S. J., and C. T. Schick. 1991b. Constancy of population parameters for life history and floral traits in *Raphanus sativus* L. II. Effects of planting density on phenotype and heritability estimates. Evolution 45: 1888–1907.

Mazer, S. J., A. A. Snow, and M. L. Stanton. 1986. Fertilization dynamics and parental effects upon fruit development in *Raphanus raphanistrum*: consequences for seed size variation. American Journal of Botany 73: 500–511.

Meade, D. E. 1996. Floral trait evolution: evidence of genetic differences among natural populations of wild radish and the role of pollinators in mediating natural selection on floral traits (Ph.D. dissertation). University of California, Santa Barbara.

Meléndez-Ackerman, E., D. R. Campbell, and N. M. Waser. 1997. Hummingbird behavior and mechanisms of selection on flower color in *Ipomopsis*. Ecology 78: 2532–2541.

Mitchell, R J. 1993. Adaptive significance of *Ipomopsis aggregata* nectar production: observation and experiment in the field. Evolution 47: 25–35.

Mitchell, R J. 1994. Effects of floral traits, pollinator visitation, and plant size on *Ipomopsis aggregata* fruit production. American Naturalist 143: 870–889.

Møller, A. P. 1995. Bumblebee preference for symmetrical flowers. Proceedings of the National Academy of Sciences, USA 92: 2288–2292.

Møller, A. P., and M Eriksson. 1995. Pollinator preference for symmetrical flowers and sexual selection in plants. Oikos 73: 15–22.

Müller, H. 1883. The Fertilisation of Flowers. Macmillan, London.

Nakamura, R. R., and M. L. Stanton. 1989. Embryo growth and seed size in *Raphanus sativus*: maternal and paternal effects *in vivo* and *in vitro*. Evolution 43: 1435–1443.

Nilsson, L. A. 1988. The evolution of flowers with deep corolla tubes. Nature 334: 147–149.

Nilsson, L. A. 1992. Orchid pollination biology. Trends in Ecology and Evolution 7: 225–259.

Paige, K. N., and T. G. Whitham. 1985. Individual and population shifts in flower color by scarlet gilia [*Ipomopsis aggregata*]: a mechanism for pollinator tracking. Science 227: 315–317.

Picker, M. D., and J. Midgley. 1996. Pollination by monkey beetles (Coleoptera: Scarabaeidae: Hopliini): flower and colour preferences. African Entomology 4: 7–14.

Pohtio, I., and I. Teräs. 1995. Bumblebee visits to different colour morphs of the Washington lupine, *Lupinus polyphyllus*. Entomologica Fennica 6: 139–151.

Schemske, D. W., and J. Ågren. 1995. Deceit pollination and selection on female flower size in *Begonia involucrata*: an experimental approach. Evolution 49: 207–214.

Schoen, D. J., and M. T. Clegg. 1985. The influence of flower color on outcrossing rate and male reproductive success in *Ipomoea purpurea*. Evolution 39: 1242–1249.

Seeley, T. D., S. Camazine, and J. Sneyd. 1991. Collective decision-making in honeybees: how colonies choose among nectar sources. Behavior, Ecology, and Sociology 28: 277–290.

Smithson, A., and M. R. Macnair. 1997a. Negative frequency-dependent selection by pollinators on artificial flowers without rewards. Evolution 51: 715–723.

Smithson, A., and M. R. Macnair. 1997b. Density-dependent and frequency-dependent selection by bumblebees *Bombus terrestris* (L.) (Hymenoptera: Apidae). Biological Journal of the Linnean Society 60: 401–417.

Stanton, M. L. 1984. Seed variation in wild radish: effect of seed size on components of seedling and adult fitness. Ecology 65: 1105–1112.

Stanton, M. L. 1985. Seed size and emergence time within a stand of wild radish (*Raphanus raphanistrum* L.): the establishment of a fitness hierarchy. Oecologia 67: 524–531.

Stanton, M L. 1987. Reproductive biology of petal color variants in wild populations of *Raphanus sativus*: I. Pollinator response to color morphs. American Journal of Botany 74: 178–187.

Stanton, M. L., J. K. Bereczky, and H. D. Hasbrouck. 1987. Pollination thoroughness and maternal yield regulation in wild radish, *Raphanus raphanistrum* (Brassicaceae). Oecologia 74: 68–76.

Stanton, M. L., and R. E. Preston. 1988. Ecological consequences and phenotypic correlates of petal size variation in wild radish, *Raphanus sativus* (Brassicaceae). American Journal of Botany 75: 528–539.

Stanton, M. L., A. A. Snow, and S. N. Handel. 1986. Floral evolution: attractiveness to pollinators increases male fitness. Science 232: 1625–1627.

Stanton, M. L., A. A. Snow, S. N. Handel, and J. Bereczky. 1989. The impact of a flower-color polymorphism on mating patterns in experimental populations of wild radish (*Raphanus raphanistrum* L.). Evolution 43: 335–346.

Stanton, M., and H. J. Young. 1994. Selecting for floral character associations in wild radish, *Raphanus sativus* L. Journal of Evolutionary Biology 7: 271–285.

Stanton, M. L., H. J. Young, N. C. Ellstrand, and J. M. Clegg. 1991. Consequences of floral variation for male and female reproduction in experimental populations of wild radish, *Raphanus sativus* L. Evolution 45: 268–280.

Thomson, J. D., and B. A. Thomson. 1992. Pollen presentation and viability schedules in animal-pollinated plants: consequences for reproductive success. Pp. 1–24 *in* R. Wyatt, ed., Ecology and Evolution of Plant Reproduction, Chapman and Hall, New York.

Waser, N. M. 1983. The adaptive nature of floral traits: ideas and evidence. Pp. 241–295 *in* L. Real, ed., Pollination Biology. Academic Press, Orlando, FL.

Weiss, M R. 1991. Floral color changes as cues for pollinators. Nature 354: 227–229.

Weiss, M. R. 1995. Floral color change: a widespread functional convergence. American Journal of Botany 82: 17–185.

Weiss, M. R., and B. B. Lamont. 1997. Floral color change and insect pollination: a dynamic relationship. Israel Journal of Plant Sciences 45: 185–199.

Willemstein, S. C. 1987. An Evolutionary Basis for Pollination Ecology. Leiden Botanical Series, vol. 10. Leiden: E. J. Brill.

Young, H. J., and M. L. Stanton. 1990. Influences of floral variation on pollen removal and seed production in wild radish. Ecology 71: 536–547.

Young, H. J., M. L. Stanton, N. C. Ellstrand, and J. M. Clegg. 1994. Temporal and spatial variation in heritability and genetic correlations among floral traits in *Raphanus sativus*, wild radish. Heredity 73: 298–308.

Zhou, S.-L., K.-Y. Pan, and K.-Y. Hong. 1996. Comparative studies on pollination biology of *Mosla hangchouensis* and *M. chinensis* (Labiatae). Acta Botanica Sinica 38: 530–540.

8

Detecting Inheritance with Inferred Relatedness in Nature

KERMIT RITLAND

In many species, the inheritance of phenotypes in the wild is difficult to demonstrate. Several obstacles stand in the way. Often it is difficult to perform artificial matings or crosses. Attributes of life history, including long generation times, mortality, and movements, can make scoring phenotypes difficult. In addition, genotype-by-environment interactions can distort estimates of genetic variation. If one is to study patterns of adaptive genetic variation for phenotypic traits, adequate demonstration of their inheritance is desirable, preferably "in nature." The purpose of this chapter is to describe how nonmanipulative methods, based on estimation of natural relatedness, might allow us to demonstrate inheritance of a trait, whether simply inherited or polygenic.

There are various manipulative ways to infer inheritance in the field that can be used for some species. An excellent approach is cross-fostering in birds (e.g., Boag and Grant 1978, van Noordwijk et al. 1980, Dhondt 1982). When offspring can be collected directly from parents as eggs or larva, another good approach is to regress laboratory-raised progeny on wild-caught parents, as done for *Drosophila* (Prout 1958, Coyne and Beecham 1987). An improvement of this method is to separately obtain estimates of additive genetic variation in the laboratory (Riska et al. 1989). However, any genotype-by-environment interaction between nature and the laboratory will distort the estimate of natural heritability (Riska et al. 1989, Groeters and Dingle 1998), and there is also considerably decreased statistical precision with the improved procedure (Ritland and Ritland 1996). With plants, studies have predominantly involved hand-planting sibships in the field (see Mitchell-Olds 1986, Shaw 1986). However, although manipulative experiments of wild populations are powerful for estimation and hypothesis testing (Mitchell-Olds and Shaw 1987), even the most careful treatments may affect the expression of traits in the field, particularly quantitative traits.

Another route for inferring heritabilities is to extrapolate from published estimates of heritabilities (e.g., Mosseau and Roff 1987) in related species or environments, or to rely on functional arguments to predict features such as genetic correlations (Endler, this volume).

However, quantitative genetic variances are dynamic (Barton and Turelli 1989), and heritabilities are often environment-specific. Extrapolation might be more valid for simply inherited traits, such as coat color polymorphisms, but such traits can have alternative genetic bases, as, for example, demonstrated with mouse coat color variants (Jackson 1994).

This chapter focuses on how molecular markers can be used to demonstrate, and possibly quantitatively estimate, the inheritance of a phenotypic trait in the field. Genetic markers are seeing an increasing spectrum of applications in evolutionary ecology (Cruzan 1998). In the 1990s, the polymerase chain reaction method created a second revolution in population genetics (isozymes causing the first, in the 1970s) by providing easily assayable, highly informative markers for a wide variety of organisms, particularly for those difficult species mentioned above, for which classical genetic approaches are impractical. In this chapter, I discuss some ways that inferred natural relationship, as measured by markers, can be used to infer adaptive patterns of variation. These ways include examining the difference of population structure between neutral genetic markers and putatively adaptive traits (Lewontin 1984, Spitze 1993) and estimating heritability using inferred relatedness (Ritland 1996a, Mousseau et al. 1998). I also introduce more recent work demonstrating simple inheritance of a coat color polymorphism in black bears.

8.1 Marker Approaches for Heritability under a Continuum of Relationships

The covariance between relatives for a quantitative trait is the basis for estimating heritability. Previous methods have assumed that the degree of relationship is known. An estimator for the case when relationship is inferred with genetic markers can be developed as follows. Consider a single population in which individuals are assayed for both genetic markers and quantitative traits, and consider all pairing of individuals (the same individual may be present in more than one pair). Let the value of a quantitative trait for the ith pair of individuals be Y_i for the first individual and Y_i' for the second, and let

$$Z_i = (Y_i - U)(Y_i' - U)/V$$

be a measure of phenotypic similarity, where U and V are the population mean and variance of Y, respectively (among all pairs, the average Z_i equals the phenotypic correlation). Taking into account the possible sharing of environments between relatives, the phenotypic similarity can be written as a linear model,

$$Z_i = 2r_i h^2 + r_e + e_i$$

where r_i is the coefficient of relationship between these individuals, r_e is the correlation of environmental effects between the individuals being compared, and e_i is random error. From this, one can derive the joint estimators for heritability and environmental correlation as

$$\hat{h}^2 = \frac{\mathrm{Cov}(Z,\hat{r})}{2\mathrm{Var}(r)} \text{ and } \hat{r}_e = \bar{Z} - \bar{r}\hat{h}^2$$

(Ritland 1996a), wherein $\mathrm{Cov}(Z,\hat{r})$ is the covariance of Z_i and r_i across all i, and $\mathrm{Var}(r)$ is the actual variance of relatedness. Actual variance of relatedness occurs when relatives of different types are intermixed (for example, full-sibs, half-sibs, and nonrelatives).

A simple modification of this estimator can also be used to find genetic correlations. For Z, one uses a different character for the second individual. Also, the sign of the genetic correlation, which is often the primary quantity of interest, is given by the sign of Cov($Z,\hat{r}$) so that actual variance of relatedness does not needed to be found. Figure 8.1 shows the result of one field study involving monkeyflowers assayed for isozymes as markers (Ritland and Ritland 1996). A positive relationship between corolla size, plant size, and estimated relatedness indicates that corolla size and plant size have a positive genetic correlation. This, and other positive genetic correlations found using this method, indicate that life history and floral size traits have positive genetic correlations, contrary to the expectations of resource allocation theories.

Figure 8.1 also shows that estimates of pairwise relatedness can have high variability, yet among a population, significant trends become apparent. Note that unbiased estimates of genetic correlations require estimation of actual variance of relatedness (see below); variance of relatedness as shown in figure 8.1 also includes estimation variance.

The heritability has the pleasing interpretation of "the (partial) regression of character similarity Z on relatedness r." It is a partial regression if the additional quantitative genetic factors, such as the environmental correlation or shared inbreeding, are also taken into account. This interpretation provides a linkage of heritability to phenotypic selection theory, wherein the phenotypic selection intensity is the regression of fitness on the character phenotype (a partial regression if several characters are included in the analysis; Lande and Arnold 1983).

The variance of $\hat{h}^2$ is determined by three primary factors: number of individuals sampled, number of marker loci assayed, and the actual variance of relatedness. However, few marker loci are needed to get adequate estimates of pairwise relatedness for the purpose of heritability estimation; generally, $n(m-1)$ should be between 25 and 100, for n loci and m alleles per locus. One could do well with a dozen triallelic isozyme loci, with 4 SSR (microsatellite) loci each with 12 alleles, or with other combinations. Beyond this number of markers, it is generally better to sample more individuals rather than more marker loci.

The third factor determining our power to infer heritability in the field is the actual variance of relatedness (e.g., the extent that different types of relatives are present). Actual variance of relatedness can be quite low and an impediment for inferring natural heritability with markers (Ritland and Ritland 1996). Although mean relatedness may be present in most populations, actual variance of relatedness may be weak, particularly in species with wide dispersal of progeny. This quantity can be intensified by using appropriate spatial sampling designs (Ritland 1996a). In contrast, philopatric species are more amenable; for example, in many Old World monkey groups (especially macaques and baboons), social groups are based on maternal lineages, so when a group is trapped, a relatively high density of relatives is obtained (J. M. Cheverud, personal communication). Another example includes the cactophilic *Drosophila*, in which colonies are founded by few gravid females, resulting in a good family structure.

The pairwise approach described here resembles an approach for heritability estimation termed "symmetric differences squared" (Grimes and Harvey 1980), which allows all levels of genetic relationships (assumed known) to be included in a single analysis. In this procedure, additive genetic variance is estimated by regressing $2(1-r_i)$ onto the squared differences among all pairs of individuals. This model was expanded in many different ways to include double co-ancestry, environmental relatedness, and so on (see Bruckner and Slanger 1986). Since then, attention has shifted to the maximum likelihood method, which can more efficiently incorporate a matrix of arbitrary relationships (e.g., Shaw 1987, 1991). However, when relationship is estimated with markers, the uncertainty that is introduced makes this

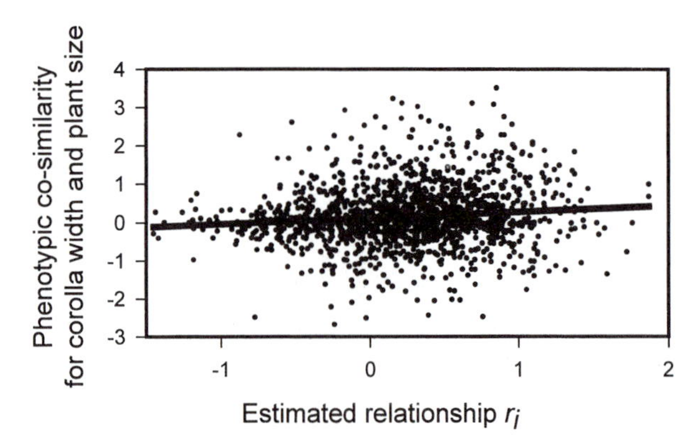

Figure 8.1 Illustration of how estimated relatedness can be used to determine the sign of the genetic correlation (in this case, positive).

matrix unstable. Further research is needed on extending the maximum likelihood method to the case of inferred relationship (T. Meagher, personal communication).

Currently, the most frequent application of markers to quantitative traits is the mapping and characterization of individual quantitative trait loci (QTLs). These methods all assume known pedigrees. The mapping of QTLs with material collected from wild populations, where pedigree is unknown, is an interesting prospect, and seems most possible using concepts borrowed from allele-sharing methods (Lander and Schork 1994). Another prospect for marker-aided inference is for those cases where inferences about heritability are made when maternity is known but paternity is not (see Konigsberg and Cheverud 1992).

8.2 Marker-Estimated Relatedness

The estimation of relationship has most often been used in studies of within-group relatedness (Reynolds et al. 1983, Queller and Goodnight 1989). However, previous methods of estimation did not behave well for pairwise relationships, where the sample size (number of marker loci) is inherently small (Weir 1996). Lynch (1988) first proposed a method for estimating pairwise relatedness based on band sharing of DNA fingerprints, where the homology of alleles is unknown. Ritland (1996b) devised a similar estimator of pairwise relatedness, which incorporates gene frequencies, and has asymptotic (large sample size) behavior with any number of marker loci. The estimator can be described as follows. Assume that at locus m there are n_m alleles, and let p_{im} be the frequency of allele i at locus m in the population. If the first relative has marker alleles $A_{im}A_{jm}$, and the second relative has marker alleles $A_{km}A_{lm}$, this estimator of relatedness is

$$\hat{r}_m = \frac{[(\delta_{ikm} + \delta_{ilm})/p_{im}] + [(\delta_{jkm} + \delta_{jlm})/p_{jm}] - 1}{4(n_m - 1)}$$

where the δs are indicator variables, e.g., $\delta_{ikm} = 1$ if $i = k$ and $\delta_{ikm} = 0$ if $i \neq k$. A multilocus estimate of relatedness is the weighted average across loci,

$$\hat{r} - \sum_m (n_m - 1)\hat{r}_m / \sum_m (n_m - 1)$$

(this is the same as eq. 5 of Ritland 1996b). A similar estimator, which is simpler and appropriate when true relatedness is higher, is given by Lynch and Ritland (in press). Both estimators assume that gene frequency is estimated from a larger population, whose size should generally be at least 50.

A formula for estimating the actual variance of relatedness was given in Ritland (1996a). The principle behind this formula is that estimates of relatedness are independent among loci; hence their cross product gives an unbiased estimated of the expected squared relatedness, from which the actual variance is obtained by subtracting the squared mean relatedness. Note that at least two loci are needed to estimate variance of actual relatedness.

8.3 Micropopulation Structure

With the advent of simple sequence repeat markers (SSR microsatellite), the characterization of local population structure is becoming easier and more powerful (e.g., Blouin et al. 1996, Bourke et al. 1997, Peacock and Smith 1997, Taylor et al. 1997). A byproduct of inferring heritabilities with markers is the attention it requires regarding local family structure, or micropopulation structure. For the ultimate purpose of characterizing heritability, we are currently surveying micropopulation structure in four British Columbia conifer species (western and mountain hemlock, silver and subalpine fir) with isozyme markers. Previous studies with conifers have indicated the presence of clustering of relatives in some conifer species (see Adams 1992). In our study, among 20 transects involving at least 100 trees, we have found highly variable levels of fine-scale structuring (this stresses the importance of preliminary screens of populations for microstructure before proceeding with intensive sampling for heritability estimates). Significant levels of relatedness between near neighbors, and its decline with distance, was apparent for at least 8 of the 20 populations, with trees within 30–50 m showing between half-sib to first cousin relationships (Travis and Ritland, unpublished data).

For the purposes of predicting population structure (to avoid marker assay of populations with little structure), it would be useful to identify ecological correlates of micropopulation structure. Two ecological correlates of microstructure we observed for conifers were patchy distributions of individuals (expected with recent colonization) and wind-protected areas (expected with wind-aided seed dispersal). In monkeyflowers, Ritland and Ritland (1996) found average relatedness to be 0.125 (SE 0.018) in a meadow habitat and 0.052 (SE 0.014) in a stream-side habitat. The actual variance of relatedness was estimated to be 0.033 (SE 0.007) for the meadow and 0.012 (SE 0.005) for the stream. Thus, probably because of greater water-mediated dispersal, streams showed lower mean and variance of relatedness compared to the meadow (Ritland and Ritland 1996).

8.4 Natural Heritabilities

In the study of Ritland and Ritland (1996), natural heritabilities were inferred in monkeyflowers using the method described above. Natural heritabilities were quite large (table 8.1), larger than those found in other studies of natural populations (see Mousseau and Roff

1987), and larger than that found by Carr and Fenster (1994) in greenhouse-raised monkeyflowers. We also found that heritability estimates based on regressing laboratory-reared offspring on wild-collected parents were substantially lower than the marker-based estimates (table 8.1; for formula, see Hoffman this volume). This was particularly dramatic for the two fitness characters, number of flowers per plant and plant size. Both Prout (1958) and Coyne and Beecham (1987) also found field-measured heritabilities in *Drosophila* to be larger than laboratory-measured heritabilities.

The lower heritabilities derived by these nonmarker methods is probably due to genotype-by-environment interactions between the field and the laboratory. Heritability in the laboratory is probably greater than in nature (as environmental conditions are more uniform), but the genetic correlation between nature and laboratory is less than one, and probably is the main reason that nonmarker heritability estimates are lower than field values. However, marker-based estimates of heritability in nature may be biased upward due to undetectable violations of assumptions such as lack of dominance variance or shared levels of inbreeding depression (see Ritland and Ritland 1996).

8.5 Partitioning of Heritability, Q_{st}, and Adaptive Variation

A feature of natural heritability not found in experimental (controlled) populations is the potential for a hierarchy of relatedness and quantitative variation. Like the partitioning of Wright's F into within (F_{is}) and between (F_{ST}) components, pairwise relatedness can be partitioned into local (r_w) and regional (r_b) components, and heritability can be partitioned into local (h_w^2) and regional (h_b^2) components. "Regional" is interpreted either as among sites within populations or as among populations. In this light, the phenotypic similarity between individuals might be modeled as

$$Z_i = 2r_{w,i}h_w^2 + 2r_{b,i}h_b^2 + r_e + e_i$$

(ignoring the potential partitioning of shared environments). The point of emphasis here is that relatedness and heritability at a regional scale would have different values than at the local scale due to different patterns of drift, migration, and selection. In particular, the among-site heritability, h_b^2, is much more likely affected by selection and may reflect adaptive genetic differentiation. Estimation of these components in natural populations has not been attempted. At least, we might consider that variation among marker-based estimates in table 8.1 may be due to variation of h_b^2, which in turn is due to the presence of selective differentiation.

At the regional scale, the relationship between marker variation and quantitative genetic variation has been used to infer selection. In his early work on isolation-by-distance, Wright (1943) introduced a quantity that measured differentiation for a quantitative trait, $\sigma_h^2/(\sigma_b^2 + 2\sigma_w^2)$, where the variances refer to within- and between-population components of genetic variance. This was subsequently termed "Q_{st}" by Spitze (1993), and it is expected to equal F_{ST} under a neutral model of evolution by drift and mutation. If selection differentiates populations with respect to the quantitative trait, Q_{st} is expected to be greater than F_{ST}. Several studies have documented excess values of Q_{st} relative to F_{ST}, indicating selective differentiation (see Podolsky and Holtsford 1995). This discrepancy has also been used to argue for the irrelevance of genetic markers for measuring genetic variation in conservation genetic applications (Karhu et al. 1996).

Table 8.1 A comparison of estimates of heritabilities in the field (meadow habitat) in the monkeyflower study (Ritland and Ritland 1996), using marker method of Ritland, regression of laboratory–grown progeny on field collected parents, and the Riska method

Trait	Markers		Regression		Riska method	
Corolla width	0.73**	(0.57)[a]	0.34*	(0.16)	0.16	(3.64)
Calyx width	0.53*	(0.54)	0.10	(0.17)	0.01	(0.11)
Stigma–anther separation	0.22	(0.48)	0.12	(0.18)	0.02	(0.05)
Corolla shape	0.32	(0.51)	0.20*	(0.09)	-0.26	(2.62)
Calyx shape	0.65	(0.70)	0.23*	(0.15)	-0.54	(1.09)
No. of flowers	0.78**	(0.59)	0.05	(0.10)	-0.02	(2.22)
Plant weight	0.63**	(0.49)	0.04	(0.11)	0.02	(0.79)

[a] Standard errors in parentheses.

*$P<.05$ based on bootstrap percentile.

However, selection on marker loci, should it exist, is more likely to be detectable at larger scales of divergence. This can be a problem with allozymes (Prout and Barker 1993), but for noncoding loci such as SSRs and anonymous restriction fragment length polymorphisms, marker-based F_{ST} would be close to neutral expectation. The remaining source of bias for marker-based F_{ST} is actual variance of F_{ST}, due to sampling variations among loci, which is particularly pronounced when populations go through bottlenecks. If population sizes are constant, actual variance of F_{ST} is quite small, and markers should give reliable measures of expected divergences.

8.6 Marker-based Inference of Simply Inherited Traits

Examples abound of apparent conspicuous polymorphisms in nature for morphological traits, particularly coat colors, for example, in beetles (Bernstein and Berstein 1996) and plants (Wolfe 1993). Because of their conspicuous nature, these polymorphisms are likely adaptive. However, rarely is the mode of inheritance directly demonstrated, for reasons mentioned above. Knowledge of their inheritance rules out the possibility that these polymorphisms are due to phenotypic plasticity and enables estimation of gene frequencies underlying the trait and, ultimately, the strength of selection affecting the polymorphism.

We have recently started a study of a coat color polymorphism in black bears. The "Kermode" bear is the white color phase of the black bear, occurring in a restricted area of the north coast of British Columbia (strictly, "Kermode" refers to a subspecies of the black bear which occurs throughout coastal British Columbia; see Byun et al. 1997). On Princess Royal and Gribbel Islands, Kermode bears are at highest frequency (about 10%). The adaptive significance of the coat color polymorphism is unknown, but possible reasons include founding events following deglaciation 10,000 years ago, genetic drift on small island populations since founding, advantages for thermoregulation or crypsis in glacial refugia (relict "glacial bear"), or relaxed hunting pressure by humans (the Kermode bear is revered by First Nations people). Regardless of the putative role of selection, demonstration of the inheritance of a polymorphism is a prerequisite for any adaptive storytelling. The white coat color does not represent the classic albino condition, so a normal recessive mode of inheritance should not be assumed. Besides shedding light on the adaptive significance of this polymorphism, information on inheritance might also aid in predicting the possible effects of forest practices on the frequency of Kermodism.

Currently, we are taking two approaches to determine the genetic basis of this polymorphism. The first is to survey candidate genes for mutations at the DNA level that correlate with coat colors. Studies with mice coat color (Jackson 1994) indicate closest similarity of the Kermode coat color with mutations at the "*p*" gene in mice. Sequence variation at the mutation might allow us to infer its age and its selective maintenance (similar to the classical *ADH* work with *Drosophila*; see Kreitman and Aguade 1986). A second approach is to infer pedigrees in mixed populations with SSRs, and then to fit models of coat color inheritance to the pedigree. In bears and other large mammals, DNA can be extracted from hairs, so that expensive and laborious trapping is not needed. Hairs can be caught in barb-wire traps or opportunistically collected from rub trees, stream banks, and so on. Paetkau et al. (1995) have developed assays for several bear SSR loci which can be used to infer relationships using hair samples.

During August and September 1997, as part of a larger study of the Kermode bear (conducted by Dawn Marshall in collaboration with John Barker of Western Forest Products), nearly 900 bear hair samples were collected from 5 islands and the adjacent mainland of northern coastal British Columbia. Assay for eight SSR loci using Paetkau's primers revealed 89 distinct genotypes in regions where Kermodism is significantly frequent [17 on Gribbel Island, 13 on Hawksbury Island, 38 on Princess Royal Island, and 21 at Terrace (mainland British Columbia)]. Because bear hair color can be reliably scored from even a single hair, this study provides an excellent opportunity to use marker-inferred relatedness to demonstrate inheritance of a discrete phenotypic trait.

The use of marker-inferred relatedness to study simply inherited traits can be developed as follows. Given marker data of two individuals of unknown relationship, the likelihood of a given relationship is (modified after Jacquard 1974):

$$\Pr(A_iA_j, A_kA_l) = p_ip_j(2-\delta_{ij})\,(2-\delta_{kl})[\Delta_9 p_kp_l + \Delta_8((\delta_{ik}+\delta_{jk})p_l \\ + (\delta_{il} + \delta_{jl})p_k)/4 + \Delta_7(\delta_{ik}\delta_{jl} + \delta_{il}\delta_{jk})/2]$$

where A_iA_j and A_kA_l are the genotypes of the two individuals at a single locus (subscripts may or may not be the same integer; e.g., may or may not indicate the same allele), and δ's are indicator variables, e.g., $\delta_{ij} = 1$ if $i = j$, and $\delta_{ij} = 0$ otherwise. The triplet of relationship coefficients ($\Delta_7, \Delta_8, \Delta_9$) are the probabilities of identity-by-descent, for respectively (1) both pairs of genes, (2) one pair of genes, and (3) no genes. They take the values of (1,0,0) for identical twins, (0,1,0) for parent–offspring, (¼, ½, ¼) for full-sibs, (0, ½ ½) for half-sibs, and (0, ¼ ¾) for first cousins (see Jacquard 1974).

At this step, it should be noted that we are now using inferred relationship (as opposed to relatedness) to demonstrate inheritance. Instead of estimating relatedness, *r*, we infer the type of relationship (full sibs, half-sibs, etc.). The problem here is that there are many possible types of relationships, and there is a limit to how many types of relatives the data can distinguish. Thompson (1986) suggested that the major types of relationships can be categorized as family (parent–offspring, full-sibs), close (half-sibs, uncle, etc.), remote (cousin, etc.) and unrelated.

Given some small number of possible relationships, the general likelihood model for inferring inheritance using markers is the mixture model

$$\sum_k \Pr(M_i|R_k)\Pr(S_i|R_k)\Pr(R_k)$$

where $\Pr(M_i|R_k)$ is the probability of markers M_i observed for pair i conditioned their relationship k (full-sibs, half-sibs, etc.), $\Pr(S_i|R_k)$ is the probability of the phenotypic similarity between the two individuals, and $\Pr(R_k)$ is the prior probability of relationship k. The total likelihood is the product over all pairs i, and the $\Pr(r_k)$ are estimated using the information from all pairs. For a continuous trait, $\Pr(S_i|R_k) = \exp[(-Z_i - .5r_k h^2)^2/2]$ where r_k is the coefficient of relatedness for relationship k. This formula was used to estimate heritability of jacking in chinook salmon (Mousseau et al. 1998).

In the case of the discrete trait of coat color, $\Pr(S_i|R_k)$ indexes the probabilities that two phenotypes are either both white, both black, or one black and one white. Unlike the continuous trait case, this quantity also depends on the mode of inheritance (e.g., single-gene recessive versus single-gene dominant, degree of penetrance). The mode of inheritance can be inferred by comparing the likelihoods of the entire data between alternative modes of inheritance (Ritland and Marshall, in preparation). Assuming a single-gene recessive mode of inheritance, we have computed the log-likelihood of data under single-gene recessive as -1476.3; the log-likelihood of the data assuming to inheritance was -1486.1. This difference is highly significant and clearly demonstrates inheritance of the coat color.

Single-gene recessive inheritance can be conclusively demonstrated by performing a goodness-of-fit test of the model expectations to the observed data. Table 8.2 shows for each of four categories of relatives (full-sibs, half-sibs, first cousins, unrelated), the observed (as probabilistically inferred by markers) and expected (as predicted by the above model, given the estimates) pattern of coat-color sharing among pairs (the gene frequency, q, of the recessive allele was estimated as 0.34). Although the data fit fairly well, the chi-square test statistic is significant (χ^2, = 46.1, 7 df, $P < .001$), mainly due to a conspicuous excess of white full-sibs (contributing 22.1 to the 46.1 value; cf. table 8.2). This can be attributed to assortative mating (Ritland and Marshall, in preparation), which is of importance in the preservation of the phenotypic frequencies, as assortative mating increases the frequency of a recessive phenotype above Hardy-Weinberg expectations.

Recent attempts to accurately identify relationships with hypervariable markers have encountered mixed success. Blouin et al. (1996) used actual allele frequencies at 20 SSRs in a population of wild mice to test the ability of this method to identify family groups. By clustering allele-matching scores, they were able to reconstruct four sets of independent maternal half sibships perfectly. However, Taylor et al. (1997) found that nine SSRs in wombats were not sufficient for parentage (pedigree) analysis and cited the problems of low genetic variation, incomplete sampling of potential parents, and paucity of concomitant observations that would suggest putative relationships. Our data indicate good success at identifying relationships; calculations of the power to exclude relationships given the polymorphism we found at the eight SSR loci indicate >95% exclusion probabilities, with the exception of first cousin versus unrelated, where resolution is poor.

I also evaluated the power, given the existing SSR variation, to infer a single-gene, recessive mode of inheritance, assuming a recessive gene frequency of 0.3, the published SSR allele frequencies of Paetkau et al. (1995), and a sample 10% full-sibs, 10% parent–offspring, 10% half-sibs, and 70% unrelated. Figure 8.2 gives the probability of accepting the model (over the alternative of no inheritance) as a function of the number of pairs sampled and the number of bear SSR loci assayed. Figure 8.2 clearly shows that the number of pairs sampled,

Table 8.2 Patterns of shared coat color in the Kermode bear, by class of inferred relationship (expected values in parentheses, assuming random mating)

Type of pair	Full sibs		Half sibs		First cousins		Unrelated	
Both white	13.4	(4.0)	2.1	(1.1)	6.5	(4.1)	20.0	(11.4)
Mixed	6.4	(17.2)	8.5	(10.3)	34.0	(41.0)	107.2	(134.8)
Both black	38.2	(36.7)	18.1	(17.3)	97.4	(92.0)	418.3	(399.4)
Total	58		28.7		137.9		545.5	

The inferred patterns are consistent with those expected under single-gene recessive inheritance of coat color, except an excess of white full-sibs, which can be explained by assortative mating.

and not the number of marker loci assayed, is the primary limiting factor for inferring inheritance. In other words, the phenotypic segregation outweighs the uncertainty of relatedness estimates, as was stressed earlier for the heritability estimator. Another field season of sampling, rather than another eight SSR loci, will best improve estimates.

8.7 Conclusions

I have outlined several approaches by which inheritance in nature can be demonstrated. The application of markers to such problems should be a fruitful area of research in the near future. In particular, linking inferred relatedness to adaptive genetic variation should be particularly challenging and innovative. Expertise with marker assays, or with statistical genetics, is not a requirement, as many service laboratories and computer programs are available for these tasks. The real issue is adopting the best statistical model that balances complexity (realism) and simplicity (statistical power). Doing this often requires insight to both the genetics and natural history of the organism.

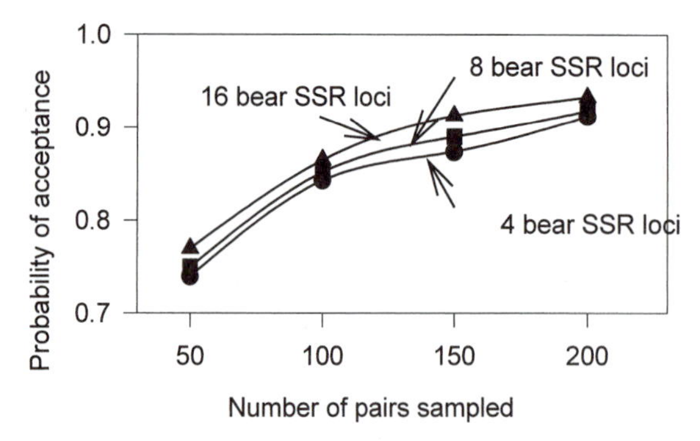

Figure 8.2 Power to accept a single-gene recessive model of inheritance as a function of number of pairs sampled and number of SSR loci assayed, assuming a certain pattern of relationships and using published allele frequencies for bear SSR loci (see text).

References

Adams W.T., 1992. Gene dispersal within forest tree populations. New Forests 6: 217–240.

Barton N.H., and M. Turelli, 1989. Evolutionary quantitative genetics: how little do we know? Annu. Rev. Genet. 23: 337–370.

Bernstein, R., and S. Bernstein, 1996. Color polymorphism and body size in males and females of *Chauliognathus basalis* (LeConte) (Coleoptera: Cantharidae). Coleopt. Bull. 50: 297–300.

Blouin M.S., M. Parsons, V. Lacaille, and S. Lotz, 1996. Use of microsatellite loci to classify individuals by relatedness. Mole. Ecol. 5: 393–401.

Boag, P.T., and P.R. Grant, 1978. Heritability of external morphology in Darwin's finches. Nature 274: 793–794.

Bourke A.F., H.A. Green, and M.A. Bruford, 1997. Parentage, reproductive skew and queen turnover in a multiple-queen ant analysed with microsatellites. Proc. R. Soc. Lond. B 264: 277–283.

Bruckner C.M., and W.D. Slanger, 1986. Symmetric differences squared and analysis of variance procedures for estimation of genetic variances and covariances for beef cattle weaning weight. 1. Comparison via simulation. J. Anim. Sci. 63:1779–1793.

Byun, S.A., B.K. Koop, and T.E. Reimchen. 1997. North Americal black bear mtDNA phylogeography: implications for morphology and the Haida Gwaii glacial refugium controversy. Evolution 51: 1647–1653.

Carr, D.E., and C.B. Fenster, 1994. Levels of genetic variation and covariation for *Mimulus* (Scrophulariaceae) floral traits. Heredity 72: 606–618.

Coyne, J.A., and E. Beecham, 1987. Heritability of two morphological characters within and among natural populations of *Drosophila melanogaster*. Genetics 117: 727–737.

Cruzan, M.B., 1998. Genetic markers in plant evolutionary biology. Ecology 79: 400–412.

Dhondt, A.A., 1982. Heritability of blue tit tarsus length from normal and cross-fostered broods. Evolution 36: 418–419.

Grimes, L.W., and W.R. Harvey, 1980. Estimation of genetic variances and covariances using symmetric differences squared. J. Anim. Sci. 50:634–644.

Groeters, F.R., and H. Dingle, 1998. Heritability of wing length in nature for the milkweed bug, *Oncopeltus fasiatus*. Evolution 50: 442–447.

Jackson, I.J., 1994. Molecular and developmental genetics of mouse coat color. Annu. Rev. Genet. 28: 189–217.

Jacquard, A., 1974. The Genetic Structure of Populations. Springer-Verlag, New York.

Karhu, A., P. Hurme, M. Karjalainen, P. Karvonen, K. Karkkainen, D. Neale, and O. Savolainen, 1996. Do molecular markers reflect patterns of differentiation in adaptive traits of conifers? Theor. Appl. Genet. 93:215–221.

Konigsberg, L.W., and J.M. Cheverud, 1992. Uncertain paternity in primate quantitative genetic studies. Am. J. Primatol. 27: 133–143.

Kreitman, M.E., and M. Aguade, 1986. Excess polymorphism at the *Adh* locus in *Drosophila melanogaster*. Genetics 114: 93–110.

Lande, R., and S. J. Arnold, 1983. The measurement of selection on correlated characters. Evolution 37: 1210–1226.

Lander, E.S., and N.J. Shork, 1994. Genetic dissection of complex traits. Science 265: 2037–2048.

Lewontin, R. C. 1984. Detecting population differences in quantitative characters as opposed to gene frequency differences. Am. Nat. 123: 115–124.

Lynch, M., 1988. Estimation of relatedness by DNA fingerprinting. Mole. Biol. Evol. 5: 584–599.

Lynch, M., and K. Ritland. 1999. Estimation of pairwise relatedness with molecular markers. Genetics, in press.

Mitchell-Olds, T., 1986. Quantitative genetics of survival and growth in *Impatiens capensis*. Evolution 40: 107–116.

Mitchell-Olds, T., and R.G. Shaw, 1987. Regression analysis of natural selection: statistical inference and biological interpretation. Evolution 41: 1149–1161.

Mousseau, T.A., K. Ritland, and D.D. Heath. 1998. A novel method for estimating heritability using molecular markers. Heredity 80: 218–224.

Mousseau, T.A., and D.A. Roff, 1987. Natural selection and the heritability of fitness components. Heredity 59: 181–197.

Paetkau, D., W. Calvert, I. Stirling, and Strobeck, C. 1995. Microsatellite analysis of population structure in Canadian polar bears. Mole. Ecol. 4: 347–354.

Peacock M.M., and A.T. Smith, 1997. Nonrandom mating in pikas *Ochotona princeps*: evidence for inbreeding between individuals of intermediate relatedness. Mole. Ecol. 6: 801–811.

Podolsky, R.H., and T.P. Holtsford, 1995. Population structure of morphological traits in *Clarkia dudleyana* I. Comparison of F_{ST} between allozymes and morphological traits. Genetics 140: 733–744.

Prout, T., 1958. A possible difference in genetic variance between wild and laboratory populations. Drosophila Information Service 32: 148–149.

Prout, T., and J.S.F. Barker, 1993. *F* statistics in *Drosophila buzzatii*: selection, population size and inbreeding. Genetics 134: 369–375.

Queller, D.C., and K.F. Goodnight, 1989. Estimating relatedness using genetic markers. Evolution 43: 258–275.

Reynolds, J., B.S. Weir, and C.C. Cockerham, 1983. Estimation of the coancestry coefficient: basis for a short-term genetic distance. Genetics 105:767–779.

Riska, B., T. Prout, and M. Turelli, 1989. Laboratory estimates of heritabilities and genetic correlations in nature. Genetics 123: 865–871.

Ritland, K., 1996a. A marker-based method for inferences about quantitative inheritance in natural populations. Evolution 50: 1062–1073.

Ritland, K., 1996b. Estimators for pairwise relatedness and individual inbreeding coefficients. Genet. Res. 67: 175–185.

Ritland, K., and C. Ritland, 1996. Inferences about quantitative inheritance based upon natural population structure in the common monkeyflower, *Mimulus guttatus*. Evolution 50: 1074–1082.

Shaw, R.G., 1986. Response to density in a natural population of the perennial herb *Salvia lyrata*: variation among families. Evolution 40: 492–505.

Shaw, R.G., 1987. Maximum likelihood approaches applied to quantitative genetics of natural populations. Evolution 41: 812–826.

Shaw, R.G., 1991. The comparison of quantitative genetic parameters between populations. Evolution 45: 143–151.

Spitze, K., 1993. Population structure and *Daphnia obtusa*: quantitative genetic and allozymic variation. Genetics 135: 367–374.

Taylor A.C., A. Horsup, C.N. Johnson, P. Sunnucks, and B. Sherwin, 1997. Relatedness structure detected by microsatellite analysis and attempted pedigree reconstruction in an endangered marsupial, the northern hairy-nosed wombat *Lasiorhinus krefftii*. Mole. Ecol. 6:9–19.

Thompson, E.A. 1986. Pedigree Analysis in Human Genetics. The Johns Hopkins University Press, Baltimore, MD.

van Noordwijk, A. J, J.H. Van Belan, and W. Scharloo, 1980. Heritability of body size in a natural population of the great tit (*Parus major*) and its relation to age and environmental conditions during growth. Genet. Res. 51: 149–162.

Weir, B., 1996. Genetic Data Analysis II. Sinauer Associates, Sunderland, MA.

Wolfe, L.M., 1993. Reproductive consequences of a flower color polymorphism in *Hydrophyllum appendiculatum*. Am. Midl. Nat. 129: 405–408.

Wright, S., 1943. An analysis of local variability of flower color in *Linanthus parryae*. Genetics 28: 139–156.

9

Laboratory and Field Heritabilities

Some Lessons from Drosophila

ARY A. HOFFMANN

The narrow-sense heritability of a trait is defined as the proportion of the phenotypic variance in the trait that is passed on to the next generation. If the phenotypic variance is given by V_P, and the additive genetic variance by V_A, then the narrow-sense heritability is defined as V_A/V_P. This parameter provides an indication of how much of the variability in a trait among individuals is genetic. It also provides an indication of how quickly a trait will respond to selection, and it is widely used to model microevolutionary change and processes maintaining variability in natural populations.

In a widely cited paper, Roff and Mousseau (1987) summarized estimates of narrow-sense heritabilities (hereafter referred to simply as "heritability") for traits in *Drosophila* falling into four categories (morphology, life-history, physiology, and behavior) from numerous laboratory studies. Their main conclusion was that life-history and behavioral traits had low heritabilities, whereas morphological and physiological traits had high heritabilities. A related survey of estimates from other organisms found the same trends (Mousseau and Roff 1987).

These patterns may also occur under field conditions. Weigensberg and Roff (1996) examined field heritabilities estimated for morphological traits from a range of organisms and found that these were similar in magnitude to estimates from the laboratory. In addition, they found that some field estimates for life-history traits were low relative to morphological traits, in agreement with the laboratory pattern for these classes of traits. However, there were few estimates for trait classes apart from morphology.

Such comparisons of heritabilities among trait classes have been used to make inferences about the effects of selection acting on them. If a trait is under intense selection and if alleles influencing it act in an additive manner, then selection is expected to favor those alleles that have a higher fitness. As the favored alleles go to fixation, the additive genetic variance of the trait is expected to decline. This will lead to a concomitant decrease in heritability if the environmental variance remains constant. Therefore, different classes of traits may have different heritabilities depending on how closely they are related to fitness. An alternative hypothesis with the same prediction (Price and Schluter 1991) is that variability in one class of traits (in particular, life-history) might be partly determined by variability in a second class of traits (morphology). Assume that both classes of traits are associated with some degree of environmental variance, V_E. Under this assumption, life-history traits will be exposed to two sources of environmental variance (from the morphological trait and the other underlying traits). This can lead to a relatively lower heritability for this trait class compared to morphological traits exposed to one source of environmental variance.

In this chapter, I take another look at the patterns of heritabilities in *Drosophila* described by Roff, Mousseau, and others. I address four main questions. First, are the laboratory patterns consistent with recent laboratory studies that have markedly increased the number of traits available? My comparisons of classes focus only on trait means (i.e., one estimate for each trait). This contrasts with Roff and Mousseau (1987), who considered patterns from all estimates in all studies and from the median estimates of each study. The reason I considered only one estimate per trait is straightforward. If inferences are being made about classes of traits, comparisons of classes should involve random samples of traits falling in each class. Although it is difficult to collect a random sample of traits, it seems more appropriate to weight all traits equally, rather than to weight studies or the number of estimates. I have therefore averaged all estimates for particular traits. Most studies used by Roff and Mousseau were included for obtaining averages, as well as more recent studies. Some early studies have been excluded because publications were unavailable or because the studies involved strain comparisons. Estimates based on the latter may be confounded by effects of deleterious alleles or dominance effects, preventing an accurate estimate of the narrow-sense heritability.

Second, are comparisons across classes confounded by differences in how accurately traits are measured? If trait classes are measured with different amounts of error, this needs to be considered when comparing them because a high degree of error can lead to heritability estimates that are too low (Henderson 1990). Measurement error may be particularly important for behavioral traits because high levels of error normally occur when an organism is measured in a behavioral assay. In addition, measurement error associated with the experimenter and apparatus needs to be separated from the effects of conditions experienced by the organism when expressing a trait in the field.

Third, do field heritabilities match laboratory heritabilities, and do they differ between trait classes? There are two problems in making these comparisons. The first is that the number of field heritability estimates for traits in *Drosophila* is still small. Several studies have considered morphological traits, but there is little information on other trait classes. The second problem is that the approach often used (regression of laboratory-reared offspring onto field-reared parents) may only provide an approximate estimate of the field heritability (Riska et al. 1989). The regression of laboratory offspring onto the midparent from nature is given by

$$\beta(o_L, p_N) = \frac{\gamma \sigma_{AL} \sigma_{AN}}{\sigma^2_{PN}} \quad (9.1)$$

where γ is the additive genetic correlation between the trait in nature and the same trait in the laboratory, σ^2_{AL} and σ^2_{AN} are the additive genetic variances in the laboratory and nature respectively, and σ^2_{PN} is the phenotypic variance in nature. Heritability estimates obtained from this equation will only equal the true field heritability ($\sigma^2_{AN}/\sigma^2_{PN}$) if $\gamma = 1$ (no genotype–environment interactions) and if the additive genetic variance is the same in the field and the laboratory. However, a lower-bound estimate can be obtained (Riska et al. 1989) from

$$\beta^2_{(O_L,P_N)}\left(\frac{\sigma^2_{PN}}{\sigma^2_{AL}}\right) = \gamma^2 h^2_N \leq h^2_N \qquad (9.2)$$

This equation describes the way the regression coefficient, the additive genetic variance in the laboratory and phenotypic variance in nature ($\sigma^2_{AL}, \sigma^2_{PN}$), and the genetic correlation between environments (γ) relate to heritability in nature when measurements on both field parents are available. Whether this is an accurate reflection of field heritability depends on how close the squared genetic correlation between laboratory and field environments (γ^2) is to 1.

So can the true heritability in *Drosophila* ever be estimated for such comparisons? If equations 9.1 and 9.2 are compared, they yield the same result when $\gamma = \gamma^2 = 1$ and when $\sigma^2_{AL} = \sigma^2_{AN}$. However, similar values can also arise when the cross-environment correlation is < 1. In this case, because $\gamma > \gamma^2$ and the minimum estimate is less than the regression estimate, similar values require σ_{AL} to be less than σ_{AN}. Nevertheless, if estimates based on equation 9.1 turn out to be generally similar to those based on equation 9.2, then both approaches may estimate field heritability, and laboratory comparisons may not be needed to compute minimum field estimates.

In this chapter I review all published *Drosophila* estimates involving field and laboratory generations (which will be referred to as "field–lab estimates") and some recent data from my laboratory. The relationship between heritability estimates computed with equations 9.1 and 9.2 are compared for data from field flies. I also compare heritability estimates in the laboratory with those obtained from field parents and consider the reasons estimates may (or may not) differ between the environments.

9.1 Heritabilties for Trait Classes

9.1.1 Behavioral traits

Roff and Mousseau (1987) found that the overall laboratory heritability for behavioral traits was 0.18, based on averaging the median value of different studies. This value was largely based on four traits: locomotion, mating activity, geotaxis, and phototaxis.

I have considered nine additional studies covering courtship signal traits (Aspi and Hoikkala 1993, Ritchie and Kyriacou 1994, 1996), temperature preference (Yamamoto 1994), mating speed (Singh and Chatterjee 1988, Stamenkovic-Radak et al. 1992), pupation height (Singh and Pandey 1993), egg insertion behavior (Kamping and Van Delden 1990), licking behavior (Welbergen and Van Dijken 1992), and copulation duration (Gromko et al. 1991). These studies cover 16 traits additional to those considered by Roff and Mousseau (1987), more than doubling the trait number considered previously. To obtain mean trait values, all estimates were first averaged for individual studies and then across studies.

Means for the individual traits are given in figure 9.1. The overall mean across the traits (0.18) is identical to the value given by Roff and Mousseau. There are nevertheless substantial differences in heritability estimates of the traits. For instance, estimates for some courtship components are high (up to 0.80), whereas estimates for other components are effectively zero (Aspi and Hoikkala 1993). In addition, the range of estimates within individual traits is high. For instance, estimates for phototaxis vary from 0.02 to 0.66.

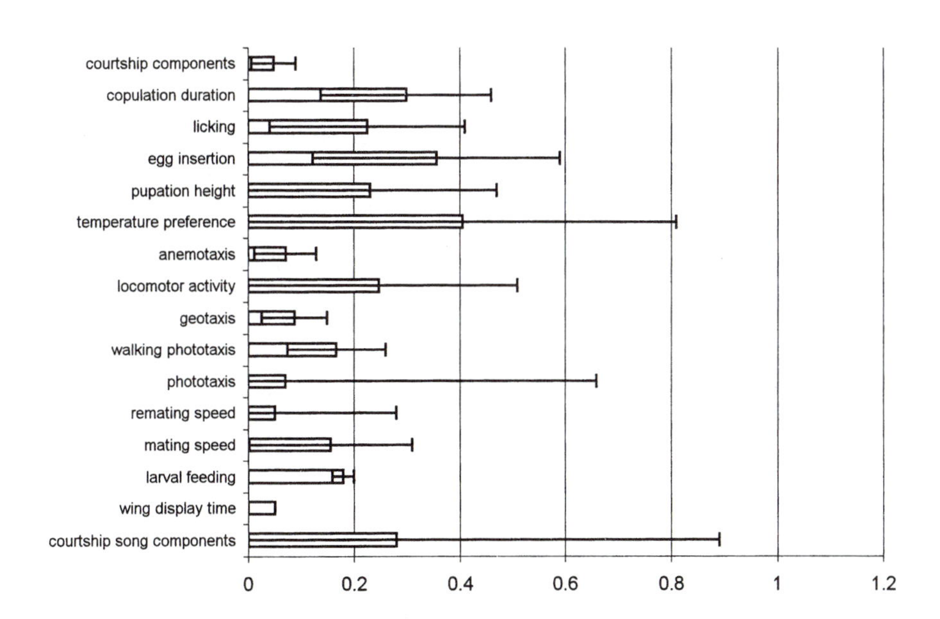

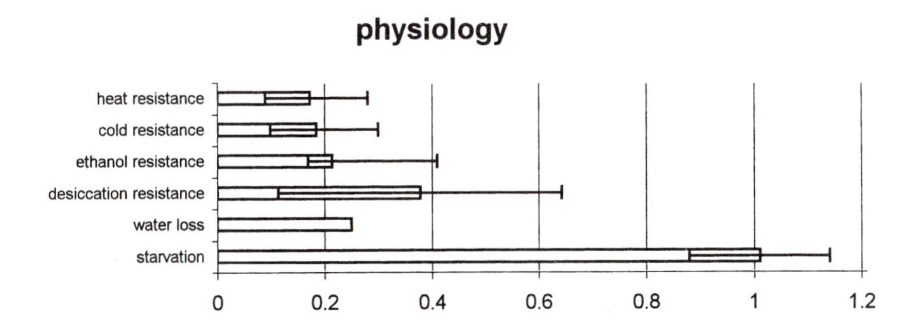

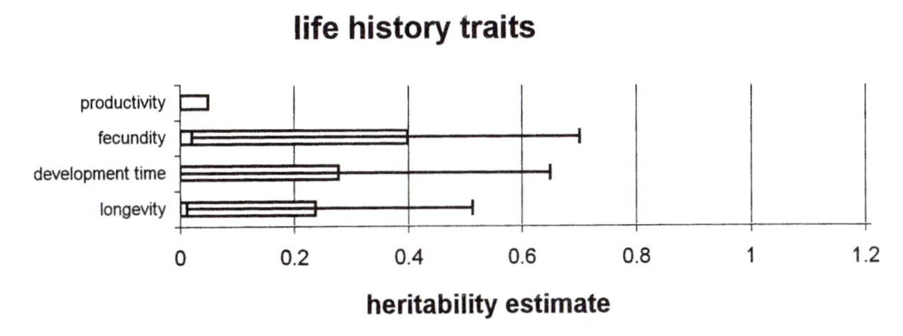

Figure 9.1 Mean estimates for narrow-sense heritabilities for traits falling into three different classes. Bars represent the range of estimates from individual studies.

One reason for this variability is that measurements of behavioral traits tend to be associated with a large amount of variability. Unfortunately, behavioral traits often have a low stability when repeat measurements are undertaken over a short time interval (Henderson 1990). If a fly is tested in a maze or along a gradient on two occasions, many flies behave differently upon retesting. For instance, for temperature preference in *D. subobscura*, Kekic and Marinkovic (1974) found that, on average, only 33% of the flies selected the same light intensity on successive days. This retesting variability can have two causes. First, learning may be involved, altering a fly's behavior. This is rarely considered, although it will lead to directional shifts in behavior. Second, minor changes in the maze, in the fly's internal state, or in testing conditions may lead to different behavior when retested. This second cause represents the measurement error associated with the behavioral assay and leads to random variability in measurements.

Measurement error is described by the measurement repeatability (r_M) of a trait. This is equivalent to the intraclass correlation (see Sokal and Rohlf 1981, Lessells and Boag 1987) and can be defined as $V_I/(V_I + V_M)$, where V_I is the variance component among individuals, and V_M is the measurement error. It therefore follows that

$$1 - r_M = \frac{V_M}{V_P} \tag{9.3}$$

where V_P is the total phenotypic variance (i.e., $V_I + V_M$). If measurement error is not identified, it becomes combined with the error variance component of a heritability estimate. The effects of low measurement repeatability on heritability can be substantial. If all measurement error was removed, the narrow-sense heritability of a trait can be defined as

$$h^2_n = \frac{V_A}{V_P - V_M} = \frac{V_A}{V_P - (1 - r_M)V_P} \tag{9.4}$$

where V_A is the additive genetic variance. The difference between a heritability estimated with measurement error and one computed without it (i.e., V_A / V_P) is therefore given by

$$\left(\frac{V_P}{V_A}\right) \frac{V_A}{V_P - (1 - r_M)V_P} \tag{9.5}$$

or $1/r_M$. This means that low measurement repeatability can lead to an underestimate of the heritability of a trait. For instance, if the measurement repeatability for a trait is 0.4, then a heritability estimate of 0.3 for this trait becomes 0.75 once measurement error is considered. As a consequence, laboratory heritability estimates that are uncorrected for measurement variation may appear to be low when the true heritability is actually high. For instance, most studies of phototaxis have yielded estimates of around 0.10 or less, but much higher estimates have been obtained from comparisons involving strains that indirectly control for measurement error (e.g., Hadler 1964).

Correcting measurement error in this way may be appropriate if a behavior is undertaken numerous times by an organism, as in phototactic or olfactory responses. The mean response of an organism is then approximated by the response that would have occurred without measurement error. However, it may be inappropriate to correct for all measurement error if a behavior occurs once or a few times in the lifetime of an organism, as in the case of mating behaviors in females that rarely remate. Behavioral assays and levels of variability measured in the laboratory also need to be related to those found under field conditions.

It should be emphasized that the measurement repeatability considered here differs from the repeatability of a trait as defined by Falconer (1989), which provides an upper limit to the heritability and refers to within-individual variance arising from local circumstances affecting the development and expression of a trait. For instance, Falconer determines the repeatability of abdominal bristles from bristle numbers on different segments of the same individual. This is different from measurement error associated with repeatedly counting the same set of bristles. Unfortunately, measurement repeatability is often viewed as a component of this form of repeatability, instead of being separated from it. When an individual repeatedly undertakes a behavior, the measurement repeatability from one behavioral event does not set an upper limit to the trait's heritability. This limit can only be determined once a correction has been made for measurement error.

Because of these concerns, published estimates of the heritability of behavioral traits in *Drosophila* (and other organisms) are too low. As a consequence, inferences based on these estimates are likely to be invalid. Behavioral traits may therefore not necessarily be closely related to fitness and do not necessarily evolve slowly under natural conditions as implied by a low heritability. As an example, consider mating speed and courtship speed in *D. melanogaster* (fig. 9.2). When heritability estimates for these traits are based on measurements of only a single event, both traits have low heritabilities (effectively 0). However, estimates increase markedly when they are based on multiple measurements of the same individuals, exceeding any of the published estimates for these behaviors (see fig. 9.1).

A low heritability for this trait class is also at odds with evidence that heritable variation in *Drosophila* behavior can easily be found in nature. Perhaps the clearest instance is larval foraging behavior studied by Sokolowski, Carton, and others (Sokolowski et al. 1986, Carton and Sokolowski 1992). Variation in foraging falls into two phenotypes, "rovers" and "sitters," depending on how far larvae move. A major gene as well as some modifiers determines the two phenotypes. This behavioral variation is closely correlated with larval pupation in culture vials (rovers pupate farther away from the medium) and in the field. Sokolowski et al. (1986) showed that pupation site selection in the field (on or off fruit) could be related to variation in the forager gene. Much of the genetic variation for pupation selection in the field as well as in the laboratory is therefore attributable to these genetic variants. Frequencies of rovers and sitters in populations can be related to the presence of parasitoids which attack one phenotype more frequently than another phenotype (Carton and Sokolowski 1992).

Another example of behavioral variation with a large genetic component in nature concerns resource response. When flies are collected from two types of fruit in the field and then reared for a generation in the laboratory, the F_1 generations show a preference for their fruit of origin when tested in a wind tunnel (Hoffmann et al. 1984). In addition, when the parental generation is reared in the laboratory and then collected from different fruit in the field, their offspring also shows a tendency to be collected from the same fruit type when released in the field (Hoffmann and O'Donnell 1992). These effects can be strong, indicating that evolution in response to fruit-related cues is potentially rapid in the field. Although the heritability of this behavior is not known, it is likely to be high when assessed without error.

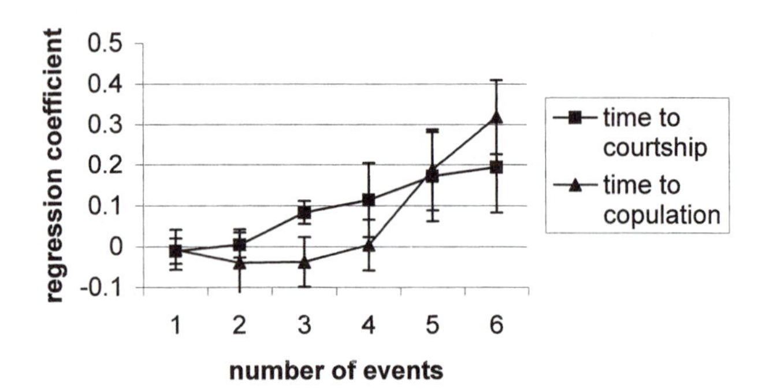

Figure 9.2 Effect of measuring repeated courtship and mating events on the regression of offspring scores onto parental scores. Males were scored for 1–6 days and regressions are based on one event (scores on day 1), the means of two events (days 1–2), the means of three events (days 1–3), and so forth. Error bars represent standard errors. Based on Hoffmann (1998).

A high heritability for behavioral traits is also consistent with the notion that behavioral traits are considered flexible at the evolutionary level because populations and closely related species show marked behavioral divergence. An illustration of the rapidity whereby behavioral traits evolve is the evolution of territoriality. *Drosophila melanogaster* males will defend food resources against other males to increase their access to females (Hoffmann 1987). This behavior is plastic in the sense that males will stop defending resources when the density of males relative to the number of resources becomes high or low (Hoffmann and Cacoyianni 1990). When laboratory cages were set up with an intermediate number of resources, levels of territorial aggression increased rapidly because territorial males had a mating advantage. After a few generations of selection, the control and selected lines were discrete in that all territories were held by males from the selected lines (Hoffmann 1988). This rapid evolution in response to environmental conditions highlights the evolutionary flexibility of such traits. Marked behavioral evolution is also evident from the diverse mating systems within the *Drosophila* genus, ranging from species that copulate as solitary pairs without any male interactions to those that show aggressive interactions among males or lekking aggregations.

Turning to field–lab heritabilities, the only estimates in *Drosophila* for behavioral traits were obtained by Aspi and Hoikkala (1993), who undertook field–lab comparisons of traits determining courtship song. Six components were measured from one species (*D. montana*) and four components from another (*D. littoralis*) under laboratory (offspring) and field (parental) conditions. Based on parent–offspring regressions, field–lab heritabilities based on equation 9.1 were significantly different from 0 for one component in *D. montana* ($h^2 = 0.43$) compared to three components for this species in the laboratory. Similar field estimates were obtained when the Riska et al. (1989) approach based on equation 9.2 was used to estimate the minimum heritabilities. None of the components in *D. littoralis* were significantly heritable in either environment. Field–lab estimates were therefore lower than in the laboratory (fig. 9.3), reflecting an increase in phenotypic variability in the field.

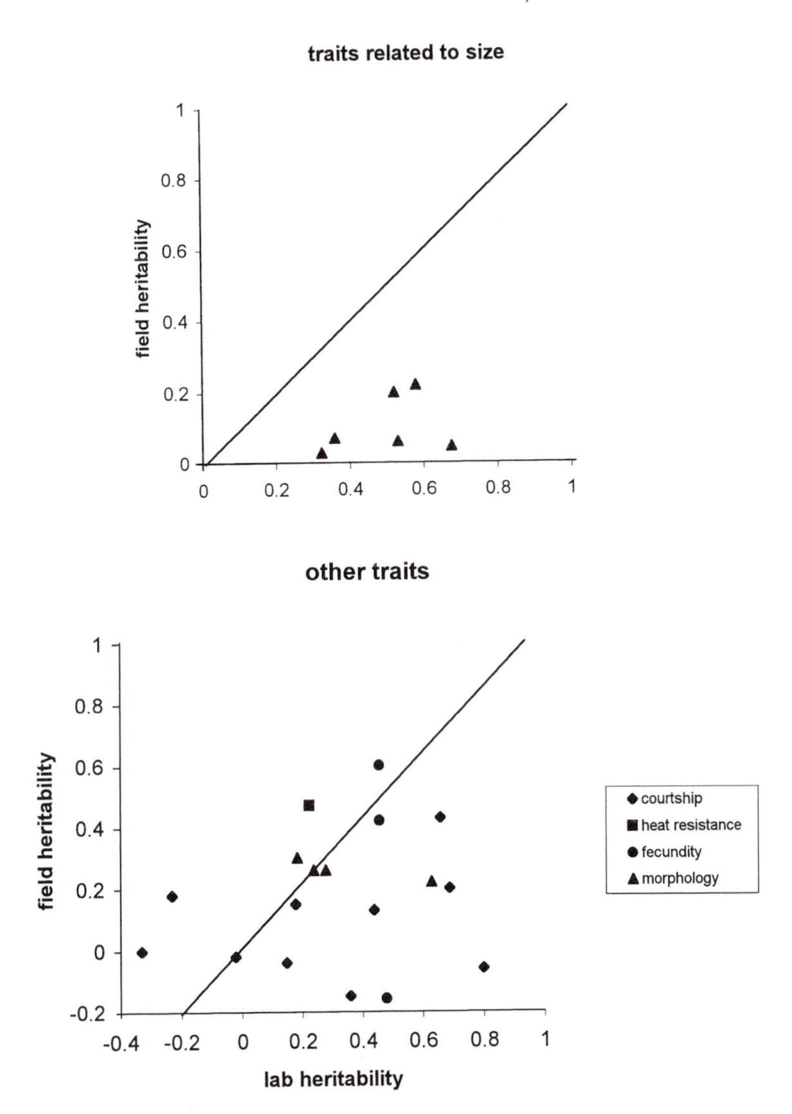

Figure 9.3 Association between two estimates of the field heritability of traits, based on data from courtship song components, thorax length, wing length and other wing measures, fecundity, and bristle counts. The two heritability estimates are defined in the text. The lines represent the expectation when estimates are equal.

In conclusion, the low heritability estimates for behavioral traits in the laboratory may be the consequence of low measurement repeatability rather than a consequence of any history of intense natural selection on these traits. This seems likely because estimates can increase twofold or more when measurement error is considered (fig. 9.2). Significant heritable variation for behavioral traits can occur under field conditions, but the only heritability estimates in field flies suggested lower values than in the laboratory.

9.1.2 Morphological traits

Heritability estimates in Roff and Mousseau (1987) mainly focused on four traits involving bristle counts and length measurements. Such traits are easy to score, but are they representative of morphology in general? One problem is that these traits have probably been chosen because they have high measurement repeatability. This is a common criterion for deciding which traits to score in such studies (for instance, see Cowley and Atchley 1988). As a consequence, there is a bias toward traits with a low measurement error, leading to heritability estimates unlikely to be confounded by this factor.

Mean estimates for 29 traits are presented in figure 9.4. Those for traits listed from basitarsus width to prementum width in figure 9.4 are from Cowley and Atchley (1988) and are each based on a single value. Other traits not considered in Roff and Mousseau (1987) but listed here are cross-vein length (Woods et al. 1998), wing tip height (Weber 1990b), orbital bristles (Woods et al. 1998), wing width (Wilkinson et al. 1990, Woods et al. 1998) and wing form (Weber 1990a, 1992). There are additional estimates for traits considered in Roff and Mousseau: weight (Clark 1990, Hillesheim and Stearns 1991), wing length (Wilkinson et al. 1990, Liebowitz et al. 1995), thorax length (Wilkinson et al. 1990, Ruiz et al. 1991), and sternopleural (Wilkinson et al. 1990) and abdominal (Coyne and Beecham 1987) bristle number. This much larger range of traits has an average heritability of 0.41 compared to the overall value of 0.32 reported by Roff and Mousseau. The proposal that morphological traits tend to have substantial heritabilities is therefore supported by this survey.

Despite this relatively high mean estimate, there are some exceptions that can be linked to measurement error. A few studies have undertaken genetic analyses by indirectly measuring morphological traits on groups of flies. These will have a relatively high degree of measurement error. For instance, Robertson (1960) undertook body size selection in *D. melanogaster* using a sieve system whereby flies were sorted into compartments through a series of slits differing in width. Realized heritabilities for size were fairly low, being 0.14 ± 0.02 for upward selection and 0.20 ± 0.02 for downward selection. Size selection is usually undertaken by carefully measuring all individuals in a population and then selecting the largest or smallest as the founders of the next generation. This approach leads to higher heritability estimates (see fig. 9.4) than found by Robertson. Baptist and Robertson (1976) estimated repeatability for the sieve size measure by running flies through the apparatus on successive days; they found a repeatability of around 0.5 based on regressing the score of the second run onto the first run. The relatively low measurement repeatability of the sieve assay is therefore likely to account for the lower heritability estimate. A second example comes from Weber (1990a) and involved an automated selection procedure for wing tip height. Groups of flies were treated so that they became torpid and flies carried their wings erect. Flies were then placed on a platform, and those with long wing tips stuck to an adhesive surface that was lowered toward the platform. This indirect selection procedure for morphological characteristics resulted in a realized heritability of 0.02 ± 0.004 for lines selected downward and 0.04 ± 0.003 for those selected upward. These values are much lower than heritabilities for wing traits and other morphological traits measured on individual *D. melanogaster* (fig. 9.4) and are likely to reflect a low measurement repeatability.

Several *Drosophila* studies have considered field–lab heritabilities of morphological traits. Not all of these report comparable laboratory estimates to allow a minimum field heritability to be determined following equation 9.2. Nevertheless, there are sufficient studies

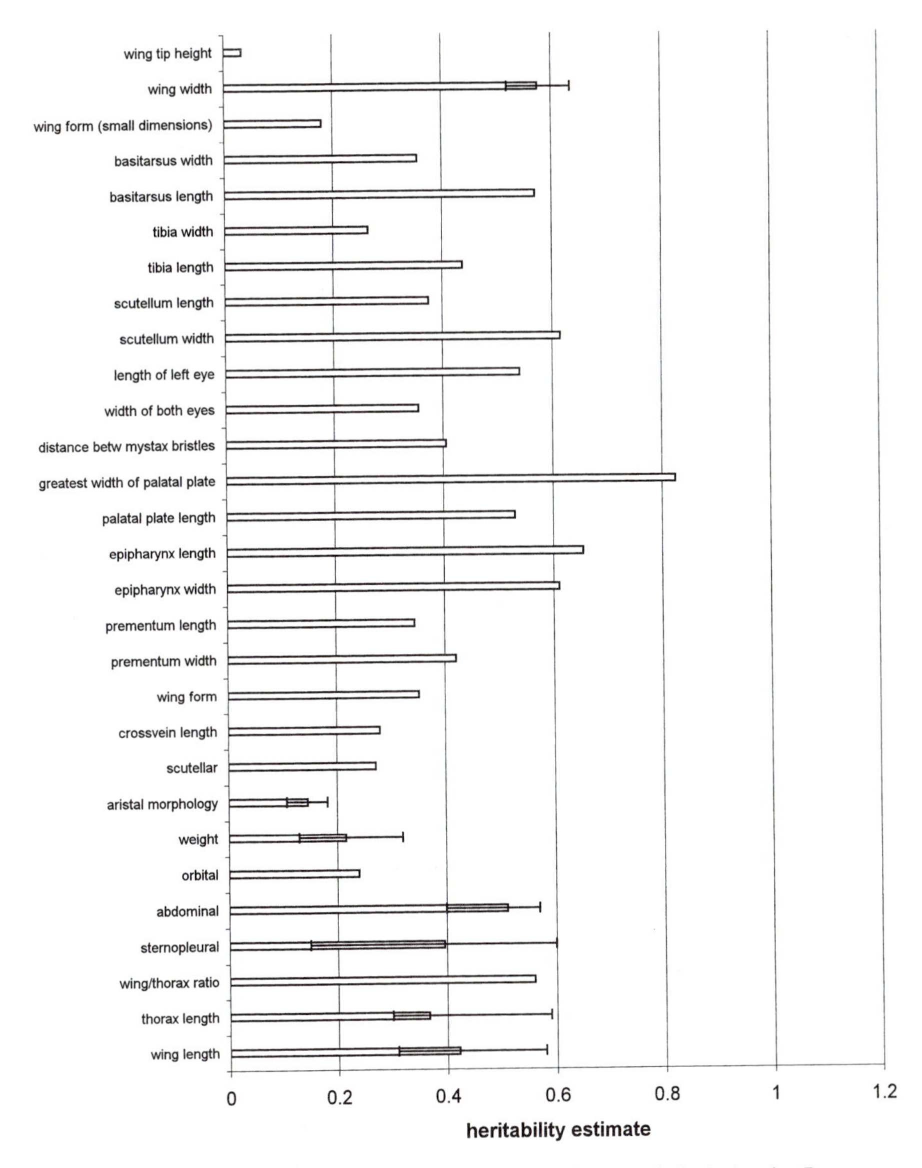

Figure 9.4 Mean estimates for narrow-sense heritabilities for morphological traits. Bars represent the range of estimates from individual studies.

giving both estimates to investigate the association between field heritabilities estimated from equations 9.1 and 9.2. As previously noted, a simple regression of laboratory offspring onto field parents will provide an estimate of field heritability if $\gamma = 1$ and if the additive genetic variance is the same in the two sets of conditions ($\sigma_{AL} = \sigma_{AN}$). The minimum heritability estimate from equation 9.2 and the regression estimate will generally yield the same result if $\gamma = \gamma^2 = 1$ and $\sigma_{AL} = \sigma_{AN}$, although similar values could also occur when the cross-environment correlation is < 1 and σ_{AL} is $< \sigma_{AN}$.

I considered the association between the heritability estimates based on data from independent estimates encompassing a range of traits: bristle traits (Coyne and Beecham 1987, Riska et al. 1989, Woods et al. 1998), wing length (Coyne and Beecham 1987, Liebowitz et al. 1995, Sgro and Hoffmann 1998b), other wing measures (Woods et al. 1998), and thorax length (Prout and Barker 1989). To examine the general relationship between these estimates, I also included courtship traits (Aspi and Hoikkala 1993) and fecundity on three occasions (Sgro and Hoffmann 1998b). The association between estimates from equations 9.1 and 9.2 closely fits a straight line with a slope of 1 (fig. 9.5), indicating similar estimates from these approaches. The poorest fit appears to be at low heritability values. This can be partly explained by the fact that the minimum estimate following equation 9.2 is obtained by squaring the regression coefficient, eliminating negative values, whereas the regression coefficient from a parent–offspring comparison can be negative. If all data points on the x-axis corresponding to values < 0 (as computed from eq. 9.1) are moved along to the zero point, then the fit of the data points to the predicted line is improved. Overall, the data therefore suggest that $\gamma = 1$ and that $\sigma_{AL} = \sigma_{AN}$; either the minimum estimate or the regression slope may provide an estimate of field heritability, at least for the traits considered here. Although the combination of conditions where $\gamma < 1$ and $\sigma_{AL} < \sigma_{AN}$ could also account for this pattern, it seems unlikely that this combination applies to all traits and field environments considered.

Because of the strong association between estimates, I have used data from all field studies including those where estimates were only based on equation 9.1 to compare field and laboratory heritabilities for morphological traits. The results indicate that the heritability for size-related traits in field-lab comparisons is always lower than in the laboratory (fig. 9.3), rarely exceeding 0.2, in comparison to laboratory values in the range 0.25–0.65. In contrast, three morphological traits measured in *D. melanogaster* (sternopleural bristles, orbital bristles, and the length of the larger cross-vein) show similar heritabilities in the field and laboratory (Woods et al. 1998). Only wing width shows the same pattern as the size traits (fig. 9.3).

To examine reasons for differences between laboratory and field heritabilities, Roff (1997) proposed that the ratios of phenotypic variances and heritabilities be examined. If lower heritabilities under field conditions are due solely to changes in the environmental variance, then the following relationship should hold:

$$\frac{h^2_{Field}}{h^2_{Lab}} = \frac{V_{P\,Lab}}{V_{P\,Field}}$$

Note that I have inverted the right side of the equation in Roff (1997) because of an apparent mistake. This relationship is subjected to considerable error due to the fact that error is associated with all the parameters that need to be estimated. Nevertheless, when these two ratios are graphed against each other for the *Drosophila* data, there is a positive association between them (fig. 9.6). The increase in the phenotypic variance typically seen in *Drosophila* field–lab comparisons (Roff 1997) therefore seems to reflect an increase in environmental variance.

In summary, laboratory heritabilities for morphological traits tend to be intermediate for a wide range of traits. Field–lab heritabilities for size-related traits are relatively lower (mean of 0.10), and this probably reflects an increased level of environmental variability under field conditions. Other morphological traits may show similar heritabilities under both field and laboratory conditions, consistent with data for other insects such as crickets (Simons and Roff 1995).

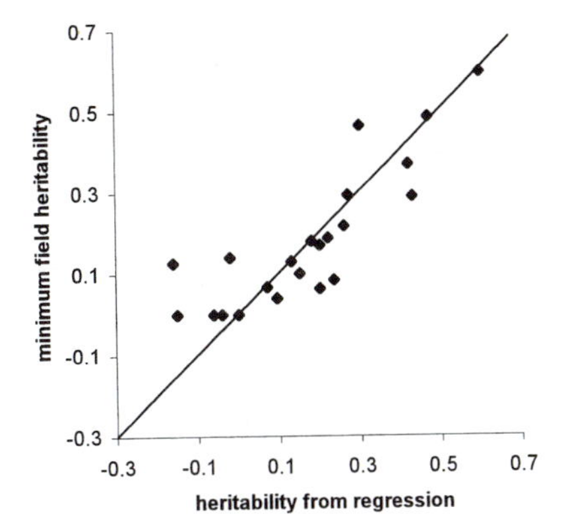

Figure 9.5 Comparisons of laboratory and field heritabilities for size traits and other traits. Points below (above) the line have higher (lower) laboratory than field heritabilities.

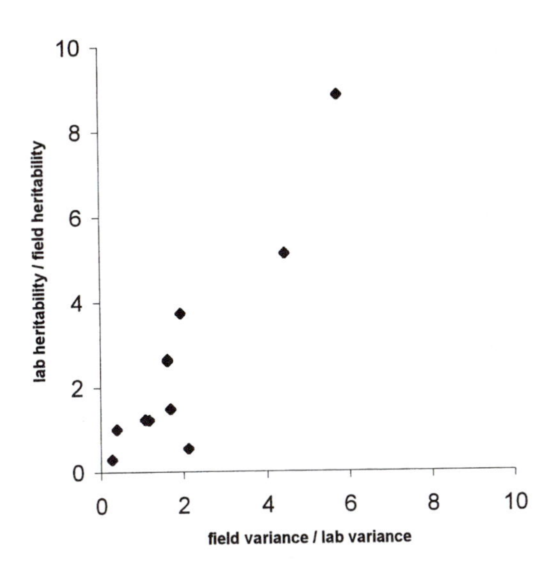

Figure 9.6 Association between the ratio of laboratory to field heritabilities and the ratio of field to laboratory variances for a range of traits.

9.1.3 Physiological traits

In Roff and Mousseau (1987), physiological traits tended to have relatively high heritabilities, comparable to values for morphological traits. This category included two physiological traits (water loss, ether resistance) as well as two studies on enzyme activity. I have included only one of these earlier studies as well as new estimates for resistance to desiccation (Hoffmann and Parsons 1989, 1993, Blows and Hoffmann 1993), heat (Huey et al. 1992), cold (Watson and Hoffmann 1995), and ethanol (Cohan and Hoffmann 1986, 1989, Weber and Diggins 1990).

Enzyme activities were excluded because these represent a different class of traits. The physiology estimates indicate high laboratory heritabilities for desiccation and starvation resistance but lower heritabilities for stress responses (heat and ethanol) measured in knockdown tubes (fig. 9.1). The lower estimates in these tubes may reflect a high measurement error; however, this would be difficult to assess because acclimation normally occurs once flies have been stressed at nonlethal levels, precluding retesting of the flies. The heritability estimates for knockdown traits are variable because for both ethanol and desiccation resistance there is evidence for heritability differences between species (Cohan and Hoffmann 1989, Hoffmann and Parsons 1993) and between conspecific populations (Cohan and Hoffmann 1986, Blows and Hoffmann 1993). Estimates for physiological traits are therefore likely to depend on how traits are measured and which species/populations are used. Nevertheless, the laboratory data suggest that physiological traits can have moderate to high heritabilities.

Heritability estimates involving field flies are only available in the case of heat resistance in *D. simulans* measured by knockdown (Jenkins and Hoffmann 1994). High values were obtained in field–lab comparisons, although there were also large maternal effects. The maternal effects were not evident under laboratory conditions, suggesting that environmental conditions can influence their transmission. Maternal effects are generally ignored in field–lab heritability studies but can be detected when field offspring are reared in the laboratory for two generations (Jenkins and Hoffmann 1994), as well as in other designs (Roff 1997).

There are now a wide range of heritability estimates available for enzyme activities, derived mainly by Clarke (1990), who considered the *in vitro* activities of 10 enzymes related to energy storage in *D. melanogaster*. Based on data from more than 1000 half-sib families, Clark showed that maximum likelihood heritabilities for the different enzymes covered a range of values, from around 0.16 for phosphoglucose isomerase to 0.59 for glycogen synthase. All estimates of V_A for the enzyme activities differed from 0 by more than two standard deviations. Enzyme activities therefore appear to show intermediate heritabilities, although these may be underestimates because of measurement error.

9.1.4 Life-history traits

Roff and Mousseau (1987) concluded that heritabilities for life-history traits in *Drosophila* are low under laboratory conditions. Unfortunately, there are only a few traits for which heritability estimates have been obtained (fig. 9.1). My overview includes recent estimates of longevity (Zwaan et al. 1995b), development time (Zwaan et al. 1995a, Sgro and Hoffmann 1998a), and fecundity (Zwaan et al. 1995b, Sgro and Hoffmann 1998a), traits which were also considered by Roff and Mousseau. Nevertheless, the mean heritability value for this class of traits (0.24) is higher than the overall value for life-history traits given by Roff and Mousseau (0.12). The range of estimates for each trait is wide (fig. 9.1), suggesting that the heritability of life-history traits may be variable, a point also noted by Roff and Mousseau (1987). The variability of life-history trait estimates is also evident from heritability changes for fecundity with female age (Rose and Charlesworth 1981) and heritability changes for development time with temperature (Sgro and Hoffmann 1998a). Measurement error may contribute to low heritability estimates for some life-history traits. To avoid this problem, traits such as egg production should probably be scored over a few days.

There is only one *Drosophila* study on heritable variation in life-history traits using field flies (Sgro and Hoffmann 1998b). *D. melanogaster* were collected on three occasions directly after eclosing from a rotting fruit heap. The fecundity of field flies was scored in the laboratory, and offspring were scored at the same age and temperature as their parents after being reared in the laboratory. Field–lab heritabilities were estimated by regression of laboratory offspring onto field parents, or by following equation 9.2 (table 9.1). These approaches yielded similar values, in agreement with the general pattern for morphological and behavioral traits (fig. 9.5). The estimates varied considerably between collections; they were similar to laboratory estimates on two occasions but lower on the other occasion. Conclusions about field and laboratory heritabilities for fecundity may therefore depend on the range of conditions being considered. It remains to be seen if particular field conditions increase heritable variation in life-history traits (Sgro and Hoffmann 1998a).

9.2 Comparisons

9.2.1 Heritabilities of trait classes

Do laboratory heritabilities for different classes of traits vary? When individual traits are considered as data points for the behavior, morphology, life-history, and physiology classes, there is a significant difference between them (Kruskal-Wallis test: $\chi^2 = 19.82$, df = 3, $P < .001$). However, post-hoc tests indicate that this is solely due to the low heritability of the behavior class, which differs significantly from both the morphology and physiology classes. Although life-history traits have a low mean heritability (0.24) comparable to that of behavioral traits (0.18), only a few data points are available for this trait class.

Do differences between trait classes really exist? Both the approach followed here and the Roff and Mousseau (1987) approach based on median values from individual studies have flaws. My approach is limited because studies that consider a range of traits are weighted much more heavily than those that provide a number of estimates for the same trait. This is particularly true for estimates of morphological traits, many of which were derived from one study. Yet the heritability estimates from this study are consistent with others for this trait class (fig. 9.4); almost all estimates for morphological traits fall in the 0.2–0.6 range. One of the outliers for this trait class (wing tip height) was measured indirectly (Weber 1990b) and almost certainly with a high degree of error, resulting in an artificially low heritability estimate. On the other hand, the Roff and Mousseau approach is limited because multiple studies of the same trait could inflate heritability estimates for an entire class. For instance, multiple estimates of the heritability of starvation resistance compared to heat resistance would markedly inflate the mean heritability of the physiology class. Yet the heritability differences between these traits may reflect the fact that knockdown resistance assays have a larger measurement error than survival assays, rather than any true heritability differences between the traits.

Measurement error may confound class comparisons regardless of the approach used. This source of error is likely to account for much of the variability in behavioral traits. The lifetime phototactic response of a fly cannot be accurately measured by a single performance in a phototactic maze. The lifetime courtship displays of a male will not be represented by his response to one female in a small mating chamber. Yet inferences about a behavior's relevance to fitness have often been made from such assays. This problem of measurement error is also relevant to other classes of traits, particularly physiological and biochemical traits.

Table 9.1 Narrow-sense heritability estimates for fecundity in *D. melanogaster* field females

Collection	Number of families	Heritability from regression	Minimum field heritability
Summer	105	0.42 ± 0.20	0.37
Autumn	135	0.60 ± 0.16	0.59
Spring	196	-0.16 ± 0.14	0.13

Adapted from Sgro and Hoffmann, 1998b

9.2.2 Variability in estimates

Apart from showing differences in means, trait classes may also differ in the variability of the heritability estimates. I have tested this by considering those traits for which estimates from at least five independent studies were available and by excluding the behavior class because of the measurement error problem. For the physiology class, two traits meet this criterion, while there are four morphological traits and three life-history traits that qualify. The life-history traits show a greater range of estimates than the other two classes (fig. 9.7); this difference is marginally nonsignificant by a Kruskal-Wallis test ($\chi^2 = 5.67$, df = 2, $P = .058$). The same conclusion holds if the variance in heritability estimates for a trait is considered instead of the range of estimates. Although more traits need to be considered, this pattern suggests that heritability estimates for life-history traits may be more labile than those for morphological and other traits.

A possible reason for this pattern is that the heritability of life-history traits is relatively more susceptible to environmental variability. Direct evidence for this hypothesis comes from a comparison of heritability estimates for fecundity, development time, and wing length in *D. melanogaster* across temperature extremes (Sgro and Hoffmann 1998a). Although heritabilities for wing length were relatively constant across different temperatures, there were significant differences in estimates for the life-history traits.

An important direction for future studies in this area is to define conditions responsible for altering heritabilities in different classes of traits. For instance, in birds, there is evidence that poor nutrition generally decreases the heritability of size-related traits (Larsson et al. 1997, Merila 1997). This can slow directional changes in such traits when periods of nutritional stress occur in natural populations. It is not yet clear if a single theoretical framework can be developed to account for such changes. It has been suggested that heritabilities may vary with levels of environmental stress (Hoffmann and Parsons 1991) or with the novelty of an environment which dictates the history of selection on a trait (Holloway et al. 1990). In addition, genetic models of phenotypic plasticity (Gavrilets and Schiener 1993) could provide a framework for interpreting changes in heritability with environmental conditions.

9.2.3 Field and laboratory heritabilities

The limited evidence available suggests that heritability estimates for traits tend to be lower in the field than in the laboratory. This applies particularly to size traits but is also apparent for courtship behavior. Nevertheless, there are traits that can have similar heritabilities in the two environments, consistent with findings in other organisms (Weigensberg and Roff 1996).

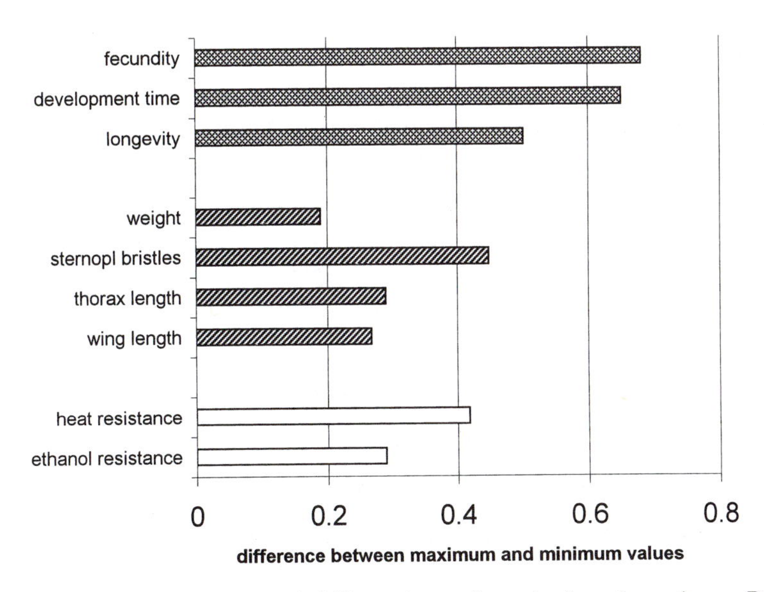

Figure 9.7 Range of laboratory heritability estimates for traits from three classes. Ranges were only computed for a trait when there were independent estimates of heritability from at least five studies.

These include bristle measurements, cross-vein length, and fecundity. There are insufficient data to compare field–lab heritabilities across trait classes. However, because environmental conditions probably have a marked influence on life-history and physiological traits, a wide range of conditions will need to be considered. Perhaps these environmental effects may be the most productive area for future studies. Can laboratory conditions that trigger changes in the heritability of a trait be linked to similar effects under field conditions? Are there particular types of field environments that increase rates of evolutionary change?

Other areas that need to be investigated are the extent to which heritable changes in laboratory conditions reflect similar changes under field conditions and whether heritable changes under one set of field conditions translate into similar changes under different conditions. The fact that heritabilities computed across field and laboratory generations are similar to minimum estimates computed following equation 9.2 (fig. 9.5) suggests that genetic correlations across laboratory and field environments are close to 1. However, there is abundant evidence from the literature that genetic correlations across some environments can be weak. In particular, agricultural work indicates that correlations for yield-related traits become weaker and even negative as increasingly stressful environments are compared to optimal ones (Hoffmann and Parsons 1991, Ceccarelli 1996). A challenge for future evolutionary studies on natural populations is to examine heritable variation under conditions when traits are likely to be under intense selection. These conditions will often be stressful. Are the same genes under selection under these conditions, compared to laboratory and optimal field conditions? Can laboratory heritabilities provide useful information when predicting likely responses under these conditions?

References

Aspi, J., and A. Hoikkala. 1993. Laboratory and natural heritabilities of male courtship song characters in *Drosophila montana* and *D. littoralis*. Heredity 70: 400–406.

Baptist, R., and A. Robertson. 1976. Asymmetrical responses to automatic selection for body size in *Drosophila melanogaster*. Theoretical and Applied Genetics 47: 209–213.

Blows, M. W., and A. A. Hoffmann. 1993. The genetics of central and marginal populations of *Drosophila serrata*. I. Genetic variation for stress resistance and species borders. Evolution 47: 1255–1270.

Carton, Y., and M. B. Sokolowski. 1992. Interactions between searching strategies of *Drosophila* parasitoids and the polymorphic behavior of their hosts. Journal of Insect Behavior 5: 161–175.

Ceccarelli, S. 1996. Specific adaptation and breeding for marginal conditions. Euphytica 77: 205–219.

Clark, A. G. 1990. Genetic components of variation in energy storage in *Drosophila melanogaster*. Evolution 44: 637–650.

Cohan, F. M., and A. A. Hoffmann. 1986. Genetic divergence under uniform selection. II. Different responses to selection for knockdown resistance to ethanol among *Drosophila melanogaster* populations and their replicate lines. Genetics 114: 145–163.

Cohan, F. M., and A. A. Hoffmann. 1989. Uniform selection as a diversifying force in evolution: evidence from Drosophila. American Naturalist 134: 613–637.

Cowley, D. E., and W. R. Atchley. 1988. Quantitative genetics of *Drosophila melanogaster*. II. Heritabilities and genetic correlations between sexes for head and thorax traits. Genetics 119: 421–433.

Coyne, J. A., and E. Beecham. 1987. Heritability of two morphological characters within and among natural populations of *Drosophila melanogaster*. Genetics 117: 727–737.

Falconer, D. S. 1989. Introduction to Quantitative Genetics, 3rd ed. Longman, London.

Gavrilets, S., and S. M. Schiener. 1993. The genetics of phenotypic plasticity. VI. Theoretical predicitons for directional selection. Journal of Evolutionary Biology 6: 49–68.

Gromko, M. H., A. Briot, S. C. Jensen, and H. H. Fukui. 1991. Selection on copulation duration in *Drosophila melanogaster*: predictibility of direct response versus unpredictability of correlated response. Evolution 45: 69–81.

Hadler, N. M. 1964. Heritability and phototaxis in *Drosophila melanogaster*. Genetics 50: 1269–1277.

Henderson, N. D. 1990. Quantitative genetic analysis of neurobehavioral phenotypes. Pp. 283–297 *in* M. E. Hahn, J. K. Hewitt, N. D. Henderson, and R. Benno, eds., Developmental Behavior Genetics: Neural, Biometrical, and Evolutionary Approaches. Oxford University Press, New York.

Hillesheim, E., and S. C. Stearns. 1991. The responses of *Drosophila melanogaster* to artificial selection on body weight and its phenotypic plasticity in two larval environments. Evolution 45: 1909–1923.

Hoffmann, A. A. 1987. A laboratory study of male territoriality in the sibling species *Drosophila melanogaster* and *D. simulans*. Animal Behaviour 35: 807–818.

Hoffmann, A. A. 1988. Heritable variation for territorial success in two *Drosophila melanogaster* populations. Animal Behaviour 36: 1180–1189.

Hoffmann, A. A., and Z. Cacoyianni. 1990. Territoriality in *Drosophila melanogaster* as a conditional strategy. Animal Behaviour 40: 526–537.

Hoffmann, A. A., and S. O'Donnell. 1992. Heritable variation in the attraction of *Drosophila melanogaster* to fruit in the field. Biological Journal of the Linnean Society 47: 147–159.

Hoffmann, A. A., and P. A. Parsons. 1989. An integrated approach to environmental stress tolerance and life-history variation: desiccation tolerance in *Drosophila*. Biological Journal of the Linnean Society 37: 117–136.

Hoffmann, A. A., and P. A. Parsons. 1991. Evolutionary genetics and environmental stress. Oxford University Press, Oxford.

Hoffmann, A. A., and P. A. Parsons. 1993. Direct and correlated responses to selection for desiccation resistance: a comparison of *Drosophila melanogaster* and *D. simulans*. Journal of Evolutionary Biology 6: 643–657.

Hoffmann, A. A., P. A. Parsons, and K. M. Nielsen. 1984. Habitat selection: olfactory response of *Drosophila melanogaster* depends on resources. Heredity 53: 139–143.

Holloway, G. J., S. R. Povey, and R. M. Sibly. 1990. The effect of new environment on adapted genetic architecture. Heredity 64: 323–330.

Huey, R. B., W. D. Crill, J. G. Kingsolver , and K. E. Weber. 1992. A method for rapid measurement of heat or cold resistance of small insects. Functional Ecology 6: 489–494.

Jenkins, N. L., and A. A. Hoffmann. 1994. Genetic and maternal variation for heat resistance in Drosophila from the field. Genetics 137: 783–789.

Kamping, A., and W. Van Delden. 1990. Genetic variation for oviposition behavior in *D. melanogaster*. Behavior Genetics 20: 645–659.

Kekic, V., and D. Marinkovic. 1974. Multiple-choice selection for light preference in *Drosophila subobscura*. Behavior Genetics 4: 285–300.

Larsson, K., K. Rattiste, and V. Lilleleht. 1997. Heritability of head size in the common gull *Larus canus* in relation to environmental conditions during offspring growth. Heredity 79: 201–207.

Lessells, C. M., and P. T. Boag. 1987. Unrepeatable repeatabilities: a common mistake. Auk 104: 116–121.

Liebowitz, A., M. Santos, and A. Fontdevila. 1995. Heritability and selection on body size in a natural population of *Drosophila buzzatii*. Genetics 141: 181–189.

Merila, J. 1997. Expression of genetic variation in body size of the collared flycatcher under different environmental conditions. Evolution 51: 526–536.

Mousseau, T. A., and D. A. Roff. 1987. Natural selection and the heritability of fitness components. Heredity 59: 181–197.

Price, T., and D. Schluter. 1991. On the low heritability of life-history traits. Evolution 45: 853–861.

Prout, T., and J. S. F. Barker. 1989. Ecological aspects of the heritability of body size in *Drosophila buzzatii*. Genetics 123: 803–813.

Riska, B., T. Prout, and M. Turelli. 1989. Laboratory estimates of heritabilities and genetic correlations in nature. Genetics 123: 865–871.

Ritchie, M. G., and C. Kyriacou. 1994. Genetic variability of courtship song in a population of *Drosophila melanogaster*. Animal Behaviour 48: 425–434.

Ritchie, M. G., and C. P. Kyriacou. 1996. Artificial selection for a courtship signal in *Drosophila melanogaster*. Animal Behaviour 52: 603–611.

Robertson, F. W. 1960. The ecological genetics of growth in Drosophila. 2. Selection for large body size on different diets. Genetical Research 1: 305–318.

Roff, D. A. 1997. Evolutionary Quantitative Genetics. Chapman and Hall, New York.

Rose, M. R., and B. Charlesworth. 1981. Genetics of life-history in *Drosophila melanogaster*. I. Sib analysis of adult females. Genetics 97: 173–186.

Ruiz, A., M. Santos, A. Barbadilla, J. E. Quezada-Diaz, E. Hasson, and A. Fontdevila. 1991. Genetic variance for body size in a natural population of *Drosophila buzzatii*. Genetics 128: 739–750.

Sgro, C. M., and A. A. Hoffmann. 1998a. Effects of temperature extremes on genetic variances for life-history traits in *Drosophila melanogaster* as determined from parent-offspring regression. Journal of Evolutionary Biology 11: 1–20.

Sgro, C. M., and A. A. Hoffmann. 1998b. Heritable variation for fecundity in field-collected *D. melanogaster* and their offspring reared under different environmental temperatures. Evolution 52: 134–143.

Simons, A. M., and D. A. Roff. 1995. The effect of environmental variability on the heritabilities of traits of a field cricket. Evolution 48: 1637–1649.

Singh, B. N., and S. Chatterjee. 1988. Selection for high and low mating propensity in *Drosophila ananassae*. Behavior Genetics 18: 357–369.

Singh, B. N., and M. B. Pandey. 1993. Selection for high and low pupation height in *Drosophila ananassae*. Behavior Genetics 23: 239–243.

Sokal, R. R., and F. J. Rohlf. 1981. Biometry. Freeman, New York.

Sokolowski, M. B., S. J. Bauer, V. Wai-Ping, L. Rodriguez, J. L. Wong, and C. Kent. 1986. Ecological genetics and behaviour of *Drosophila melanogaster* larvae in nature. Animal Behaviour 34: 403–408.

Stamenkovic-Radak, M., L. Partridge, and M. Andjelkovic. 1992. A genetic correlation between the sexes for mating speed in *Drosophila melanogaster*. Animal Behaviour 43: 389–396.

Watson, M. J. O., and A. A. Hoffmann. 1995. Acclimation, cross-generation effects, and the response to selection for increased cold resistance in Drosophila. Evolution 50: 1182–1192.

Weber, K. E. 1990a. Selection on wing allometry in *Drosophila melanogaster*. Genetics 126: 975–989.

Weber, K. E. 1990b. Increased selection response in larger populations. I. Selection for wing-tip height in *Drosophila melanogaster* at three population sizes. Genetics 125: 579–584.

Weber, K. E. 1992. How small are the smallest selectable domains of form? Genetics 130: 345–353.

Weber, K. E., and L. T. Diggins. 1990. Increased selection responses in larger populations. II. Selection for ethanol vapor resistance in *Drosophila melanogaster* at two population sizes. Genetics 125: 585–597.

Weigensberg, I., and D. A. Roff. 1996. Natural heritabilities: can they be reliably estimated in the laboratory? Evolution 50: 2149–2157.

Welbergen, P., and F. R. Van Dijken. 1992. Asymmetric response to directional selection for licking behavior of *Drosophila melanogaster* males. Behavior Genetics 22: 113–124.

Wilkinson, G. S., K. Fowler, and L. Partridge. 1990. Resistance of genetic correlation structure to directional selection in *Drosophila melanogaster*. Evolution 44: 1990–2003.

Woods, R., M. Hercus, and A. A. Hoffmann. 1998. Estimating the heritability of fluctuating asymmetry in field *Drosophila*. Evolution, 52: 816–824.

Yamamoto, A. H. 1994. Diallel analysis of temperature preference in *Drosophila immigrans*. Japanese Journal of Genetics 69: 77–86.

Zwaan, B., R. Bijlsma, and R. F. Hoekstra. 1995a. Artificial selection for developmental time in *Drosophila melanogaster* in relation to the evolution of aging: direct and correlated responses. Evolution 49: 635–648.

Zwaan, B., R. Bijlsma, and R. F. Hoekstra. 1995b. Direct selection on life span in *Drosophila melanogaster*. Evolution 49: 649–659.

10

Intra- and Interpopulation Genetic Variation

Explaining the Past and Predicting the Future

TIMOTHY A. MOUSSEAU

There are two principal reasons for interest in the genetics of traits associated with adaptation: the first concerns the elucidation of past evolutionary responses to environmental change, and the second concerns our ability to predict future evolutionary responses to selection. If we can understand the causes and mechanisms underlying adaptive phenotypic variation that have occurred in the past, this will provide clues to how populations, communities, and ecosystems may evolve in response to future environmental change. For example, much of the latitudinal variation currently observed in north temperate insect life histories (e.g., Tauber et al. 1986, Danks 1987) is genetically based and has evolved since the last glacial period. Because similar patterns of life-history variation have evolved repeatedly, independently, and even on different continents, this suggests that these patterns of genetically based phenotypic variation are generally adaptive (Endler 1987) and likely to arise again in response to similar patterns of environmental change. Examples of such patterns include Bergmann's and Allen's rules for endotherms (Futuyma 1998) and the converse to Bergmann's Rule in ectotherms (Mousseau and Roff 1989, Mousseau 1997).

The second, perhaps more compelling reason for understanding the genetic mechanisms responsible for adaptation is that it may permit the prediction of future response to selection. More important, an understanding of the genetic basis of individual variation can allow for prediction of *rates* of response. The combination of descriptions of past evolutionary response and the ability to predict future response to selection could play a fundamental role in attempts to predict population, community, and ecosystem response to global climate change, urbanization, and other anthropogenic influences, in addition to improving our understanding of general evolutionary principles. Although predictive evolutionary genetics has been used with great success in agriculture, animal husbandry, and plant cultivation (Falconer and Mackay 1996), there have been few applications to natural populations (Roff 1997). Among the few examples of application of quantitative genetic approaches for the prediction of evolutionary response in nature include studies of annual variation in development in the fall web worm, *Hyphantria cunea* (Morris and Fulton 1970, Morris 1971), changes in bill shape in Galapagos finches (Grant and Grant 1995), and male morphology and mating strategy in side-blotched lizards (Sinervo and Lively 1996). This is clearly an area worthy of further pursuit.

My research program has sought to address explanation and prediction of evolutionary response in a diversity of natural systems using a variety of population-level approaches (especially common garden experiments) to test for the genetic basis to geographic variation in insect life histories. I have also used a variety of techniques to test for and quantify the nature of heritable genetic variation within natural populations. This chapter summarizes several studies conducted with crickets, grasshoppers, beetles, and trees that my laboratory has conducted during the past decade.

10.1 Population-level Genetic Variation: Common Garden Approaches

One of the simplest methods for examining the genetic basis of adaptive variation in the wild is to compare populations using either common garden or reciprocal transplant techniques. We have used these approaches to estimate the genetic and environmental basis to life-history variation in model cricket and grasshopper systems that show phenotypic variation along clines in latitude and altitude. The method is simple: populations spanning the geographic range are sampled, and individuals are reared in a common garden for two to three generations. A minimum of two generations are required for this approach because nongenetic maternal effects on phenotypic variation can persist for multiple generations (Roff 1997), although their influence is expected to decrease rapidly with temporal distance between environmental cue and developmental response (Mousseau and Dingle 1991b, Heath and Blouw 1998). Any differences among populations that persist for multiple generations in a common garden are presumed to reflect genetic differentiation. Further, a comparison of phenotype between field-collected and laboratory-reared populations will reveal the extent to which phenotypic variation is genetically versus environmentally induced (Mousseau and Roff 1989).

Reciprocal transplant methods extend the common garden approach one step further by examining phenotypic variation in multiple environments. As with simple common garden approaches, individuals should be reared in a common environment for two generations before transplantation to minimize the potentially confounding influences of maternal effects. Reciprocal transplant methods have the potential to provide much greater information concerning genetic differentiation because the expression of variation may be environment specific (i.e., genetic differences may be expressed in some environments, but not in others). This is particularly important if the reaction norm of environment-specific phenotypic expression is of interest. For example, in crickets the reaction norm of adult-male calling song (in response to ambient temperature) is dramatically influenced by the rearing environment of developing nymphs (Olvido and Mousseau 1995).

10.2 Adaptation to Seasonality in Crickets: Geographic Variation in Body Size, Cell Size, Ovipositor Length, Egg Depth, Diapause, and Egg Size

10.2.1 Patterns of phenotypic variation in the wild

Possibly the greatest challenge faced by most insects is the need to synchronize growth and reproduction with favorable seasons and the evolution of physiological and behavioral adaptations for the endurance of unfavorable conditions (Tauber et al. 1986). The striped ground cricket, *Allonemobius fasciatus* (fig. 10.1), like many insects, has a wide geographic distribution and is found in North America from Florida to Canada, from coast to coast, and from sea level to > 3000 m altitude (Alexander and Thomas 1959, Mousseau and Roff 1989, 1995). This cricket has adapted to a wide range in seasonal variation by adjustments in development time (with corresponding changes in body size), ovipositor length (with consequent changes in the depth at which eggs are deposited into the soil), egg size, and propensity to diapause under various environmental conditions.

In eastern North America, this cricket shows marked clinal variation in body size that reflects corresponding variation in the length of the growing season (fig. 10.2). In the northern part of this species' distribution and at high altitudes, this species is univoltine, producing a single generation per year, whereas in the south it is bivoltine (two generations per year), and may even be trivoltine at the southernmost end of its range. Body size (and development time) show a "sawtooth" distribution, with size gradually increasing from north to south, followed by a rapid decrease in mean population body size corresponding to the region of transition from univoltine to bivoltine life cycles. South of the transition zone, body size once again gradually increases with increasing season length. The observed clinal variation in size has resulted from balancing selection mediated by positive directional selection favoring increased fecundity (which pulls body size up as a result of the positive genetic and phenotypic correlations between body size and fecundity that exist in most organisms) and negative directional or truncation selection favoring decreased development time (which pulls body size down due to the genetic and phenotypic correlation between development time and body size). The net effect is an optimal body size at a given location, which varies positively with the length of the growing season (Roff 1980, 1983, Mousseau and Roff 1989). The drop in body size that occurs at the transition from univoltine to bivoltine populations results from the division of the growing season into two parts to accommodate the two generations. This pattern of size variation is known as the "Converse to Bergmann's Rule," and has been reported for many insects (Park 1949, Mousseau 1997).

It is worth noting that the detection of body size clines requires samples from many populations, as well as knowledge concerning the type of life cycle found in a given population (i.e., univoltine versus bivoltine). When too few samples are taken or when there is incomplete understanding of the life cycle, it is easy to suggest that a cline follows Bergmann's Rule (i.e., big in north, small in south) when in fact the size differences are simply due to the different number of generations among populations.

10.2.2 Genetics of population differences in cricket body size

In most instances it is not known if the observed phenotypic patterns of clinal variation reflect genetic differentiation among populations or simply environmentally modulated phenotypic plasticity. To test the hypothesis that the patterns of variation observed among wild populations reflect genetic differentiation, a series of common garden experiments was conducted. It was predicted that genetically differentiated populations should continue to show phenotypic variation even when reared under identical environmental conditions in the laboratory. Further, it is possible to quantify the relative contribution of genetics and environment to phenotypic variation by comparing field-collected parents to common-garden–reared descendants.

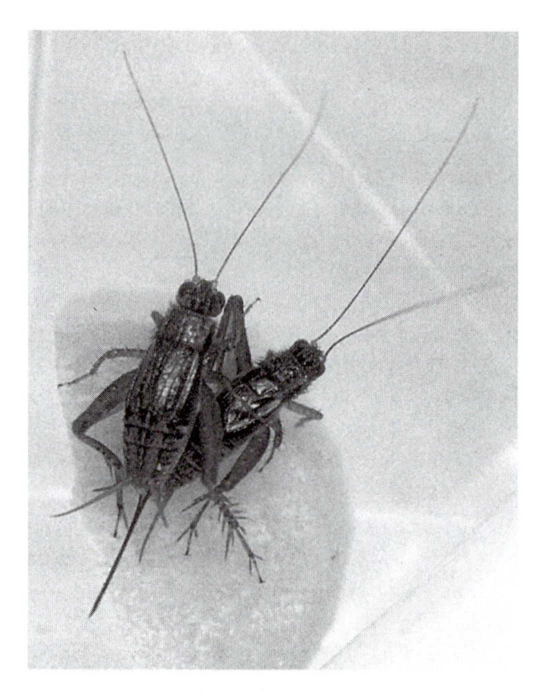

Figure 10.1 A pair of striped ground crickets, *Allonemobius fasciatus*. Note that the female has a long, needlelike ovipositor which is used to insert eggs into the ground.

Figure 10.3 shows the results of an experiment demonstrating the genetic basis to body size variation in the striped ground cricket. These third-generation common garden crickets show highly significant variation in body size among populations that reflects genetic differentiation. Although the time frame for this differentiation is largely unknown, differentiation among northern and high altitude populations has most certainly occurred since the start of the current interglacial period, circa 14,000 mya (Smith 1996), as these regions were covered in snow before this time. It is worth noting that the magnitude of variation observed among populations is considerable, especially that seen between southern univoltine and northern bivoltine populations (Mousseau and Roff 1989), and this variation exceeds that often observed among closely related species.

Although common garden experiments can provide evidence for genetic differentiation among populations, the interaction between genotype and environment can either mask or enhance genetic variation, especially when common garden experiments are conducted in only one environment. For example, if reaction norms for a given trait vary among populations, it possible for common garden approaches to erroneously conclude that no genetic differences exist among populations when in fact they are simply not expressed in the common garden environment (fig. 10.4). One way to avoid this problem is to conduct common garden experiments in multiple environments so that population-level reaction norms can be characterized. Indeed, it may well be that the reaction norms themselves are under selection (e.g., Gibert et al. 1998). Alternatively, comparisons between laboratory-reared populations and their field-collected ancestors (fig. 10.5) can provide hints concerning the relative contribu-

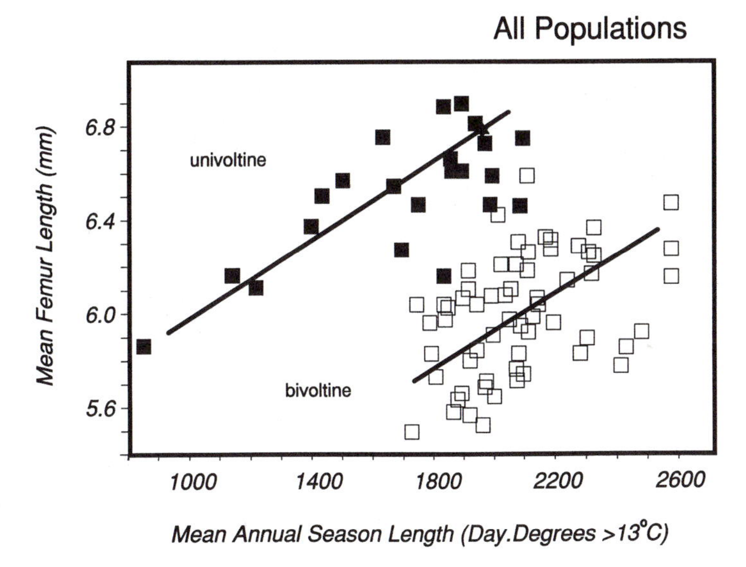

Figure 10.2 The pattern of body size variation (as indicated by femur length) among wild populations of striped ground cricket plotted against summer season length (day.degrees > 13°C) for populations ranging from Montreal, Canada, to southern Georgia, USA. Populations that exhibited a univoltine life cycle (i.e., one generation per year) are denoted by filled symbols; open symbols denote bivoltine populations (i.e., two generations per year). Note the sawtooth pattern of body size variation, whereby size is small in the north then gradually increases, then drops in the transition zone where populations shift from univoltine to bivoltine life cycles. Size then increases again with increasing season lengths. This pattern of size variation (small in the north, large in the south) is referred to as the "Converse to Bergmann's Rule" (Park 1949, Mousseau and Roff 1989, Mousseau 1997).

tion of genetics and environment to phenotypic expression in the wild. Our analysis indicates that about half of the total variation observed in body size in the wild reflects genetically based adaptation to the variation in season length experienced by these populations.

10.2.3 Cell size and its relationship to body size

To determine whether the observed variation in cricket body size reflects a corresponding variation in cell size or cell number (i.e., number of cell divisions before final ecdysis), we measured cell size for wild-caught and laboratory-reared crickets. Figure 10.6 shows the pattern of cell size variation for wild-caught crickets ranging from Montreal, Canada, to Georgia, USA. There is highly significant variation in cell size, with northern crickets having larger cells, on average, compared to their southern counterparts. Given that body size is smaller in the north, this implies that body size behaves either independently of cell size or in opposition to it. This suggestion is further emphasized by split-brood laboratory experiments where full-sib families were divided into two groups: half were reared in cool temperatures, and the other half were reared in a hot environment (fig. 10.7). High temperatures resulted in large crickets with small cell size, whereas cool temperatures yielded the reverse.

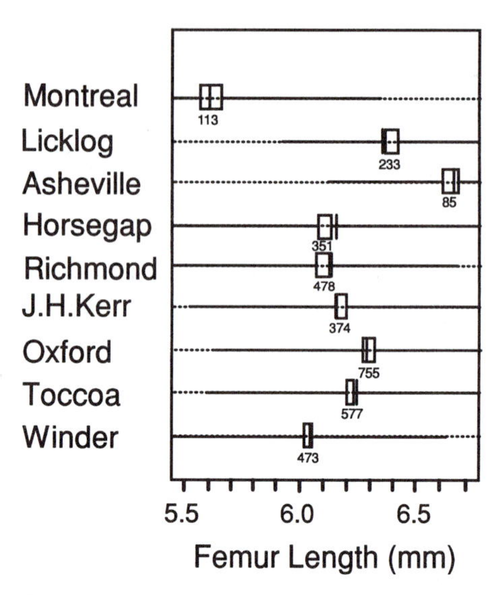

Figure 10.3 Body size variation among populations of striped ground cricket following two generations of rearing in a common garden (30°C, 14:10 h light:dark). Boxes denote 95% confidence intervals. Numbers below boxes indicate sample sizes. There were highly significant differences among populations in the common garden that paralleled the variation observed in nature (fig. 10.5). Data from Mousseau and Roff (1989).

Thus, although cell sizes are indeed larger in cool environments, as has been suggested by van Voorhies (1996), cell size variation per se cannot be responsible for the converse to Bergmann's Rule observed in this and many other orthopterans (Mousseau 1997). At present it is not known if cell size variation is entirely ecophenotypic or if there is genetic variation for this trait.

10.2.4 Ovipositor length in the laboratory and field

Female crickets, and many other insects, have an ovipositor that is used to deposit eggs into a substrate suitable for embryonic development. In *Allonemobius socius* and *A. fasciatus*, the depth at which eggs are deposited into the ground is positively correlated to the length of the ovipositor. As with cell size, ovipositor length follows a clinal size pattern opposite to that observed for body size (Mousseau and Roff 1995; fig. 10.8). Crickets from high latitudes and altitudes have longer ovipositors than crickets from warmer locations, both in the wild and in the laboratory after two generations of common garden rearing (fig. 10.9). Also, there is a good relationship between mean ovipositor length in the wild and crickets reared in the common laboratory environment (fig. 10.10; Mousseau and Roff 1995). This supports the hypothesis that there are genetically based differences for ovipositor length in this species. However, there are also strong environmental influences on ovipositor development (fig. 10.11). When full-sib families from a southern bivoltine population were split between two rearing environments (hot temperatures and long photoperiod versus cooler tempera-

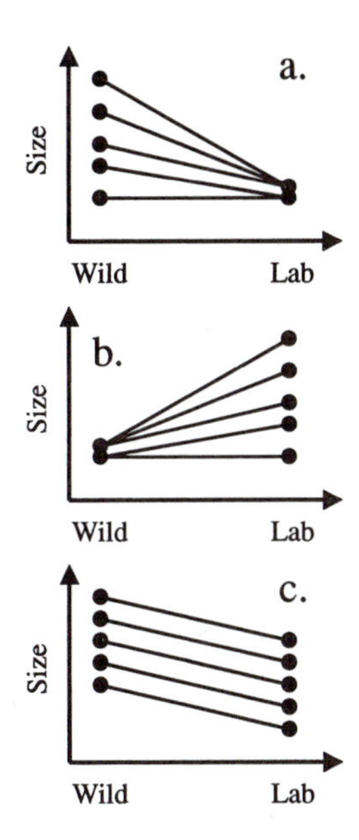

Figure 10.4 Hypothetical reaction norms for the effect of rearing environment on body size. (a) Variance among genotypes/populations is much lower in the laboratory than in the field. In this scenario, the simple common garden approach (i.e., organisms are reared in only one environment) would erroneously suggest that phenotypic variation in the wild is due to environmental effects. (b) Variance in the laboratory is much higher than that observed in the wild. This would suggest genetic variation among genotypes/populations results in convergent size in nature. Alternatively, genetic variation is masked by environmental effects in the wild. (c) Variance in size is similar under laboratory and field conditions, but the means have changed. This indicates environmental effects, but no genotype × environment interactions.

tures and shorter photoperiod), crickets reared in the hot environment developed a relatively short ovipositor compared to crickets reared at cooler temperatures (Mousseau and Roff 1995). The reaction norm observed in the laboratory is paralleled in the wild in bivoltine populations (fig. 10.8).

10.2.5 The effects of egg depth on overwintering survival in a cricket

It has been suggested that a long ovipositor (and deep egg deposition in the soil) is adaptive for crickets that lay overwintering eggs that need to endure long periods of cold and desiccation (Masaki 1979, 1986, Mousseau and Roff 1995). However, this hypothesis has never been rigorously tested. In addition, Bradford et al. (1993) have clearly demonstrated that short ovipositors (and shallow egg deposition) are adaptive when eggs will be directly developing (i.e., nondiapausing) because of the difficulties newly hatched nymphs face while excavating themselves from the soil. Indeed, Bradford et al. (1993) were unable to detect any advantages to deep egg deposition (or long ovipositors) for the range of species and conditions they tested.

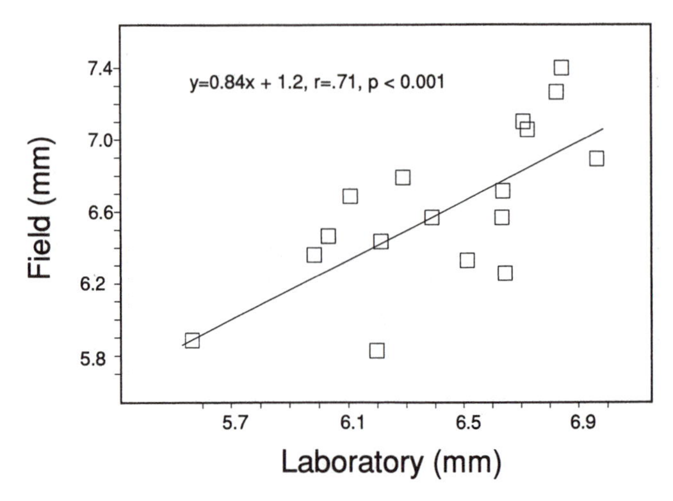

Figure 10.5 Body size variation for wild-caught versus laboratory-reared populations. This plot suggests that much of the variation in size observed in the wild results from genetic differentiation among populations (about 50%). The slope does not differ significantly from 1, indicating no genotype × environment interaction. However, wild crickets had femurs that were on average 1.2 mm larger than their laboratory-reared descendants. Redrawn from Mousseau and Roff (1989).

To test the hypothesis that geographic variation in ovipositor length and egg depth have adaptive significance in the striped ground cricket, we conducted a series of egg transplant experiments in the wild. Diapausing eggs from a laboratory stock were collected and placed at varying depths within open-bottomed plastic tubes (see fig. 10.12 for a schematic of the experimental design). These tubes were then transplanted to five sites chosen to span the range of environments this cricket's eggs might experience in nature (Columbia, South Carolina, to Montreal, Canada; see fig. 10.12), plus a controlled environment in the laboratory (constant 4°C). Tubes containing eggs were planted in the ground in October, left to overwinter, and were retrieved the following April. Tubes were then placed at 28°C in the laboratory, and all hatchlings were counted for estimates of overwintering survival. When eggs were placed in the wild (e.g., Montreal, Canada, or at high altitudes along the Blue Ridge Mountains of North Carolina), only the deepest eggs survived (fig. 10.13). However, in controlled laboratory conditions, survival was not related to egg depth. This result supports the hypothesis that the variation observed for ovipositor length in the wild is adaptive.

10.2.6 Geographic variation in diapause

Many insects and plants overwinter in a state of arrested development known as diapause (for insects) or dormancy (for plants). The physiological state of diapause protects the individual from inclement conditions and coordinates development with the favorable season(s) (Tauber et al. 1986, Danks 1987). In the striped ground cricket, the expression of diapause in the wild follows a stepwise pattern with season length. In northern and high altitude univol-

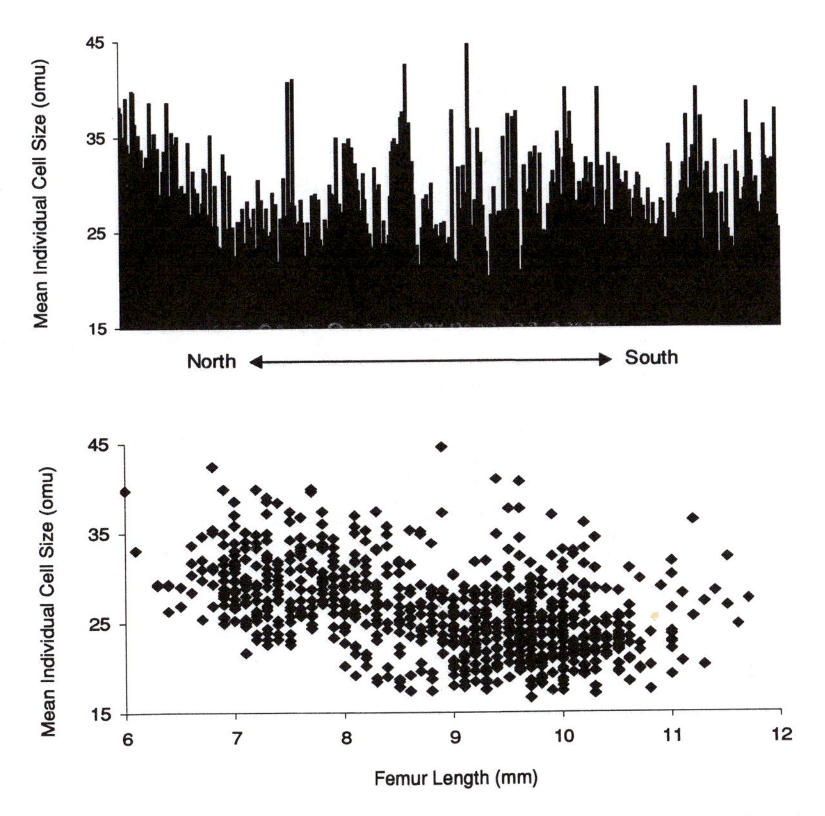

Figure 10.6 Cell size variation for 849 wild crickets collected from Montreal, Canada, to southern Georgia, USA. Although crickets from the coldest sites had significantly larger cell sizes (cell sizes based on the mean diameter of 10 epithelial cells on the femur), there was no overall pattern of geographic variation (top panel). However, there was a tendency for cell size and femur length to be inversely related (bottom panel), indicating that greater body size is achieved through greater numbers of cell division (i.e., cell number) rather than as a consequence of cell size.

tine populations, all individuals diapause, whereas in southern bivoltine populations, the first-generation adults produce direct-developing eggs that do not diapause, while second-generation adults produce eggs that diapause and overwinter. The pattern of variation in the wild reflects a combination of both genetic and ecophenotypic control that acts both via the mother (i.e., maternal effects; Mousseau and Fox 1998) and the developing embryo (Mousseau and Roff 1989, Olvido et al. 1998). When reared in a common garden, the expression of diapause shows continuous clinal variation, indicating genetic differentiation among populations (fig. 10.14). However, this genetically based variation in propensity to diapause is modulated by direct environmental effects experienced both by the mother during oviposition and by the embryos during early development (fig. 10.15; Olvido et al. 1998). Thus, both genetic and environmental effects act in concert to generate the stepwise patterns of expression observed in nature.

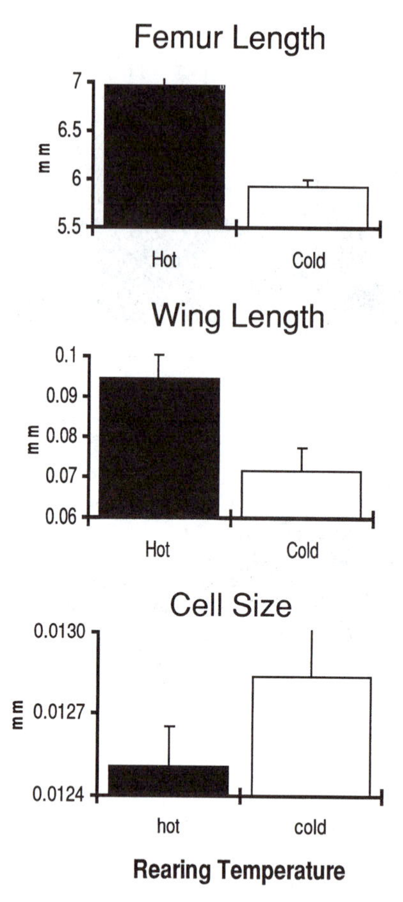

Figure 10.7 The effects of rearing temperature on body size and cell size in the striped ground cricket. The nymphs from 51 full-sib families (424 individuals) were split-brood reared in two environments. The hot environment was 30°C; the cold environment was 24°C. Crickets reared in the hot environment were dramatically larger than their cold-reared siblings, but had significantly smaller mean cell diameters. This indicates that that increased body size is a consequence of increased cell numbers, not cell size, as was found for wild-caught crickets (fig. 10.6).

10.2.7 Geographic variation in egg size

As mentioned above, northern and high altitude striped ground crickets possess long ovipositors used to place eggs deep in the soil. Although being deep in the soil has benefits with regard to overwintering survival (fig. 10.13), it does impose a cost in terms of nymphal emergence from the soil after hatching; nymphs emerging from eggs at greater depths suffer higher mortality than shallow eggs (Bradford et al. 1993). However, when comparing among cricket species, Bradford et al. (1993) found that depth effects were considerably less for crickets with large eggs (e.g., *Gryllus firmus*). Given the large variation in ovipositor length

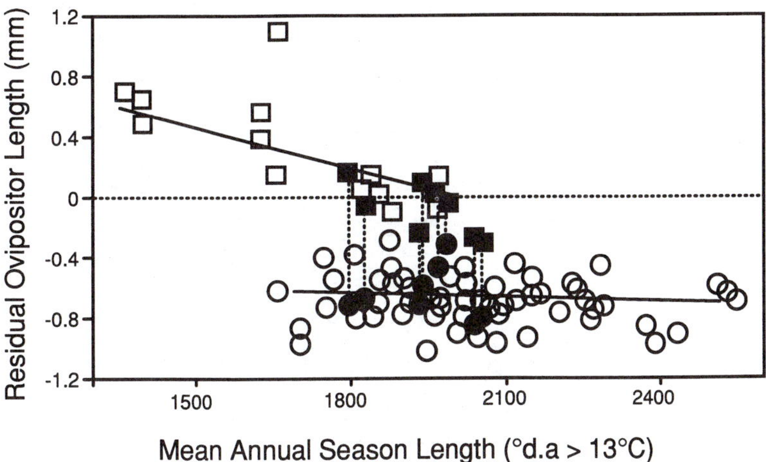

Figure 10.8 Clinal variation in mean ovipositor length for wild-caught striped ground crickets collected from the mountains of North Carolina to southern Georgia, USA. Plotted is the mean ovipositor length after body size effects are removed by regressing ovipositor length on femur length and determining the residuals from a common allometric relationship. Circle symbols refer to first–generation bivoltine populations (i.e., crickets that will be producing direct-developing embryos); squares denote univoltine and second–generation bivoltine crickets (i.e., crickets that will be laying diapausing–overwintering eggs). Filled symbols refer to populations for which both first– and second–generation crickets were collected. There is no pattern of geographic variation for first–generation bivoltine crickets, whereas crickets that produced overwintering offspring displayed a negative relationship between ovipositor length and season length of the collection site (i.e., crickets from cold sites had relatively long ovipositors). Season length at each site was estimated from climatological data obtained from the National Climatic Data Center as the average annual total day°degrees > 13°C (the lower threshold temperature for development in *Allonemobius fasciatus*). Note that in bivoltine populations the second generation develops a significantly longer ovipositor than their first-generation parents. Data from Mousseau and Roff (1995).

(and egg depth) that occurs in the striped ground cricket, it was predicted that egg size should be larger in northern populations as a result of selection mediated via greater egg depth. To test this hypothesis, six populations were reared in a common garden. Three populations were from north of the hybrid zone between *A. fasciatus* and *A. socius*, three were from the south, and one population originated from within the hybrid zone. As predicted, northern crickets had significantly larger eggs and hatchling sizes than their southern counterparts, while crickets from the hybrid zone population produced eggs and nymphs of intermediate size (Mousseau, unpublished data). This common garden experiment supports the hypothesis that genetically based differences in egg size have evolved as an adaptation to the differing depths at which mothers place their eggs in different environments.

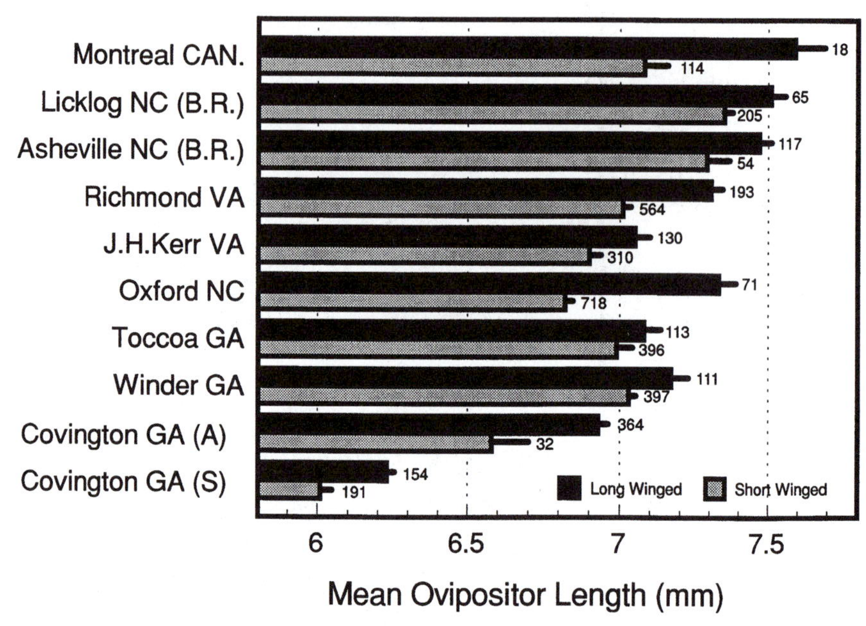

Figure 10.9 Mean ovipositor length for nine populations of crickets reared in the laboratory. All populations except Covington, GA, represent individuals measured following three generations in a common garden (30°C, 14:10 h light:dark photoperiod), and thus the significant differences observed among populations primarily reflect genetic variation, rather than maternal or grandmaternal environmental effects. Covington (A) crickets were reared under similar conditions to the other populations (30°C, 15:9 h light:dark), while Covington (S) were reared under springlike conditions (27°C day, 23°C night; 11:13 h light:dark). Standard errors and total sample sizes are indicated next to each bar, and long- and short-winged morphs are plotted separately because of differences in body size. Populations are ranked from top (Montreal, Canada) to bottom (Covington, Georgia) in ascending order with respect to mean annual season length in the wild. Crickets from cold locations (top) have significantly longer ovipositors than crickets from warmer sites.

10.2.8 Summary: geographic variation in the striped ground cricket

The use of common garden approaches has permitted the elucidation of genetically based interpopulation variation in development, morphology, and life-history in the striped ground cricket that has evolved in response to the geographic range in climate that this cricket experiences. In the north and at high altitudes, crickets develop quicker into smaller crickets that have long ovipositors, deeper egg placement, larger eggs, and a greater propensity to diapause than their southern counterparts. It is worth noting that for most traits it is the interaction between genetic differentiation and direct environmental effects (i.e., phenotypic plasticity) that regulates trait expression in the wild. For example, in bivoltine populations,

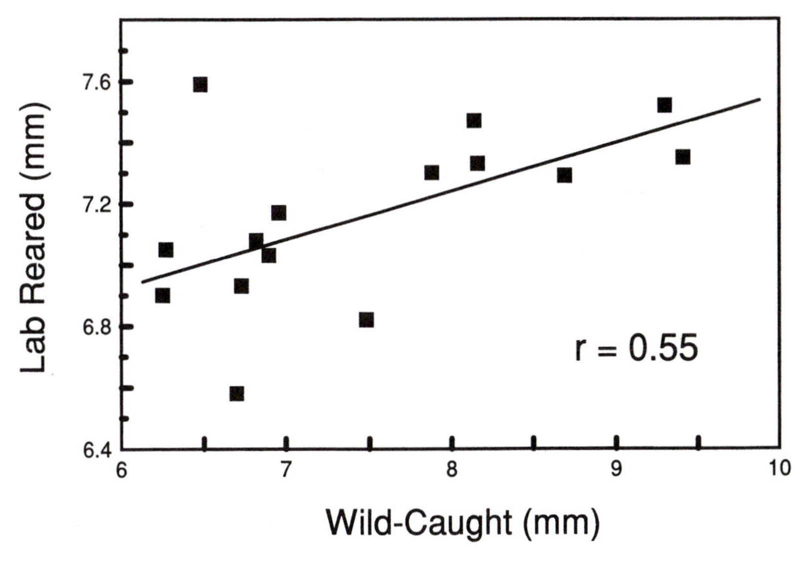

Figure 10.10 The relationship between mean ovipositor lengths of laboratory-reared populations and their field-collected progenitors. This relationship suggests that a significant proportion ($r^2 = .3$, $P < .035$) of the variation in ovipositor length observed in the wild can be attributed to genetic differentiation among populations.

first-generation crickets have a short ovipositor and direct-developing eggs, whereas second-generation crickets have long ovipositors and diapausing eggs.

Similar patterns of variation have been observed in many other crickets, most notably a variety of species studied by Masaki and colleagues in Japan (e.g., Masaki 1967, 1973, 1978, Tauber et al. 1986), suggesting that these patterns of interpopulation variation reflect general adaptive strategies common to a large number of independently evolving species.

10.3 Adaptation to Altitudinal Variation in the Lessor Migratory Grasshopper

In many ways, altitudinal and latitudinal clines are analogous. Season length and mean temperature are both negatively related to altitude and latitude, and at first glance one might expect life-history variation to evolve in a parallel manner. However, there is one significant difference between these two types of clines (all else being the same) with respect to evolutionary response to selection along a climatic gradient: gene flow among geographically close but ecologically divergent populations may act as an impediment to local adaptation and interpopulation genetic differentiation. This is more likely to be the case along altitudinal clines where climate can change dramatically over short distances. In many ways this idea is similar to that developed by Kirkpatrick and Barton (1997) concerning the limits to the latitudinal range expansion of a species. In essence the argument is that the flow of alleles from adjacent but ecologically different populations (i.e., higher or lower altitude or latitude) is

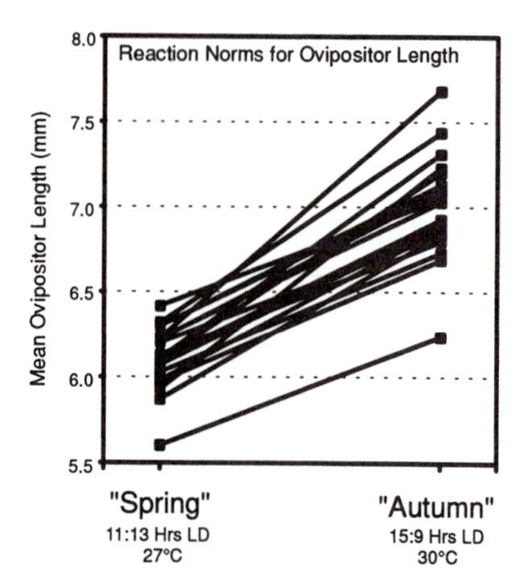

Figure 10.11 The effect of rearing environment on mean ovipositor length in 20 full-sib families that had been split between springlike (27°C, 11:13 h light:dark photoperiod) and autumn-like conditions (30°C, 15:9 h light:dark photoperiod). This figure suggests that crickets hatching in the spring (i.e., first-generation bivoltine) would develop a short ovipositor, whereas crickets hatching in the middle of the summer (i.e., second-generation bivoltine) would develop a significantly longer ovipositor. These reaction norms are analogous to those observed for field-collected crickets (fig. 10.8) and suggest the importance of adaptive, environmentally modulated phenotypic plasticity in bivoltine life cycles.

maladaptive because these alleles evolved under different climatic/ecological conditions. Two predictions stem from this thesis: (1) under conditions of high gene flow in ecologically heterogeneous environments, dispersal rates should decline. It is interesting to note that high altitude insects are often flightless, although other hypotheses have been proposed to explain this observation (Roff 1992). (2) In the face of gene flow, environmentally cued phenotypic plasticity is expected to evolve. Adaptive plasticity allows the individual to adjust its phenotype in response to the local environmental conditions in which it finds itself. Under these conditions it could be said that plasticity itself is an adaptation to the deleterious effects of gene flow in highly heterogeneous environments. This idea is somewhat novel and as yet largely untested. However, it provides a rationale for comparative studies of life-history variation simultaneously along latidudinal and altitudinal clines.

10.3.1 Population variation in a model grasshopper system

The lessor migratory grasshopper, *Melanoplus sanguinipes*, is an excellent model for studies of adaptive genetic variation. It is found throughout North America (Alexander and Hilliard 1969) and as a consequence has adapted to a wide range of climates (Dingle et al. 1990). Of particular interest here is this species' range in California; it is found from the deserts of southern California to the upper reaches of the Sierra Nevada mountains (>3000 m altitude), and it is possible to find very rapid elevational gradients in many parts of the state.

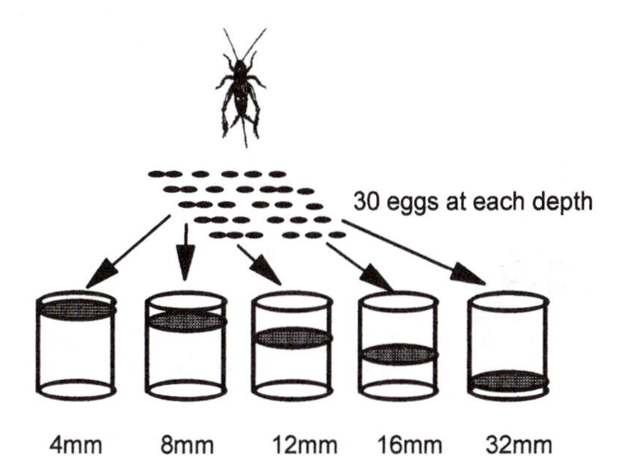

Figure 10.12 A schematic for an experiment designed to test for the adaptive significance of ovipositor length and egg depth in the striped ground cricket. Approximately 12,000 eggs were placed at varying soil depths (4, 8, 12, 16, and 32 mm) in open-bottomed 7-dram plastic vials (30 eggs per vial). In October 1995, 50 vials (10 replicates of each depth) were inserted flush with the ground at each of five geographic locations ranging from Montreal, Canada (high latitude), Mt. Mitchell, NC (2000 m), Asheville, NC (1100 m), and two sites near Columbia, SC (low latitude and altitude). An additional 150 vials (30 from each depth) were placed at 4°C in a cold room as controls. Vials were removed from the ground the following April and incubated for 30 days (or until hatching stopped) at 28°C in the laboratory.

Because this grasshopper has wings and is a capable flyer, the opportunity exists for high levels of gene flow among geographically close but ecologically diverse populations. We have used this grasshopper system to test for genetically based life-history, morphological, and physiological variation among populations along elevational gradients. Given the opportunity for gene flow among populations, we expected there to be little genetic differentiation among populations.

10.3.2 Geographic variation in nymphal development time and adult body size

There is a large range among populations in the length of the growing season available for grasshopper development, with essentially a year-long season for southern coastal populations (e.g., San Diego), and a 6- to 8-week growing season at high altitudes (e.g., Lake Winnemucca, near Carson's Pass) (Dingle et al. 1990, Dingle and Mousseau 1994). This variation in available season length is reflected in differences in development rates among grasshopper populations. At low altitudes grasshoppers take longer to develop and attain a larger size than

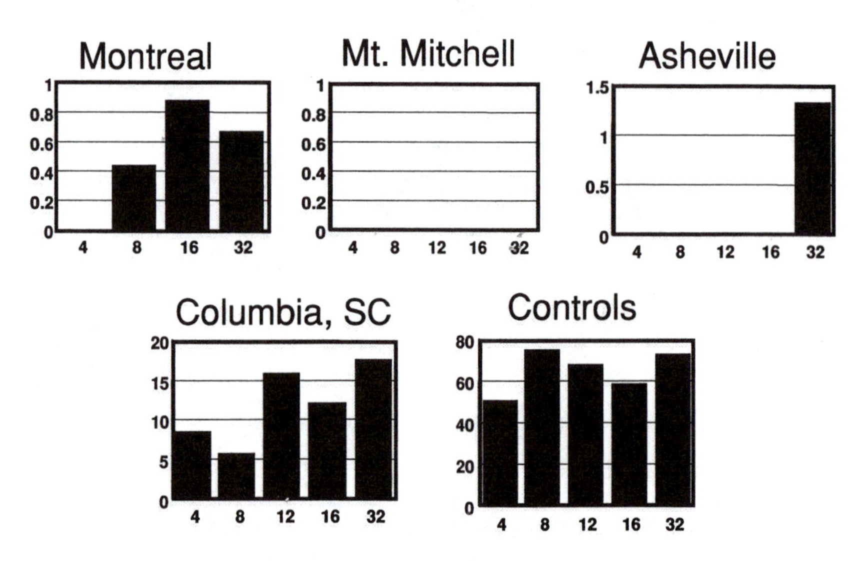

Figure 10.13 Egg survival versus overwintering depth for the study sites. Survival for the controls averaged 62%. Survival at all three cold sites (Montreal, Mt. Mitchell, and Asheville) was low; no eggs survived the winter from the Mt. Mitchell site. There was increased survival at greater depth for all sites. However, only the deepest eggs survived for the Montreal and Asheville sites. These results support the hypothesis that overwintering survival is enhanced by deeper egg placement by females where winter climates are harsh.

at high altitudes (table 10.1 and fig. 10.16). Because these experiments were conducted in a common garden, this indicates highly significant genetically based variation among populations. This pattern of altitudinal variation in body size and development time, whereby size is smaller at high elevations, I refer to as Dingle's Rule.

10.3.3 Geographic variation in embryonic development time and stage of embryonic diapause

As with the striped ground cricket, *M. sanguinipes* overwinters as a diapausing embryo. Using a common garden approach, it was found that embryonic development times and the stage at which embryonic diapause occurs vary dramatically among 23 populations of this species in California. Grasshoppers were collected from a wide range of latitudes (32°57' N to 41°20' N) and altitudes (10–3031 m), spanning much of the variation in climatic conditions experienced by these insects in California. Total embryonic development times in a common garden were positively correlated to the mean annual temperature of the habitat from which the grasshoppers were collected and ranged from <19 days for some high altitude populations to >30 days for low altitude populations (Dingle and Mousseau 1994). These

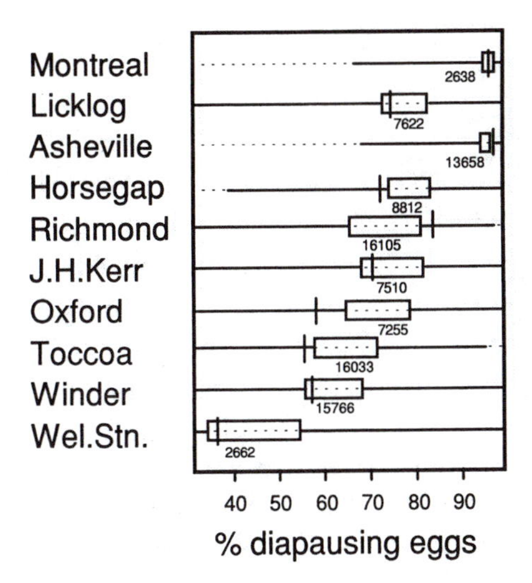

Figure 10.14 The mean proportion of eggs in diapause for striped ground cricket populations reared in the laboratory (30°C, 14:10 h light:dark). The populations are ranked according to mean annual season length. Boxes represent ± 1 SE about the mean; vertical lines denote the median proportion of diapausing eggs. The total number of eggs examined is denoted below each box. This result suggests genetically based clinal variation in the propensity for crickets to enter diapause in a common garden environment. Data from Mousseau and Roff (1989).

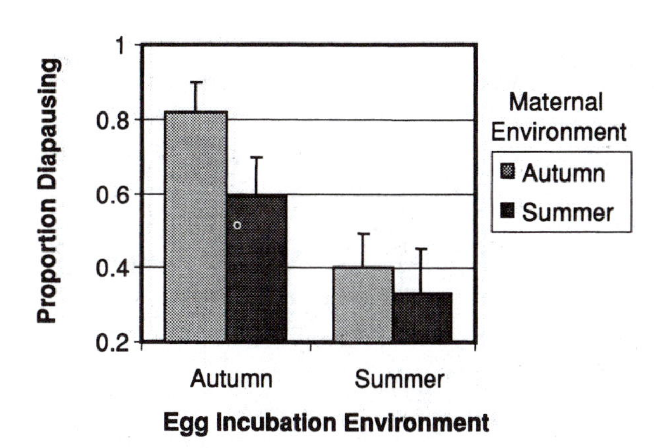

Figure 10.15 The effects of maternal and embryonic environment on the expression of diapause in the striped ground cricket. The summer treatment was 31°C, 15:9 h light:dark; the autumn treatment was 24°C, 11:13 h light:dark. There were significant maternal environmental effects on diapause; mothers laying in the autumn environment produced a significantly higher proportion of diapausing offspring than did mothers experiencing the summer treatment. However, there were also highly significant direct environmental effects on developing embryos that could partially negate or reinforce the maternal effect. Data from Olvido et al. (1998).

Table 10.1 Nymphal development times (hatch to adult eclosion) for *Melanoplus sanguinipes* grasshoppers collected from four populations in the Sierra Nevada Mountains of California, USA

Altitude (m)	Development time (+ SE)	Season length (°.days > 15°C)	*n*
90	27 ± 0.08	486	40
1500	25.1 ± 0.08	452	40
2150	21.5 ± 0.10	387	32
2650	21.2 ± 0.15	382	26

All grasshoppers were reared in a common garden environment of 14:10 h light:dark and 33°C. Data from Dingle et al. (1990).

grasshoppers overwinter as diapausing eggs, and the proportion of embryonic development completed before diapause was significantly higher in populations collected from cool habitats (>70% development completed before diapause) than in populations collected from warm environments (<26%). The length of prediapause development time is determined by the stage of embryonic development at which diapause occurs, which varied considerably among populations of these grasshoppers. Grasshoppers from warmer environments tended to diapause at early stages of embryogenesis, whereas grasshoppers from cooler environments diapaused at late stages (fig. 10.17). The combined effect of variation in embryonic development time and variation in the stage at which diapause occurs resulted in a dramatic reduction in the time needed to hatch in the spring. Populations from warm environments required up to 20 days (at 27°C) to hatch; populations from cool environments required as few as 5 days to complete embryonic development before hatching (fig. 10.18). We interpret

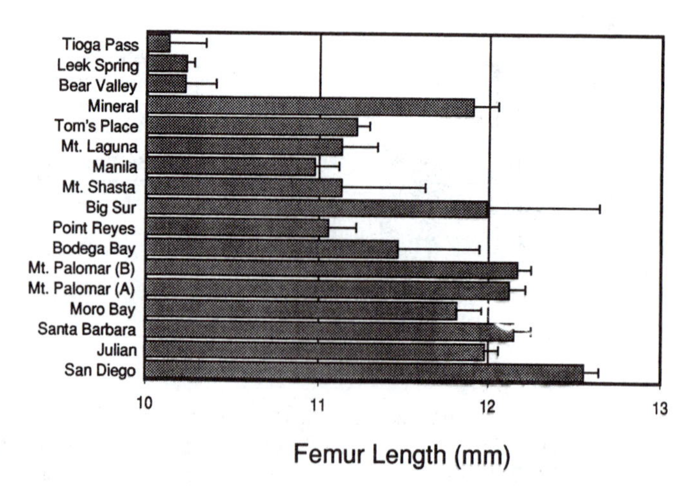

Figure 10.16 Geographic variation in body size (as indicated by femur length) of the lessor migratory grasshopper, *Melanoplus sanguinipes*. Means + 1 SE are plotted, and populations are ranked according to mean annual temperature (cold at top). All grasshoppers were reared in a common garden at 31°C, 14:10 h light:dark. Grasshoppers from cold, high altitude sites were dramatically smaller than grasshoppers from southern, low altitude sites.

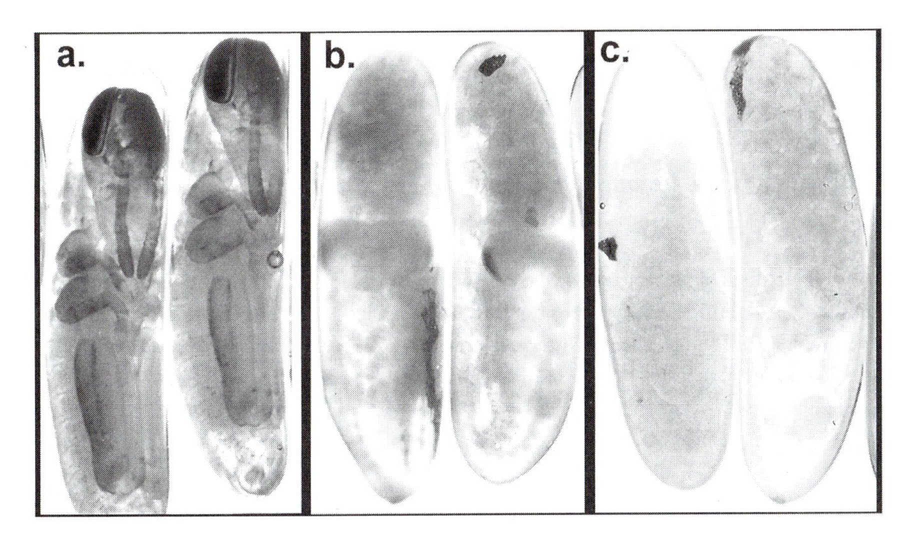

Figure 10.17 *Melanoplus sanguinipes* eggs diapausing at three different states of embryonic development. Eggs from high altitude populations most often diapaused at a late stage of embryonic development (a), whereas low altitude populations tended to diapause at early stages (c). Some intermediate populations showed evidence of eggs diapausing at multiple stages of development. Late-stage diapausers required dramatically less time to hatch than did early-stage diapausers (fig. 10.18). Data from Dingle and Mousseau (1994).

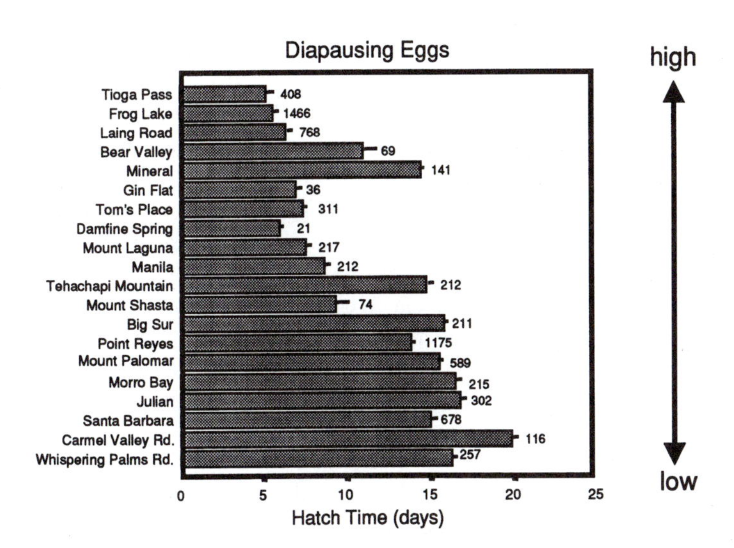

Figure 10.18 Mean hatch times for diapausing eggs in a common garden (27°C) from 20 populations ranked with respect to mean annual temperature (coldest on top). Error bars denote ±1 SE, and sample sizes are indicated next to each bar. There was a significant relationship between hatch time and mean annual temperature ($r = .86$, $P < .001$).

these findings as supporting the hypothesis that embryonic development time and the stage of embryonic diapause have evolved as adaptations to prevailing season lengths in the study populations.

10.3.4 Geographic variation in the biophysical properties of M. sanguinipes *cuticle lipids*

Due to their small size and large surface-to-volume ratio, insects are vulnerable to desiccation, especially in dry desert environments. Epicuticular lipids are the primary barrier to water loss in insects and other terrestrial arthropods. However, cuticular lipids are only effective when they are in a solid phase; water loss increases dramatically when lipids melt at high temperatures. We tested for adaptive variation in cuticle-lipid melting temperatures across this species' range in California. When reared in a common garden, grasshoppers from hot desert environments had lipid-melting temperatures that were dramatically higher than grasshoppers from cooler and/or humid environments (Gibbs et al. 1991), suggesting significant genetic variation among populations for this trait. However, there were also highly significant environmental influences on lipid-melting temperature (fig. 10.19; Gibbs and Mousseau 1994); individual grasshoppers were capable of changing lipid-melting temperatures in response to ambient temperatures during development. These findings suggest that both genetic and developmental effects act in concert to produce epicutilar lipids appropriate for the environment in which an insect develops (Gibbs et al. 1998).

10.3.5 Summary: Geographic variation in M. sanguinipes

The lessor migratory grasshopper shows considerable genetically based adaptive variation in life-history, morphology, and physiology among ecologically diverse (but geographically close) populations in California. This suggests that, despite the potential for high gene flow among populations, genetic differentiation is maintained. This differentiation no doubt reflects the high intensity of natural selection that occurs for these traits. At high altitudes the growing season available for insect development is extremely restricted, with snow-free periods often shorter than 2 months at high altitudes. This grasshopper has evolved both a short development time (and a consequent smaller body size) and a short postdiapause embryonic development time as mechanisms to dramatically increase development rate in cold, high altitude environments. The interpopulation variation observed for these traits appears to have a large genetic basis. In contrast, cuticular lipids, the primary mechanism by which this insect reduces water loss in hot, dry, desiccating environments, show evidence of both genetic differentiation and adaptive phenotypic plasticity (in response to ambient thermal conditions).

10.4 Testing for Heritable Genetic Variation within Populations: The Foundation for Predictive Evolutionary Ecology

Robert Henry Peter's life-long mantra was that for a science to have general utility, it must be predictive (Peters 1983). To generate testable evolutionary hypotheses in natural populations, it is necessary to first determine whether a trait that is under selection is heritable. Knowledge

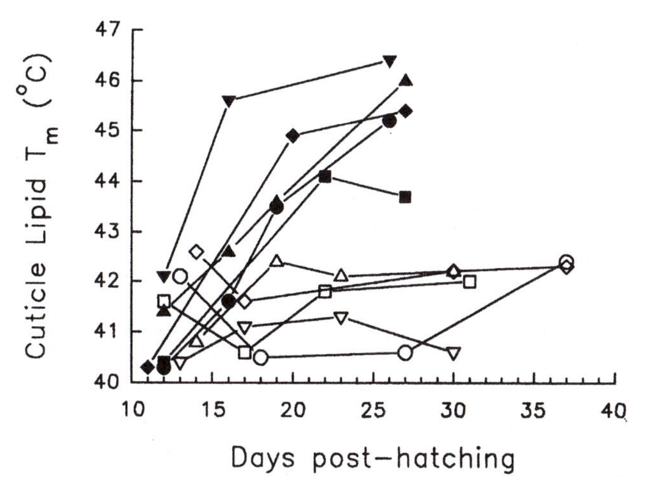

Figure 10.19 Temperature-induced phenotypic plasticity (acclimation) in the mean melting temperature (T_m) of epicuticular lipids for individual grasshoppers. Nymphs were hatched at 27°C and then transferred to either 29°C (open symbols) or 34°C (filled symbols). Each line represents a single nymph, and all individuals were full-siblings. Data from Gibbs and Mousseau (1994).

of the heritability of a trait and how it is genetically correlated to other traits can provide the basis for predicting population response to changing environmental conditions. Heritability, the proportion of total phenotypic variance attributable to the additive effects of genes, is a measure of the potential response to selection. Although this approach has been extremely productive in animal and plant breeding programs, it has rarely been applied to the study of natural populations. Natural populations have for the most part escaped classical quantitative genetic study because of the difficulty in using traditional experimental designs in the wild. Typically, quantitative genetic designs use controlled breeding regimes to estimate the covariance among individuals of known (and manipulated) relatedness (e.g., parents and offspring, full-sibs, half-sibs; Roff 1997). With the exception of a dedicated few (e.g., Grant 1986, this volume, Sinervo, this volume), most researchers have conducted their quantitative studies in the laboratory (e.g., most studies listed in Mousseau and Roff 1987, Roff and Mousseau 1987) with the hope that laboratory-based estimates would reflect the expression of genetic variation under natural conditions. Although there is some support for the argument that genetic variation expressed in a controlled laboratory setting can sometimes be extrapolated to the wild (Weigensberg and Roff 1996), there are good reasons to suspect that this is not generally the case. For example, it is widely believed that heritability will generally be lower in the wild because of the obscuring effects of natural environmental variation (Price and Schluter 1991, Houle 1992) and maternal effects (Mousseau and Dingle 1991a, Mousseau and Fox 1998). However, field-based heritabilities may be higher when selection and/or inbreeding is high in the laboratory. Similarly, the strength and direction of genetic correlations among traits are expected to be sensitive to environment (Service and Rose 1985). The remainder of this chapter explores several methods that can be used to estimate quantitative genetic variation in wild populations.

10.4.1 Heritability of calling song in the striped ground cricket

The males of many insects generate a calling song used to attract conspecific females. In crickets, males rub their forewings (the tegmina) together to generate the calling song. Calling song typically varies greatly among species and is often used as a taxonomic indicator. Because calling song, and female preference for a given song, is also often assumed to be the primary mechanism for prezygotic reproductive isolation among closely related species (e.g., Otte 1989), understanding the genetic basis for variation in calling song is a prerequisite for predictions concerning rates of speciation mediated via song/preference coevolution. This is especially true if reproductive character reinforcement for calling song differences in a hybrid zone is the presumed mechanism for the evolution of song differences (e.g., Benedix and Howard 1991).

The striped ground cricket species *Allonemobius fasciatus* and *A. socius* meet and hybridize along a zone that occurs from southern New Jersey in the east to at least southern Illinois in the west (Howard and Furth 1986). Although there has been discussion as to whether these two taxa are truly separate species (they hybridize readily both in the laboratory and in the field), there is evidence for some degree of genetic differentiation and reproductive isolation between these taxa. In addition, there is evidence that calling song differences between the two "species" show an abrupt change across the hybrid zone (Mousseau and Howard 1998), and it has been suggested that there may be evidence for character displacement for song within the hybrid zone (Benedix and Howard 1991). However, there is little evidence that females prefer conspecific over heterospecific song (Doherty and Howard 1996). Also, despite strong genetically based differences for songs among species when reared in a common garden (Mousseau and Howard 1998), large environmental effects have also been documented (Olvido and Mousseau 1995) that could easily obscure species-level variation.

To test whether heritable genetic variation for calling song is expressed under natural conditions, I have used a relatively new technique that compares wild caught parents with their laboratory-reared offspring to estimate lower bounds for heritabilties in nature (Lande in Coyne and Beecham 1987, Riska et al. 1989). This technique has been successfully used in several study systems during the past decade (e.g., Coyne and Beecham 1987, Prout and Barker 1989, Riska et al. 1989, Hoffmann 1991, Aspi and Hoikkala 1993, Groeters and Dingle 1996, Hoffmann, this volume) and shows great promise for organisms with short generation times that can be reared in a controlled environment. In essence, a regression is performed for laboratory-reared offspring on field-collected parents. (See Hoffmann, this volume, for more details.)

In 1990 wild parents were collected from within the hybrid zone between *A. fasciatus* and *A. socius* in southern New Jersey. These crickets were brought to the laboratory, and 118 mating pairs were generated. The offspring from these matings were reared in a common garden, and calling songs were measured as described in Mousseau and Howard (1998). Phenotypic variances were similar in the laboratory and field environments, allowing the use of a standard linear regression to estimate the genetic covariance among generations. Figure 10.20 shows the relationship between parents and offspring for four calling song components often associated with species recognition in crickets. In this hybrid zone population, heritabilities are moderately high and significant. Although the data are not presented here, song components had the highest heritabilities in the hybrid zone population, presumably reflecting the effects of allelic introgression from adjacent source species populations (Mousseau and Howard 1998, Roff et al. 1999). Similar effects of hybridization have been noted by Grant (this volume).

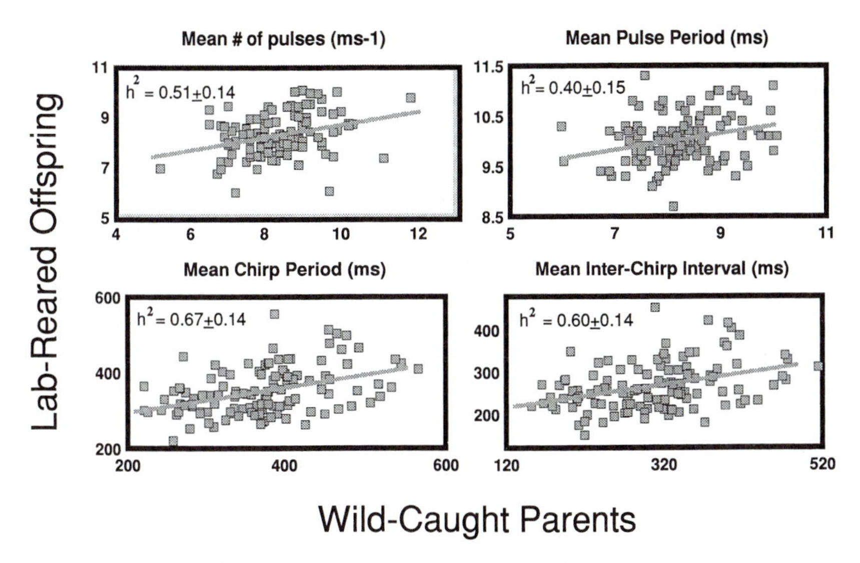

Figure 10.20 The relationship between 118 wild-caught fathers and their laboratory-reared sons for calling song components in a population of striped ground crickets within the hybrid zone between *Allonemobius fasciatus* and *A. socius*. These data suggest that these calling song components are moderately heritable in nature. Data from Mousseau and Howard (1998).

10.4.2 Heritability of larval performance in a leaf-mining beetle

The most direct approach for obtaining heritability estimates in the wild is to generate controlled pedigrees by directly manipulating mating and reproductive behavior. Offspring that are generated are subsequently tracked through their development, and the traits of interest are measured in much the same manner as a laboratory-based study. This approach is dependent on the ability to capture, manipulate, and track individuals of the study species. In general, it also requires that the organism complete its development (or the relevant portion of it) in a reasonable time frame. Thus, this approach may work for many insects, annual plants, and short-lived vertebrates, but it is not effective for trees or long-lived vertebrates (unless the investigator is not dependent on federal funding sources to support the work). The main advantage to this approach is that the investigator is able to optimize a breeding program to estimate quantitative genetic parameters (e.g., full-sibs, half-sibs, diallel, offspring–parent, etc.) using standard methods of analysis (e.g., Falconer and Mackay 1996, Lynch and Walsh 1998). One disadvantage is that this approach imposes patterns of mating that may not be "natural," which can lead to artifacts or biases similar to those observed for laboratory-based experiments. For example, it is possible that a natural population may exhibit a genetic correlation for a pair of traits that is generated as a consequence of nonrandom mating patterns that would be lost if the investigator imposed a random design. However, this approach permits genetic studies of organisms in their natural habitat, experiencing the full range of natural environmental variation. Essentially, this approach is analogous to a typical laboratory-based study except that organisms are reared in the wild.

We have used this approach to study the genetic basis to maternal oviposition behavior and larval performance in the leaf-mining buprestid beetle, *Brachys tessellatus*. In South Carolina, USA, this beetle shows a marked preference for several species of oak trees and is considered a specialist on the turkey oak, *Quercus laevis* (Turnbow and Franklin 1981, Waddell and Mousseau 1996). Adults emerge from the leaf litter in the spring and eat young foliage of the oaks until late May, at which time mating and oviposition starts. Eggs are laid singly on the upper surface of the leaf, and females hyperdisperse their eggs among leaves. Hyperdispersion has presumably evolved as an adaptation to reduce competition among siblings; most leaves will only support a single successful larva (Waddell 1996). The eggs hatch within 2 weeks, and the larvae burrow into the mesophyll of the leaf. Over the course of the summer and early fall a larva will excavate between 30% and 100% of the leaf tissue (hence the name "leaf miner"). Larvae remain within senescing leaves and pupate within the leaf (on the ground) during late fall and winter.

Because this beetle requires living foliage to survive, it cannot be easily manipulated in the laboratory. However, adults can be contained within mesh bags placed over clusters of leaves, and they will mate and oviposit within these bags. Females can then be rotated among host trees to test for host effects on beetle behavior and life-history. Similarly, male beetles can also be rotated among females, facilitating the generation of full-sib and half-sib family groups. Once an egg is deposited on a leaf, it can be marked and the leaf tagged with a unique identifier. Because larvae are restricted to the interior of a single leaf (and never move from this leaf until they emerge as adults), their performance (growth and survival) can be tracked through the season. Our ability to easily manipulate these beetles has permitted experimental designs in the wild that can be used to test for and estimate the amount of heritable genetic variation for traits associated with adaptation to host plant variation in nature.

We have found heritable genetic variation for several performance traits in this beetle (table 10.2). Based on 51 half-sib families, heritabilties ranged from 0.12 for lifetime fecundity to 0.43 for early larval survival. Heritability estimates generated using 169 full-sib families were similar for offspring weight at pupation and lifetime fecundity, but were significantly higher for early larval survival, suggesting the role of maternal effects on this trait (table 10.2).

Using this same approach we have also tested for the genetic basis to variation in oviposition preferences, larval performance, and their genetic correlations within and among host plants in this beetle. Our objective is to test the hypothesis that maternal host preferences have evolved in response to selection imposed by variation in offspring performance on different host plants. As with many organisms, this hypothesis is not addressable or meaningful in the laboratory. The ability to manipulate and track individuals in the wild has permitted us to tackle these important evolutionary questions in an ecologically relevant context.

10.4.3 Using molecular markers to infer heritability in the wild

The holy grail for many evolutionary ecologists is to examine the expression of genetic variation in naturally occurring, nonmanipulated populations because it is only under these circumstances that the interactions between genotype and environment and the role of nonrandom mating and selection on phenotypic variation can be fully appreciated. In recent years a number of approaches have been developed that permit the use of molecular markers to infer patterns of relatedness in natural populations, and from this to estimate the heritability of quantitative traits in naturally occurring populations (Ritland 1996, this volume, Ritland

Table 10.2 Heritabilities (h^2) for performance traits in experimentally manipulated beetles

	h^2	
Trait	Half-sib	Full-sib
Weight	0.18	0.18**
Lifetime fecundity	0.12*	0.10***
Early survival	0.43***	0.60***

Estimates are based on 51 half-sib, 169 full-sib families (51 sires, 169 dams), and 511 offspring.

$P<.05$; ** $P<.01$; *** $P<.001$.

and Ritland 1996, Mousseau et al. 1998). In the simplest case molecular markers can be used to infer a pedigree for a natural population, and this inferred pedigree can be analyzed using standard quantitative genetic models (e.g., Mousseau et al. 1998). Alternatively, if a large number of molecular markers are available, the exact degree of genetic relatedness can be estimated and used to test for heritable variation (e.g., Ritland's linear estimator method; Ritland 1996), and it may even be possible to test for quantitative trait loci (QTLs) if a very large number of markers are available. We have used this approach to examine the genetic basis to quantitative variation in trees and fish.

Detecting naturally occurring clones to test for heritable resistance to herbivory in an oak tree A common observation in many forests is the degree of heterogeneity among individual trees in the degree of damage inflicted by herbivores. In managed systems, where plantations are established using cultivars of known genetic background, it is relatively simple to test for the genetic basis to resistance. However, in natural systems genetic studies can be difficult because of the unknown genetic history and the long time required for tree growth and development. The use of molecular markers shows great promise for the assessment of quantitative trait variation in long-lived plants (and animals) because they can permit inferences concerning patterns of genetic relatedness within naturally occurring populations.

The turkey oak, *Quercus laevis*, propagates both vegetatively along underground runners, and sexually. Like most oaks, it is an obligate outcrosser. Most populations that have been examined are a mixture of clones (100% related) and less related individuals in a mosaic spatial pattern (Berg and Hamrick 1995, Waddell and Mousseau, unpublished data). Using a combination of molecular markers (RAPDs) and morphological characters, we have mapped more than 200 pairs of trees at a study site located within Sesquicentennial State Park, South Carolina, USA. One hundred of these pairs were clones; the other 100 pairs were either unrelated or siblings. At present our markers do not have sufficient resolution to differentiate between unrelated individuals and siblings, although we are now developing microsatellite markers for this purpose. Similarly, our current molecular markers do not permit mapping of clones at broader spatial scales. Because individuals within a given pair of trees were physically adjacent to each other (usually within 3 m of each other), environmental differences were minimal relative to spatially distant individuals. Using this mosaic of related and relatively unrelated pairs of trees has allowed us to test the hypothesis that level of herbivory has a genetic basis in this population. If traits associated with resistance have no genetic basis, then levels of herbivory should be random, and clones will be no more similar than nonclones. However, if traits have a significant genetic basis, then clones (within pairs) should be relatively more similar to each other than nonclones (within pairs). Figure 10.21 shows a simple schematic for this design.

Figure 10.22 shows the degree of similarity between nonclones and clones for various aspects of herbivory by the turkey oak leaf-mining beetle, *Brachys tessellatus*. These data are consistent with the hypothesis that resistance has a genetic basis because clones (within pairs) were much more similar with respect to levels of herbivory then were nonclones (within pairs). Because clonal pairs are genetically identical, differences between trees presumably reflected environmental effects on these traits. Analysis of variance indicated that genetic relatedness (clone, nonclone) had a highly significant effect on all aspects of herbivory on the trees ($P<.0001$ for all characters). Variance component analyses found that broad-sense heritabilities ranged from 0.05 to 0.11 for these traits (table 10.3). Although these heritability estimates include the effects of dominance, maternal effects, and other nonadditive genetic effects, they likely underrepresented the true heritability for these traits, as many of the nonclone pairs were likely sibs (or other related individuals), which would tend to decrease the difference between trees (within pairs), assuming a genetic basis to the traits.

Overall, this study suggests that several aspects of resistance to herbivory in this oak are genetically based and would likely respond to selection. At this point, the mechanisms underlying resistance (e.g., leaf morphology, tree phenology, leaf biochemistry, etc.) are unknown. However, the results from this preliminary study indicate that further studies of this sort are both feasible and likely to provide valuable insights to the potential rates of evolutionary response to selection imposed by herbivores in a natural system.

Testing for heritable variation in a captive salmon population using DNA fingerprints Many animals of economic and ecological interest have life cycles that make them intractable for quantitative genetic study. Long generation times and/or complex (and often unknown) life histories often preclude manipulation of the study population in any controlled way. And even for species (e.g., many fish) that can be crossed and reared in a hatchery environment, it is often not economically or logistically feasible to rear the large numbers of families required for rigorous quantitative genetic studies.

As with trees, the use of molecular markers can provide novel opportunities for genetic studies by allowing the inference of genetic relatedness and subsequent tests for quantitative genetic variation in naturally occurring populations. Mousseau et al. (1998) developed a method that uses molecular markers to estimate heritability and genetic correlation for quantitative traits in populations that have simple family structure (e.g., mixtures of full-sibs and unrelated individuals). This novel method was tested on a population of captive-bred Pacific chinook salmon, *Oncorrhynchus tsawaytscha*, which were crossed and reared in a controlled hatchery environment. The goal was to test this new method on a population for which heritabilities had been estimated using conventional approaches.

The basic approach is as follows. First, a captive population containing a mixture of full-sibs and unrelated individuals was generated. Wild-caught males and females collected from Robertson Creek, British Columbia, Canada, were mated to generate 15 full-sib families. During the first year, offspring were reared in large freshwater holding tanks. In the second year offspring were transferred to large ($10 \times 10 \times 10$ m) sea pens. One hundred and seventy 2-year-old fish were collected and measured for phenotypic traits. Second, fish were genotyped using

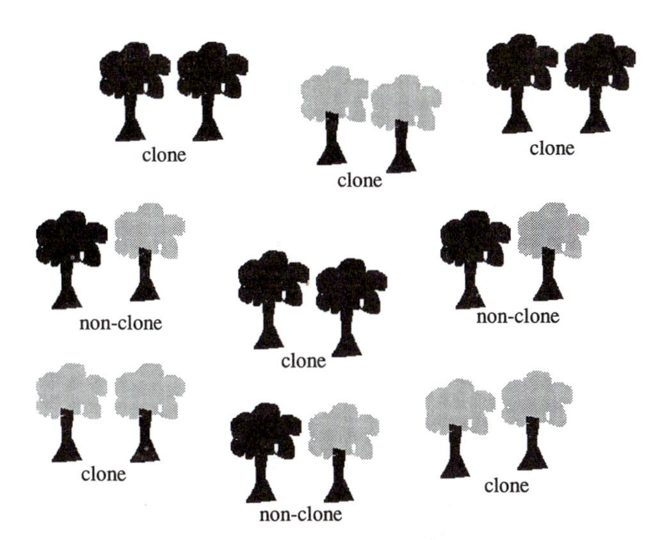

Figure 10.21 A simplified schematic of the experimental design used to study genetic variation for resistance traits in a wild turkey oak (*Quercus laevis*) population. One hundred pairs of trees were genetic clones (based on a combination of molecular markers and morphological traits); another 100 pairs were nonclones. Pairs of trees were scattered randomly throughout the study site in Columbia, South Carolina. Because individuals within each pair were spatially close to each other, environmental differences were minimized.

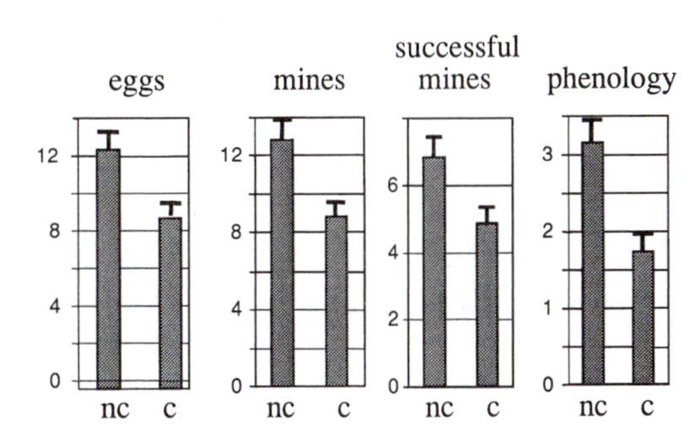

Figure 10.22 The mean difference between clones (c) and nonclones (nc; within pairs of trees) for resistance traits in a wild turkey oak population. Eggs refers to the average number of eggs laid by wild leaf-mining beetles (*Brachys tesselatus*) per 100 leaves, mines were the number of leaf-mining beetle larvae that initiated development, successful mines were the number larvae that successfully pupated, and phenology refers to the median date of leaf expansion on the trees. On average, the mean difference was significantly less for clone pairs (i.e., clones were more similar to each other than nonclones), suggesting a genetic basis to phenotypic variation (table 10.3).

Table 10.3 Heritabilities (h^3) of traits in a wild turkey oak (*Quercus laevis*) population based on similarities within pairs of clones and nonclone trees (100 clone and 100 nonclone pairs; total of 400 trees)

Trait	h^2
Mean number of eggs per leaf	0.08
Mean number of mines (larvae) per leaf	0.08
Mean number of surviving larvae per leaf	0.05
Tree phenology in the spring	0.11

All estimates are significant at P<.01. See figure 10.21 for a schematic of the study design.

two hypervariable minisatellite DNA fingerprint probes (Heath et al. 1994, 1995). Next, a maximum likelihood model was developed using the molecular marker data that tested for relatedness among all pairs of individuals in the study population (Mousseau et al. 1998). Pairs of individuals were determined to be either full-sibs or unrelated based on the degree of similarity for markers. Heritabilities and genetic correlations were then estimated using the equations outlined in Mousseau et al. (1998), and standard errors were generated by bootstrapping the data.

The results from this analysis are presented in table 10.4. Although limited sample sizes precluded the detection of small heritabilities, in general, estimates were in accordance with those generated using classical approaches for salmon (e.g., Gjerde and Gjedrem 1984, Heath et al. 1994) and for animals in general (e.g., Mousseau and Roff 1987, Roff and Mousseau 1987, Houle 1992). And, despite the limitations of this preliminary study, it does effectively demonstrate the potential ability to examine quantitative genetic variation in wild, undisturbed populations.

10.5 Conclusions

If we accept that the principle objectives for studies of genetic variation are to understand past evolutionary events and to predict future responses to natural and anthropogenically induced selection, then greater effort must be directed toward studies conducted in naturally occurring populations. Although common garden approaches have been effective at elucidating past and probable future responses to environmental change, our ability to predict *rates* of future response are premised on knowledge of the genetic architecture underlying the traits responsible for adaptation. A first step in this direction is to estimate the heritability for and the genetic correlations among traits involved in adaptation. Given the importance the environment plays in modulating the expression of genetic variation, such studies must be conducted in wild, naturally occurring populations. In this chapter and in the other chapters in this book, a wide variety of techniques have been outlined that can be brought to bear on this problem. Principal among these methods are (1) comparisons of wild parent and laboratory-reared offspring, a method promoted by Coyne and Beecham (1987), Riska et al. (1989), and Prout and Barker (1989), which is effective for species that can be reared in the laboratory; (2) the generation of artificial populations in the wild (e.g., Mitchell-Olds 1986, Shaw 1986), whereby controlled matings are conducted and offspring are reared and tracked in a natural setting. This method is effective for species that can be manipulated and followed

Table 10.4 Estimates of heritabilties in a seminatural salmon population based on the molecular marker method of Mousseau et al. (1998)

Trait	Jacking	Length	Weight	Color
Jacking	0.67 +0.20	0.35	0.35	0.08
Length		0.38 +0.41	0.88	-0.18
Weight			0.20 +0.30	0.00
Color				0.76 +0.44

Standard errors were calculated by boostrapping. "Jacking" refers to precocious male maturation. Numbers along the diagonal indicate heritability; other numbers refer to the genetic correlations. Only the heritability of jacking was statistically greater than zero.

through their life cycle; (3) the tagging and tracking of naturally occurring populations using conventional ecological methods (e.g., Sinervo, this volume, Grant, this volume). This method requires that observers be able to determine paternity, either through behavioral observations or by using molecular genetic paternity analyses. It also generally requires that the offspring be tracked through their life cycle; (4) The advent of molecular genetic tools now makes it feasible for ecological geneticists to infer patterns of genetic relatedness in naturally occurring populations (e.g., Ritland, this volume). Developments in modern statistical methods make it possible to use molecular data to estimate the genetic basis of quantitative traits, even in organisms that cannot be easily manipulated (e.g., trees, long-lived vertebrates). Given the rapid progress being made in molecular genetic methods, these hybrid molecular-ecological approaches will likely lead to analyses of QTLs in wild organisms in the near future.

Acknowledgments Many individuals have contributed to the research presented in this chapter, including Derek Roff, Mike Bradford, Yves Carriere, Sejal Shah, Amy Desai, Rebbeca Heil, Alex Olvido, Hugh Dingle, Allen Gibbs, Dan Howard, John Doherty, Kim Waddell, Chuck Fox, Fran Groeters, Eilleen Lawson, Elizabeth Mack, Kermit Ritland, and Daniel Heath. Special thanks goes to Heather Preston. This work has been supported at various times by NSERC (Canada), NSF, and the USDA.

References

Alexander, G., and J.R. Hilliard. 1969. Altitudinal and seasonal distribution of Orthoptera in the Rocky Mountains of northern Colorado. Ecol. Monogr. 39:385–431.

Alexander, R.D., and E.S. Thomas. 1959. Systematic and behavioral studies of the crickets of the *Nemobius fasciatus* group. (Orthoptera: Gryllidae: Nemobiinae). Ann. Entomol. Soc. Am. 52: 591–605.

Aspi, J, and A.A. Hoikkala. 1993. Laboratory and natural heritabilities of male courtship song characters in *Drosophila montana* and *D. littoralis*. Heredity 70:400–406.

Benedix, J.H. Jr., and D.J. Howard. 1991. Calling song displacement in a zone of overlap and hybridization. Evolution 45:1751–1759.

Berg, E.E., and J.L. Hamrick. 1995. Fine-scale genetic structure of a turkey oak forest. Evolution 49:110–120.

Bradford, M.J., P.A. Guerette, and D.A. Roff. 1993. Testing hypotheses of adaptive variation in cricket ovipositor lengths. Oecologia 93:263–267.

Coyne, J.A., and E. Beecham. 1987. Heritability of two morphological characters within and among natural populations of *Drosophila melanogaster*. Genetics 117:727–737.

Danks, H.V. 1987. Insect Dormancy: an Ecological Perspective. Biological Survey of Canada Monograph Series, no. 1. Ottawa, Canada.

Dingle, H., and T.A. Mousseau. 1994. Geographic variation in embryonic development time and stage of diapause in a grasshopper. Oecologia 97:179–185.

Dingle, H., T.A. Mousseau, and S.K. Scott. 1990. Altitudinal variation in life cycle syndromes of the California grasshopper *Melanoplus sanguinipes*. Oecologia 84:199–206.

Doherty, J.A., and D.J. Howard. 1996. Lack of preference for conspecific calling songs in female crickets. Anim. Behav. 51:981–989.

Endler, J.A. 1987. Natural Selection in the Wild. Monographs in Population Biology, no. 21. Princeton University Press, Princeton, NJ.

Falconer, D.S., and T.F.C. Mackay. 1996. Introduction to Quantitative Genetics, 4th ed. Longman, Essex, UK.

Futuyma, D.J. 1998. Evolutionary Biology, 3rd ed. Sinauer Associates, Sunderland, MA.

Gibbs, A.G., A.K. Louie, and J.A. Ayala. 1998. Effects of temperature on cuticular lipids and water balance in a desert *Drosophila*: Is thermal acclimation beneficial? J. Exp. Biol. 201:71–80.

Gibbs, A., and T.A. Mousseau. 1994. Thermal acclimation and genetic variation in cuticular lipids of the lesser migratory grasshopper (*Melanoplus sanguinipes*): effects of lipid composition on biophysical properties. Physiol. Zool. 67:1523–1543.

Gibbs, A., T.A. Mousseau, and J. Crowe. 1991. Genetic and acclimatory variation in biophysical properties of insect cuticle lipids. Proc. Natl. Acad. Sci. USA 88:7257–7260.

Gibert, P., B. Moreteau, J.R. David, and S.M. Scheiner. 1998. Describing the evolution of reaction norm shape: body pigmentation in *Drosophila*. Evolution 52:1501–1506.

Gjerde, B., and T. Gjedrem. 1984. Estimates of phenotypic and genetic parameters for carcass traits in Atlantic salmon and rainbow trout. Aquaculture 36:97–110.

Grant, P.R. 1986. Ecology and evolution of Darwin's finches. Princeton University Press, Princeton, NJ.

Grant, P.R., and B.R. Grant. 1995. Predicting microevolutionary responses to directional selection on heritable variation. Evolution 49:241–251.

Groeters, F.R., and H. Dingle. 1996. Heritability of wing length in nature for the milkweed bug, *Oncopeltus fasciatus*. Evolution 50:442–447.

Heath, D.D., N.J. Bernier, and T.A. Mousseau. 1995. A single-locus minisatellite discriminates chinook salmon (*Oncorhynchus tshawytscha*) populations. Mole. Ecol. 4:389–393.

Heath, D.D., and M. Blouw. 1998. Are maternal effects in fish adaptive or merely physiological side-effects? Pp. 178–201 *in* T.A. Mousseau, and C.W. Fox, eds., Maternal Effects as Adaptations. Oxford University Press, New York.

Heath, D.D., G.K. Iwama, and R.H. Devlin. 1994. DNA fingerprinting used to test for family effects on precocious sexual maturation in two populations of *Oncorhynchus tshawytscha* (Chinook salmon). Heredity 73:616–624.

Hoffmann, A.A. 1991. Heritable variation for territorial success in field-collected *Drosiphila melanogaster*. Am. Nat. 138:668–679.

Houle, D. 1992. Comparing evolvability and variability of quantitative traits. Genetics 130:-195–204.

Howard, D.J., and D.G. Furth. 1986. Review of the *Allonemobius fasciatus* (Orthoptera: Gryllidae) complex with the description of two new species separated by electrophoresis, songs, and morphometrics. Ann. Entomol. Soc. Am. 76:1014–1021.

Kirkpatrick. M., and N.H. Barton. 1997. Evolution of a species' range. Am. Nat. 150:1–23.

Lynch, M., and B. Walsh. 1998. Genetics and analysis of quantitative traits. Sinuaer Associates, Sunderland, MA.

Masaki, S. 1967. Geographic variation and climatic adaptation in a field cricket (Orthoptera: Gryllidae). Evolution 21:725–741.

Masaki, S. 1973. Climatic adaptation and photoperiodic response in the band-legged ground cricket. Evolution 26:587–600.

Masaki, S. 1978. Seasonal and latitudinal adaptations in life cycles of crickets. Pp. 72–100 *in* H. Dingle, ed., Evolution of Insect Migration and Diapause. Springer-Verlag, New York.

Masaki, S. 1979. Climatic adaptation and species status in the lawn ground cricket. III. Ovipositor length. Oecologia 43:207–219.

Masaki, S. 1986. Significance of ovipositor length in life cycle adaptations of crickets. Pp. 20–34 *in* F. Taylor and R. Karban, eds., The Evolution of Insect Life Cycles. Springer-Verlag, New York.

Mitchell-Olds, T. 1986. Quantitative genetics of survival and growth in *Impatiens capensis*. Evolution 40:107–116.

Morris, R.F. 1971. Observed and simulated changes in genetic quality in natural populations of *Hyphantria cunea*. Can. Entomol. 103:893–906.

Morris R.F., and W. Fulton 1970. Heritability of diapause intensity in *Hyphantria cunea* and correlated fitness responses. Can. Entomol. 102:927–938.

Mousseau, T.A. 1997. Ectotherms follow the "Converse to Bergman's Rule". Evolution 51:630–632.

Mousseau, T.A., and H. Dingle. 1991a. Maternal effects in insect life histories. Annu. Rev. Entomol. 36:511–534.

Mousseau, T.A., and H. Dingle. 1991b. Maternal effects in insects: examples, constraints, and geographic variation. Pp. 745–761 *in* E.C. Dudley, ed., The Unity of Evolutionary Biology. Dioscorides Press, Portland, OR.

Mousseau, T.A., and C.W. Fox. 1998. Maternal Effects as Adaptations. Oxford University Press, New York.

Mousseau, T.A., and D.J. Howard. 1998. Genetic variation for calling song across a hybrid zone between two sibling cricket species. Evolution 52:1104–1110.

Mousseau, T.A., K. Ritland, and D.D. Heath. 1998. A novel method for estimating heritability using molecular markers. Heredity 80:218–224.

Mousseau, T.A., and D.A. Roff. 1987. Natural selection and the heritability of fitness components. Heredity 59:181–197.

Mousseau, T.A., and D.A. Roff. 1989. Adaptation to seasonality in a cricket: patterns of phenotypic and genotypic variance in body size and diapause expression along a cline in season length. Evolution 43:1483–1496.

Mousseau, T.A., and D.A. Roff. 1995. Genetic and environmental contributions to geographic variation in the ovipositor length of a cricket. Ecology 76:1473–1482.

Olvido, A.E., S. Busby, and T.A. Mousseau. 1998. Oviposition and incubation environmental effects on embryonic diapause in a ground cricket. Anim. Behav. 55:331–336.

Olvido, A., and T.A. Mousseau. 1995. The effect of rearing environment on calling song plasticity in the striped ground cricket. Evolution 49:1271–1277.

Otte, D. 1989. Speciation in Hawaiian crickets. Pp. 482–526 *in* D. Otte and J.A. Endler, eds., Speciation and Its Consequences. Sinauer Associates, Sunderland, MA.

Park, O. 1949. Application of the converse Bergmann principle to the carabid beetle, *Dicaelus purpuratus*. Physiol. Zool. 22:359–372.

Peters, R.H. 1983. The Ecological Implications of Body Size. Cambridge University Press, New York.

Price, T., and D. Schluter. 1991. On the low heritability of life-history traits. Evolution 45:853–861.

Prout, T., and J.S.F. Barker. 1989. Ecological aspects of the heritability of body size in *Drosophila*. Genetics 123:803–813.

Riska, B., T. Prout, and M. Turelli. 1989. Laboratory estimates of heritabilities and genetic correlations in nature. Genetics 123:865–871.

Ritland, K. 1996. A marker-based method for inferences about quantitative inheritance in natural populations. Evolution 50:1062–1073.

Ritland, K., and C. Ritland. 1996. Inferences about quantitative inheritance based upon natural population structure in the common monkeyflower, *Mimulus guttatus*. Evolution 50:1074–1082.

Roff, D.A. 1980. Optimizing development time in a seasonal environment: the 'ups and downs' of clinal variation. Oecologia 45:202–208.

Roff, D.A. 1983. Phenological adaptation in a seasonal environment: a theoretical perspective. Pp. 253–270 *in* V.K. Brown and I. Hodek, eds., Diapause and Life Cycle Strategies in Insects. W. Junk, The Hague.

Roff, D.A. 1992.The Evolution of Life Histories; Theory and Analysis. Chapman and Hall, New York.

Roff, D.A. 1997. Evolutionary Quantitative Genetics. Chapman and Hall, New York.

Roff, D.A., and T.A. Mousseau. 1987. Quantitative genetics and fitness: lessons from *Drosophila*. Heredity 58:103–118.

Roff, D.A., T.A. Mousseau, and D.J. Howard. 1999. Variation in genetic architecture of calling song among population of *Allonemobius socius*, *A. fasciatus* and a hybrid population: drift or selection? Evolution 53:216–224.

Service, P.M., and M.R. Rose. 1985. Genetic covariation among life-history components: the effects of novel environments. Evolution 39:943–945.

Shaw, R.G. 1986. Response to density in a natural population of the perennial herb *Salvia lyrata*: Variation among families. Evolution 40:492–505.

Sinervo, B. and C.M. Lively. 1996. The rock-paper-scissors game and the evolution of alternative male strategies. Nature 380:240–243.

Smith, R.L. 1996. Ecology and Field Biology, 5th ed. Benjamin Cummings, New York.

Tauber, M.J., C.A. Tauber, and S. Masaki. 1986. Seasonal Adaptations of Insects. Oxford University Press, New York.

Turnbow, R.H. Jr., and R.T. Franklin. 1981. Bionomics of *Brachys tessellatus* in coastal plain scrub oak communities. Ann. Entomol. Soc. Am. 74: 351–358.

Van Voorhies, W.A. 1996. Bergmann size clines: a simple explanation for their occurrence in ectotherms. Evolution 50:1259–1264.

Waddell, K.J. 1996. Variables that influence the relationship between oviposition preference and offspring performance in a leafmining beetle (PhD dissertation). University of South Carolina, Columbia.

Waddell, K.J., and T.A. Mousseau. 1996. The oviposition preference hierarchy of a leafmining beetle, *Brachys tessellatus* (Coleoptera: Buprestidae). Environ. Entomol. 25:63–67.

Weigensberg, I., and D.A. Roff. 1996. Natural heritabilities: can they be reliably estimated in the laboratory? Evolution 50:2149–2157.

11

Adaptive Genetic Variation in the Wild

JOHN A. ENDLER

After many decades of neglect, we are finally considering seriously the extent and causes of adaptive genetic variation in natural populations. One of the reasons for the neglect until recently is that, in order to address this problem, we need to use extensive, complex, and time-consuming techniques of several poorly integrated fields: ecology, genetics, functional morphology, physiology, and behavior, and we need to use methods ranging from difficult field work to advanced molecular laboratory work. It is a rare person who can do all of these and integrate disparate fields. The chapters in this book are a good sample of the elegant, clever, and original approaches to understanding adaptive genetic variation, and they show that significant progress is now being made in understanding it. In this chapter I summarize the general patterns that are emerging and make some tentative suggestions about where we should go from here. To avoid long-standing arguments about the meaning of "adaptation," I use the term "adaptive genetic variation" in the sense of genetic variation that is correlated with variation in lifetime or total fitness of individuals.

11.1 Genetic Variation for Adaptive Traits

Evidence from a wide variety of plants and animals suggests that there is heritable variation in a all kinds of adaptive traits, ranging from physiological through morphological, life-history to behavioral traits (Endler 1986, and the chapters in this volume).

There may be differences in average heritabilities (h^2) among different classes of traits (Hoffman, this volume). In *Drosophila*, life-history and behavioral traits have lower heritabilities than morphological and physiological traits (Roff and Mousseau 1987, Weigensberg and Roff 1996; Hoffman, this volume). One explanation for this is the rate-of-loss hypothesis. If the environment remains constant, advantageous alleles get fixed and narrow-sense h^2 gets reduced. Traits that are more tightly related to fitness have less genetic variation because advantageous alleles get fixed more rapidly in traits with stronger contributions to fitness (Falconer 1989, Hoffman, this volume). This prediction is problematic because, as anyone who has worked with natural populations for more than a breeding season knows, environments are anything but constant, and there are fluctuations at time scales ranging from months to millennia and more (Davis 1986, McCauley 1993, Bell 1997); we consequently expect fluctuations in the direction and magnitude of selection over all time scales. The consequences of fluctuating selection are summarized by Grant and Grant (this volume). For example, there may not be enough time between changes in direction of selection for the temporarily advantageous alleles to be fixed. Nevertheless, if some traits respond more quickly to the environment than others, then they may be responding more to shorter time-scales of selective changes than traits with a longer response time. This will probably have predictable consequences on the variation in heritability within and among trait classes, but this has not been studied theoretically or empirically.

Life-history traits show a greater range of h^2 estimates per trait than other traits, but there are still too few data on which to form strong conclusions (Hoffman, this volume). If true, this is probably due to the greater influence of the environmental component (V_E) of life-history traits compared to other traits. Behavior is also more sensitive to V_E than morphology, so we would also expect behavior to have a greater range of h^2, but this was not found. The rate-of-loss hypothesis awaits more data for confirmation or rejection.

There is an alternative explanation for life-history and behavioral traits having lower h^2 than morphological and physiological traits. Life history and behavior are partially influenced by morphology. Consequently, they have two sources of V_E: the direct effects of the trait V_E and the indirect effects of the associated morphology V_E. The effect of the additional V_E would be to lower h^2 of these traits (Price and Schluter 1991, Hoffman, this volume).

There are also more prosaic explanations for differences in h^2 among trait classes. Low h^2 may be erroneously recorded due to low repeatability, as is common for behavioral and, to a lesser extent, life-history traits. For example, life-history traits depend strongly on conditions of measurement and rearing (Merila 1997). Morphological traits may have higher h^2 because traits are often chosen for study precisely because they have high repeatability. It is hard to know whether to take these trait differences seriously because the sample sizes of these comparisons are still small (each datum requires a tremendous amount of work), and most h^2 estimates come from the laboratory rather than from the field. For excellent discussions of the problems and biases of h^2 estimates, see Hoffman (this volume), Grant and Grant (this volume), and Meade and Mazer (this volume).

These problems may not merely be nuisances but may have fascinating evolutionary consequences. For example, lower h^2 due to poor nutrition (as in pied flycatchers; Merila 1997) might slow the response to selection during stressful years, with faster response in good years. The good years may have lower selection intensities, counteracting the faster response per standardized selection, yielding slower long-term evolution than expected from the average selection over many generations.

We might well ask why classes of similar traits should have similar h^2? Given huge fluctuations in the magnitude and direction of selection and differences among different traits within trait classes and among taxa, it may not be reasonable to look for general patterns in this way. Perhaps we should instead look for differences among groups of species with particular combinations of generation lengths relative to the temporal scale of selection. The problem of temporal scale of selection relative to generation length is a long-standing one in studies of phenotypic plasticity and may determine whether plasticity, polymorphism, or low phenotypic variation (V_P) has evolved in any given trait (Bradshaw 1965, Nager et al. this volume).

Nager et al. (this volume) present an interesting perspective on phenotypic plasticity. An important point is that different proximal causes of nongenetic components of phenotypic variation may have different evolutionary consequences. Nager et al. placed this in the context of phenotypic plasticity in egg-laying dates in great tits. How does *Parus major* respond to variability during the egg-laying period, and how does the environment affect fitness? The data are fascinating because they seem to challenge the conventional viewpoint that phenotypic plasticity is adaptive in the sense that fitness is maintained despite environmental changes. The results bring up the possibility of a different kind of adaptive "nongenetic" variation. Many traits may not be conventionally plastic (which implies a functional or adaptive genotype–environment interaction), but just poorly regulated. For example, the mere randomness of the pattern of breeding dates ensures that some birds accidentally hit the right laying date to synchronize their laying with high insect density when the young hatch. This is logically equivalent to plants producing thousands of seeds in the near certainty that at least some of them will reach an appropriate place for germination. The "plasticity" of timing is only adaptive in the sense that some birds breed successfully some seasons, and the same pair breeds successfully some seasons. This is a bet-hedging form of developmental noise favored by natural selection, not plasticity in the conventional sense. Thus there are two forms of environmental influences on the phenotype, random developmental noise and phenotypic plasticity. Either form may be adaptive and selected for, but the evolutionary consequences are different.

One source of genetic variation that has only recently begun to be considered seriously is introgression from other species and other populations (Arnold 1997, Smith et al. 1997, Grant and Grant, this volume). Alleles, allele combinations, traits, and trait combinations that appear in one population or species can spread into others by gene flow, introgression, or hybridization. Grant and Grant (this volume) provide two lines of evidence that this has happened in Darwin's finches: there is more genetic variation than expected from effective population size and the models of maintenance of quantitative genetic variation, and certain gene combinations could only have been produced by hybridization and introgression. Grant and Grant reviewed the relatively unappreciated literature on the maintenance of genetic variation by spatial and temporal variation. Temporal and spatial heterogeneity can maintain variation within populations, and spatial variation can also maintain variation among populations (see also Smith et al. 1997, Meade and Mazer, this volume, Smith and Girman, this volume), and longer time-scales of temporal variation can mean that the same populations may have different kinds of genetic variation at different times. This makes comparisons among species more difficult, so more work needs to be done to see if there are predictable patterns in spite of these confounding factors.

11.2 The Diversity of Adaptive Traits

As the chapters in this volume show, the diversity of adaptive traits with a genetic basis is impressive. But considering their function, this is not really surprising; there seem to be an infinite diversity of ways to successfully pass genes between generations. It is an excellent trend that more and more people are working on the function and ecology of adaptive traits because the functional approach allows one to predict variation in fitness and evolutionary change rather than simply proving its existence. A particular advantage of knowing the function of a trait in detail is that it allows specific predictions about fitness and the direction of evolution (Arnold 1983, Endler 1986, Bell 1997, Smith and Girman, this volume, Mazer and Meade, this volume, Robinson and Schluter, this volume, Nager et al., this volume). The functional approach also allows us to explain and predict genetic correlations, covariance selection, and complex relationships among traits (Sinervo, this volume, Smith and Girman, this volume, Nager et al., this volume). An additional value of the functional approach is that it can illuminate constraints to variation, a subject which has heretofore been plagued with speculation and strong opinions in the absence of evidence. An excellent example is Sinervo's (this volume) work on egg size and offspring size in lizards. Stabilizing selection for egg size is caused by fecundity selection favoring smaller size (for a given body size), offspring survival favoring larger size, and pelvic size limiting the upper end of egg size because it increases egg-bound mortality. One still has to avoid being overconfident about the functional approach; factors or functions that are important to the trait, but that are not included in the study, can result in confusing or even misleading results (Endler 1986, Meade and Mazer, this volume, Sinervo, this volume).

Diversity in adaptive traits is also found within species, and there is increasing evidence for geographical variation in these traits. Geographical variation in traits, function, and environment allows inferences about the relationships between structure, function, fitness, and maintenance which are not possible in geographically uniform traits (see chapters by Meade and Mazer, Smith and Girman, Grant and Grant, Mousseau, and Mopper et al.; Endler 1977, Sandoval 1994, Endler and Houde 1995, Foster and Endler 1998); but incomplete knowledge of the geographically varying selective factors can mislead and confuse the interpretations (Endler 1977, Meade and Mazer, this volume, Mopper et al., this volume). A study of the adaptive traits and selectively neutral genetic markers can yield interesting insights into local adaptation and selection in general, as well as suggest better ways to interpret other studies involving only neutral traits (Smith and Girman, this volume; Mopper et al., this volume). Mopper et al.'s (this volume) study of leaf miners and adaptations to plant secondary compounds (plant defenses) is a fascinating example. There is not always any obvious relationship between the ecological and functional traits and F_{ST} of neutral markers, but this reflects the interesting dynamics of the population structure within and among trees. The model (Mopper et al., this volume) is that young oaks are colonized and founder effects occur and cause population structure, which is eliminated as new migrants arrive. As densities rise, fitness of new migrants relative to natives will decline owing to growing local adaptation as well as competition, eventually resulting in the (now older) tree populated by locally adapted forms. Miners experimentally transferred from younger to older trees experience stronger selection, but not the converse. Selection may have removed the nonadaptive forms, but nevertheless there has been a longer history of gene flow on older trees, so the neutral markers have the opposite pattern: smaller F_{ST}, on older trees. The study of the relationship between neutral and adaptive genetic variation is just beginning; no doubt further surprises await us. One significant new insight is that population structure evolves relative to adaptation, not in the simplistic way assumed by those working on neutral markers (Mopper et al., this volume).

11.3 Emerging Patterns Suggested by Studies of Adaptive Genetic Variation

There are a number of interesting patterns beginning to emerge from studies of adaptive genetic variation in natural populations. There is increasing evidence for genetic correlations among adaptive traits, both positive and negative. This is particularly common in life-history traits (Sinervo, this volume), but this might be because those studying life-history traits regularly measure many traits, whereas those studying other traits tend to concentrate on a few. A notable exception is bill size and shape in birds, where genetic covariances between different measurements are regularly found (Smith and Girman, this volume, Grant and Grant, this volume). Genetic correlations are often assumed to result from correlational selection or selection that favors particular combinations of trait values over others. This can arise either from positive correlations between different environmental factors or from strong functional relationships among different traits. Evidence for this is particularly strong in birds, where biomechanical properties required for seed crushing successfully predict the shape as well as the size of bills; certain sizes and shapes are required for seeds of particular hardness and diameter (Smith and Girman, this volume, Grant and Grant, this volume). Optimum combinations of various life-history characters are also predicted from functional considerations of life history, but the results are not as clear cut (Sinervo, this volume). The difference probably arises because there are fewer selective factors operating on bill shape than on life-history traits. Still, we can probably generally say that traits that function together are subject to correlational selection and will probably evolve genetic correlations. This prediction will only be true if the joint function of the associated traits continues with the same sign over many hundreds of generations. It would be interesting to know if traits subject to correlational selection of fluctuating form (e.g., sign) have weaker genetic correlations than traits subject to continuous and consistent correlational selection even in the face of fluctuating environments.

Genetic correlations can result from pleiotropy or linkage disequilibrium. The former is stable (aside from mutation), and the latter breaks down at a rate proportional to the recombination rate. Linkage disequilibrium must therefore be maintained either by population structure (such as gene flow between populations with different gene combinations) or correlational selection. Correlational selection results from correlated function or correlated environmental factors that favor particular combinations of trait values over others. For example, if we plot fitness as a function of two traits, correlational selection shows as a ridge on the surface. Such selection will select for and refine the orientation of genetic correlation between the traits until the two match on the trait axes. Sinervo (this volume) gives an interesting example based on the function of the traits: follicle-stimulating hormone, estrogen, and vitelloprotein levels in the female can be modified individually and in concert to produce the appropriate genetic correlation to match the correlational selection induced by life-history selection. Selection by multiple factors on multiple traits, or parts of traits (such as life-history components) often gives rise to confusing, if not counterintuitive, results. The net affect may be stabilizing, directional, or disruptive, and this is not related simply to what happens for each trait. Multivariate selection may be nonlinear and complex, and the outcome dependent on positive and negative genetic correlations. It would be fascinating and productive to model the evolution of genetic correlations under specified multivariate correlational selection, but it is not clear where to start. One possibility would be to start with suites of traits with known interrelated adaptive functions (as did Sinervo, this volume) which allow explicit predictions of what form correlational selection would take. When we have enough of these functions, we may be able to generalize about what classes of correlational selection exist and what their consequences are.

One has to be careful in interpreting some forms of genetic correlations as arising or being maintained by correlational selection. Tight associations may arise from pleiotropy. Pleiotropy may be observed as a result of pleiotropy in the formal sense (one gene having multiple effects), but it may also arise as a result of mechanical or other physical limits. A gene may have an effect on development and morphogenesis, and the physical relationships between the structures assembled by the same gene may cause additional effects that are secondary to the gene. In this situation it is not so much a case of pleiotropy as measuring the effects of the same gene at varying places on its cascade of phenotypic effects. Some of these effects are so nonlinear that the correlations may be weak; morphological pattern generation is a good example (Murray 1989). On the other hand, tight genetic correlations must not be treated as cases of pleiotropy. Strong functional relationships, such as that between the upper and lower jaws of vertebrates (Smith and Girman, this volume), will induce such strong correlational selection that genetic correlations will become very strong. In these cases one might well ask if the genetic system will evolve tighter physical genetic linkage to make the system behave as if it were in fact pleiotropy and allow a tighter fit between phenotypes and the environment.

The idea of evolving tighter linkage between loci in response to selection for particular allele combinations is an old one (Sheppard 1958, Dobzhansky 1970). It has been extended to the evolution of sex chromosomes (Bull 1983) and could easily be extended to quantitative traits. The hypothesized sequence (for any kind of trait) is that (1) strong correlational selection (2) results in linkage disequilibrium, (3) translocations of the genes themselves or their gene control mechanisms, (4) evolution of tighter linkage and supergenes, and (5) perhaps eventual conversion to single genes. We will not be able to understand or demonstrate this sequence until we know more about the form of correlational selection, but these data are finally starting to accumulate. Effects of the environment and phenotypic plasticity further complicate interpretation of selection and genetic variation (Meade and Mazer, this volume, Mousseau, this volume, Nager et al., this volume), but a predictive functional approach to the phenotypes should help unravel these factors. It would be productive to model the conditions under which correlational selection favors tighter linkage or under which it fluctuates so much that it favors phenotypic plasticity.

If correlational selection is common, and there is widespread evolution of tighter linkage to allow genetic correlations to match selective correlations, then under what conditions does this favor single phenotypes or multiple discontinuous phenotypes such as morphological or trophic polymorphism? The original version of the evolution of supergenes (Sheppard 1958, Dobzhansky 1970) explained polymorphisms rather than unimodal phenotypic distributions. Natural selection was thought to favor two or more distinctly different combinations of alleles that were much superior to all other combinations. These superior combinations became tightly linked (for example, color and banding in *Cepaea nemoralis*; Sheppard 1958) or held together in inversions (in *Drosophila*, Dobzhansky 1970). The tightly linked groups were called "supergenes" because they consisted of multiple genes but segregated as units (formally they probably should have been called "superalleles"). Additional selective factors were required to explain why more than one of these superalleles or morphs were maintained in a population: heterozygous advantage (Dobzhansky 1970), frequency-dependent and density-dependent natural selection (Clarke et al. 1988), and multiple-niche polymorphisms and selection-gene flow balance (Endler 1977, Smith et al. 1997, Smith and Girman, this volume, Meade and Mazer, this volume). Until recently evidence for these mechanisms has been scant, probably because both ecological and genetic methods were required in the same study, and until recently people were not trained in both fields.

Adaptive genetic variation is found in both continuous (quantitative) and discontinuous (polymorphic) traits (Smith and Girman, this volume, Grant and Grant, this volume, Robinson and Schluter, this volume). In addition, the mode of genetic variation may even vary within the same trait for different species. For example, bill size and shape is polymorphic in some birds (Smith and Girman, this volume) and continuous in others (Grant and Grant, this volume). What causes some traits to be discontinuous and others to be continuous? Smith and Girman (this volume) suggest that discontinuous variation in bill shape and size may have evolved to exploit a discontinuous distribution of seed hardness and size classes. On the other hand, if there are large temporal fluctuations in the distribution and abundance of seed hardness and size, as in Darwin's finches (Grant and Grant, this volume), then the resulting disruptive and correlational selection may not last enough generations for strong genetic correlations or supergenes (Sheppard 1958) to evolve, resulting in a continuous distribution. It would be interesting to know what conditions favor polymorphisms and what conditions favor continuous variation, and how closely related this is to the temporal variation in intensity and form of correlational selection.

There has now been good success in explaining polymorphisms due to ecological differences between morphs, and the chapters on resource and trophic polymorphisms in this volume are excellent examples (Smith and Girman, this volume, Grant and Grant, this volume, Robinson and Schluter, this volume). Resource polymorphisms have been found in insects, fishes, and birds. Interestingly, resource polymorphisms are not necessarily based on supergenes, and many are polygenic. Robinson and Schluter suggest six attributes of systems in which resource polymorphisms are caused by natural selection: (1) repeated patterns of niche differentiation and associated body shape; (2) excessive variation among ecotypes — trophic or resource polymorphisms (these are most dramatic where competitors and predators are rare. Do predators and competitors limit differentiation?); (3) polymorphisms may be examples of character release; (4) ecotypes persist in single populations in the face of gene flow; (5) parallel patterns independently evolved many times, and in different systems; (6) ecologically intermediate forms exhibit reduced fitness compared to more specialized ecotypes. This has been observed within species and among hybrids. Finally, niche-based variation may sometimes be "converted" from within to between species, as shown in sticklebacks (Robinson and Schluter, this volume).

One of the interesting ideas suggesting why variation should be discontinuous is that there are "adaptive valleys" between the morphs — combinations of traits and trait values that have low fitness. The best data come from birds and do not explicitly depend on correlational selection. Bird bills have discontinuous sizes because the seed base that they crack is also discontinuous (Smith and Girman, this volume). Smith and Girman suggest that finches crossed the adaptive valley to form new morphs when the new seed size became available, but there is another possibility. The seeds may have evolved with the finches, so there was never any adaptive valley to cross. The plants may evolve a discontinuous distribution of seed size and hardness in response to the disruptive selection caused by the birds, and vice-versa. Consider initially a continuous distribution of seed hardness/size and bill size. By chance, either seeds or bill distributions may form a multimodal distribution. This would immediately cause nonuniform selection on the other species, causing it to become partially multimodal in the appropriate parts of its trait distribution. The two systems induce natural selection on each other, and this may result in their respective distributions, eventually splitting into discontinuous distributions of bill size and seed size/hardness. Formal modeling of the evolution of trophic polymorphisms would be interesting. It would also be interesting to know if other systems in which two species interact with each other also result in polymorphisms, or at least generate more variation than when only one species is involved. For example, can this happen with other predator–prey systems, parasite–host systems, or floral characters and pollinators? Invasions of new selecting species may also allow the study of natural selection of traits found in systems with interacting species (for example, Meade and Mazer, this volume, Mopper et al., this volume). There may be a host of new phenomena in evolution having to do with multispecies interactions, just as the study of interactions among species yields new insights into ecology not possible from studies of single species.

For whatever reason, the existence of the gaps in seed size and hardness cause disruptive selection which maintains but may not have caused the origin of the morphs. Similar arguments have been made about other trophic polymorphisms (Smith and Girman, this volume, Robinson and Schluter, this volume). A polymorphism is not necessarily expected for resource-based variation. If there is no discontinuous variation in resource, but the resources vary in time, so too does the selection, and there may be no disruptive selection. This may be true even though its strength and direction varies in time (Grant and Grant, this volume). One pattern that has emerged is that the larger the continuous habitat, the more likely the polymorphism (Robinson and Schluter, this volume). Another pattern is that recently deglaciated regions contain a higher incidence of resource-based polymorphisms (Robinson and Schluter, this volume). Perhaps they are more likely in recently expanded geographical distributions among species that expanded before niches were occupied by other species. Recent expansions of geographic range are superb but still understudied systems in which to study expansion into formerly unoccupied niches (Mopper et al., this volume) and may eventually allow us to explore these problems in more detail. We know more about maintenance than origin of ecologically relevant (adaptive) phenotypic variation.

The subdivision of populations and species into trophic and other polymorphisms brings up the possibility that such multiple-niche polymorphisms are the stepping stones for speciation; all that is needed is reproductive isolation (Robinson and Schluter, this volume; Smith and Girman, this volume). There is abundant evidence that divergent selection in different niches or selective regimes is the primary mechanisms that sorts heritable variation into different ecological niches at all levels, from phenotypes to species (Endler 1977, 1986, Bell 1997, Robinson and Schluter, this volume, Smith and Girman, this volume, Grant and Grant, this volume). But divergent variation must be associated with positive assortative mating for speciation to occur. Speciation may not occur if trophic variation is not associated with other genetic variation, as is the case in the finches (Smith and Girman, this volume, Grant and Grant, this volume). Does the association take a long time to form, or does it only occur when there is active selection associating the mating system with divergent phenotypic variation? Such studies are still in their infancy.

11.4 Conclusions

Although we have been nominally interested in adaptive genetic variation for a century, the serious study of the problem has only recently become common. The result is an explosion of new patterns, explanations, and ideas, indicating that we have barely scratched the surface of the problem. Like all rapidly expanding fields, this has generated a large number of interesting questions. Here are a few which now seem practical:

1. Does heritability vary with kinds of traits or with temporal fluctuation scale relative to generation length? What are the conditions that cause h^2 to vary?
2. Might there be selection for a particular change of heritability with changing environmental conditions?
3. What are the effects of various scales of temporally varying selection? Are there scale-dependent effects? How does this interact with spatially varying selection?
4. Genetic covariance may be induced by covariance selection — are there particular patterns? What are the mechanisms? Is it more likely when the sign of the selective correlation remains constant for a long time?
5. Covariance selection should be common because certain sets of environmental parameters covary — this needs documentation.
6. Is there a pattern of genetic covariance that follows covarying environmental factors? Do traits subject to correlational selection of fluctuating form (e.g., sign) have weaker genetic correlations than traits subject to continuous and consistent correlational selection even in the face of fluctuating environment?
7. What causes some traits to be discontinuous and others to be continuous? If two species are interacting and this affects traits in both species, will this likely lead to discontinuities in both sets of traits?
8. How widespread is the maintenance of adaptive genetic variation by multiple-niche polymorphisms and the balance between environmental heterogeneity and gene flow? Do scales of heterogeneity in time and space yield predictable patterns in this variation?
9. Environmental fluctuations may favor plasticity if they are short relative to the generation length. If so, are there traits that are more likely to be subject to randomly fluctuation selection than others, and are these in fact more plastic? Are traits that respond to shorter time-scale events more plastic than traits that respond to longer time-scale events?
10. If environments are predictably fluctuating, does this favor particular V_{GE}? Or does particular V_{GE} evolve in cases where the selective grain is smaller than the scale of gene flow?
11. Can randomly varying V_E (i.e., no V_{GE} component) be adaptive or selected?
12. If lineages invading new environments diversify more rapidly than noninvading lineages, do such lineages have suites of characteristics which make them expand rapidly particularly easily (genetically set for radiation) or are ecologically predisposed to invade (generalists, high r, fugitive species, etc.)?
13. Is there any correlation between adaptational changes in morphology, physiology, or development and speciation? Too often this is assumed. How often does trophic polymorphism lead to speciation, or just faster rates of morphological evolution and divergence? This is the old question of divergence versus speciation.
14. How often can we predict the direction of evolution from a knowledge of the mechanisms of natural selection? Is this more difficult for multivariate correlational selection?

15. How often will multivariate evolution and speciation be stopped by awkward or counteracting genetic correlations? Does the presence of these negative correlations imply former correlational selection in different directions, or just mechanisms of development and physiology?
16. Question 15 leads to the question about how development and physiology evolved. Do development and physiology evolve explicitly to adapt to changing environments? This seems likely and goes back to question 6 and the question of responses to environmental fluctuations of different temporal and spatial scales. It would be beneficial to seriously integrate development and physiology into evolution, now that ecology is finally being integrated.

References

Arnold, M. L. 1997. Natural Hybridization and Evolution. Oxford University Press, Oxford.

Arnold, S. J. 1993. Morphology, performance and fitness. American Zoologist 23:347–361.

Bell, G. 1997. Selection: The Mechanism of Evolution. Chapman and Hall, New York.

Bradshaw, A. D. 1965. Evolutionary significance of phenotypic plasticity in plants. Advances in Genetics 13:115–155.

Bull, J. 1983. *Evolution of Sex Determining Mechanisms*. Benjamin/Cummings, Menlo Park, CA.

Clarke, B., L. Partridge, and A. Robertson. 1988. Frequency-Dependent Selection. Cambridge University Press, London.

Davis, M. 1986. Climatic instability, time lags, and community disequilibrium. Pp. 269–284 *in* J. Diamond and T. Case, eds., *Community Ecology*. Harper & Row, New York.

Dobzhansky, T. 1970. *Genetics of the Evolutionary Process*. Columbia University Press, New York.

Endler, J. A. 1977. *Geographic Variation, Speciation and Clines*. Princeton University Press, Princeton, NJ.

Endler, J. A. 1986. *Natural Selection in the Wild*. Princeton University Press, Princeton, NJ.

Endler, J. A., and A. E. Houde. 1995. Geographic variation in female preferences for male traits in *Poecilia reticulata*. Evolution 49:456–468.

Falconer, D. S. 1989. *Introduction to Quantitative Genetics*, 3rd ed. Longman, London.

Foster, S. M., and J. A. Endler. 1998. Geographic Variation in Behavior, Perspectives on Evolutionary Mechanisms. Oxford University Press, Oxford.

McCauley, D. E. 1993. Evolution in metapopulations with frequent local extinction and recolonization. Oxford Surveys in Evolutionary Biology 9:109–133.

Merila, J. 1997. Expression of genetic variation in body size of the collard flycatcher under different environmental conditions. Evolution 51:526–36.

Murray, J. 1989. Mathematical Biology. Springer-Verlag, New York.

Price, T., and D. Schluter. 1991. On the low heritability of life-history traits. Evolution 45:-853–861.

Roff, D. A., and T. A. Mousseau. 1987. Quantitative genetics and fitness: lessons from *Drosophila*. Heredity 58:103–118.

Sandoval, C. P. 1994. The effects of the relative geographic scales of gene flow and selection on morph frequencies in the walking-stick *Timema cristinae*. Evolution 48:1866–1879.

Sheppard, P. M. 1958. *Natural Selection and Heredity*. Hutchinson University Library, London.

Smith, T. B., R. K. Wayne, D. J. Girman, and M. W. Bruford. 1997. A role for ecotones in generating rainforest biodiversity. Science 276:1855–1857.

Weigensberg, I., and D. A. Roff. 1996. Natural heritabilities: can they be reliably estimated in the laboratory? Evolution 50:2149–2157.

Index